Air Pollution XVII

WIT*PRESS*

WIT Press publishes leading books in Science and Technology.
Visit our website for the current list of titles.
www.witpress.com

WIT*eLibrary*

Home of the Transactions of the Wessex Institute.
Papers presented at Air Pollution XVII are archived in the WIT eLibrary in volume 123 of WIT Transactions on Ecology and the Environment (ISSN 1743-3541).
The WIT eLibrary provides the international scientific community with immediate and permanent access to individual papers presented at WIT conferences.
http://library.witpress.com

Seventeenth International Conference on Modelling, Monitoring and Management of Air Pollution

AIR POLLUTION XVII

Conference Chairman

C. A. Brebbia
Wessex Institute of Technology, UK

International Scientific Advisory Committee

Organised by
Wessex Institute of Technology, UK

Sponsored by
WIT Transactions on Ecology and the Environment

WIT Transactions

Transactions Editor

Carlos Brebbia
Wessex Institute of Technology
Ashurst Lodge, Ashurst
Southampton SO40 7AA, UK
Email: carlos@wessex.ac.uk

Editorial Board

Air Pollution XVII

Editors

C.A. Brebbia

&

V. Popov

Wessex Institute of Technology, UK

WIT*PRESS* **Southampton, Boston**

Editors

C.A. Brebbia
Wessex Institute of Technology, UK

V. Popov
Wessex Institute of Technology, UK

Published by

WIT Press
Ashurst Lodge, Ashurst, Southampton, SO40 7AA, UK
Tel: 44 (0) 238 029 3223; Fax: 44 (0) 238 029 2853
E-Mail: witpress@witpress.com
http://www.witpress.com

For USA, Canada and Mexico

WIT Press
25 Bridge Street, Billerica, MA 01821, USA
Tel: 978 667 5841; Fax: 978 667 7582
E-Mail: infousa@witpress.com
http://www.witpress.com

British Library Cataloguing-in-Publication Data

A Catalogue record for this book is available
from the British Library

ISBN: 978-1-84564-195-5
ISSN: (print) 1746-448X
ISSN: (on-line) 1734-3541

The texts of the papers in this volume were set individually by the authors or under their supervision. Only minor corrections to the text may have been carried out by the publisher.

Printed in Great Britain by Athenaeum Press Ltd.

Preface

This volume contains the reviewed papers accepted for the Seventeenth International Conference on Modelling, Monitoring and Management of Air Pollution held in Tallinn, Estonia in July 2009. This meeting similarly to the previous meetings has attracted outstanding contributions from leading researchers in the field. All the presented contributions are permanently stored in the WIT eLibrary as Transactions of the Wessex Institute (see http://library.witpress.com/).

Air pollution is highly topical nowadays due to the increase in the number of emission sources and the significance that good quality air has on human health. The complexity of this topic is increased by the fact that air pollution generated locally can have an impact on a regional and in some cases even on a global scale. The contaminants emitted in one place can quickly disperse through the atmosphere and industrial activities in one country can affect the air quality in another. More accurate and reliable predictive models are necessary, which can be used to assess the influence of one or several sources of pollution on various end points. The improvements are possible through achieving better quantification of emission rates and more accurate information on composition of pollutants of various sources, improving transport models on a regional scale which can include accurate predictions on local scale where necessary, enhancing the knowledge on the chemical reactions transforming existing and creating new pollutants, deposition of particles, and improving the understanding of impact of separate pollutants and combinations of pollutants on human health and the environment.

The technology constantly brings new products to the consumers and with this comes the possibility of creating new contaminants. Perhaps a good example is the emerging industry producing nanoparticles. Although nanoparticles have always existed in the environment, only recently have people become concerned with their effect upon human health due to the increasing quantities being produced. The history of new products, e.g. asbestos, shows that sometimes whether one product represents a threat to human health, or commodity, depends only upon our level of understanding about the product properties. This indicates that technological development will constantly demand research in the field of air pollution in order to better understand and prevent, or bring to acceptable levels, new pollution sources. The process of defining the acceptable levels is constantly demanding new research to understand better the impact of long term exposure to various pollutants, or mixtures where separate components can have synergistic effects. The improved knowledge on the effect of pollutants on human health forces periodic review of the regulations for air quality and emissions. Further research for improving, monitoring and detection technology is required in order to be able to verify that the current regulations for air quality are satisfied, and to identify areas where further improvements are required.

The papers in this volume address a wide range of topics including air management and policy, aerosols and particles, air pollution effects and environmental health, air pollution modelling, emission studies, global and regional studies, monitoring and measuring and pollution effects and reduction.

The Editors are grateful to all the participants for the quality of their contributions as well as the eminent members of the International Scientific Advisory Committee and other colleagues who reviewed the papers published in this volume.

The Editors
Tallinn, 2009

Contents

Section 4: Monitoring and measuring

Section 5: Aerosols and particles

Section 6: Air pollution effects and environmental health

Section 1
Air pollution modelling

Air quality in street canyons: a case study

F. Patania[1], A. Gagliano[1], F. Nocera[2] & A. Galesi[1]
[1]*Energy and Environment Division of D.I.I.M.,*
Engineering Faculty of University of Catania, Italy
[2]*Department of Analysis, Representation and*
Project in Mediterranean Areas,
Architectural Faculty of University of Catania, Italy

Abstract

Keeping the air quality acceptable has become an important task for decision makers as well as for non-governmental organizations. Particulate and gaseous emissions of pollutants from auto-exhausts are responsible for rising discomfort, increasing airway diseases, decreasing productivity and the deterioration of artistic and cultural patrimony in urban centers. Air quality limit values, which are aimed at protecting public health, are frequently exceeded especially in streets and other urban hotspots. Within these streets, pedestrians, cyclists, drivers and residents are likely to be exposed to pollutant concentrations exceeding current air quality standards.

In order to give the right support to decision makers for air pollution control, a suitable microscale dispersion model must be used to investigate phenomenon The paper presents the results obtained by utilizing a three dimensional numerical model based on Reynolds-averaged Navier–Stokes equations to simulate the fluid-flow development and pollutant dispersion within an isolated street canyon. Finally, the authors tested the reliability of the same code examined resemblances and differences between the measured data coming from a survey measurement within the canyon and the data coming from the code.
Keywords: urban canyon, air pollution, traffic emissions, CFD.

1 Introduction

Despite significant improvements in fuel and engine technology, urban environments are mostly dominated by traffic emissions. Most of the substances

WIT Transactions on Ecology and the Environment, Vol 123, © 2009 WIT Press
www.witpress.com, ISSN 1743-3541 (on-line)
doi:10.2495/AIR090011

directly emitted by vehicles in the atmospheric or indirectly produced through photochemical reactions could represent a serious hazard for human health.

In urban environments and especially in those areas where population and traffic density are relatively high, human exposure to hazardous substances is expected to be significantly increased [1]. This is often the case near busy traffic axis in city centres where urban topography and microclimate may contribute to the condition of air stagnation rising the contamination hotspots. High pollution levels have been observed in "Urban Canyons", which is a term frequently used for urban streets bounded by buildings on both sides [2].

The dimensions of an urban canyon are usually expressed by its aspect ratio [3, 4], which is the height (H) of the canyon divided by the width (W). A canyon might be called regular, if it has an aspect ratio approximately equal to 1 and not big openings along the walls. An avenue canyon may have an aspect ratio below 0.5, while a value of 2 may be representative of a deep canyon. Finally, the length (L) of the canyon usually expresses the road distance between two major intersections, subdividing street canyons into short ($L/H\approx3$), medium ($L/H\approx5$); and long canyons ($L/H\approx7$). Urban Canyons might be also classified in symmetric canyons if the buildings bounding the street have approximately the same height. On the contrary they may be classified asymmetric, if there are significant differences in the building heights forming the canyon.

The study on wind flow and pollutant transport inside and over urban street canyons, have attracted great concern during the past two decades mainly due to increasing urban pollutants and their adverse impacts on human health.

Field measurements [5], laboratory-scale physical modelling [6] and computational fluid dynamics (CFD) techniques [7] are the common tools used to study the wind flow and pollutant distributions in street canyons.

A three-dimensional numerical model based on Reynolds-averaged Navier–Stokes equations coupled $k–\varepsilon$ turbulence models was developed to simulate the fluid-flow development and pollutant dispersion within an isolated street canyon using the FLUENT code.

The field experiment used to validate the model outputs is described in Section 2 followed by the model descriptions and results in Section 3 and 4. The flow and concentrations predicted by the FLUENT code are then compared with the measurements from the field experiment in Section 4 along with a discussion of the discrepancies.

2 Experiment

2.1 Canyon geometry and sampling methods

The Vincenzo Giuffrida Street canyon (Fig. 1) is located in a residential part of Catania city. Regular net of perpendicularly intersecting street canyons forms this urban area. Information about geometry of the urban canyon was obtained from Catania Council

The studied canyon has a NNW orientation and its main axis is 345 degrees from the North. The canyon is 250 m length, 10 m width and the average height of the buildings is 21 m. The Balconies facades were taken into account in the model because they significantly influence airflow in the façade boundary layer. The authors, also, considered the roof of the buildings flat.

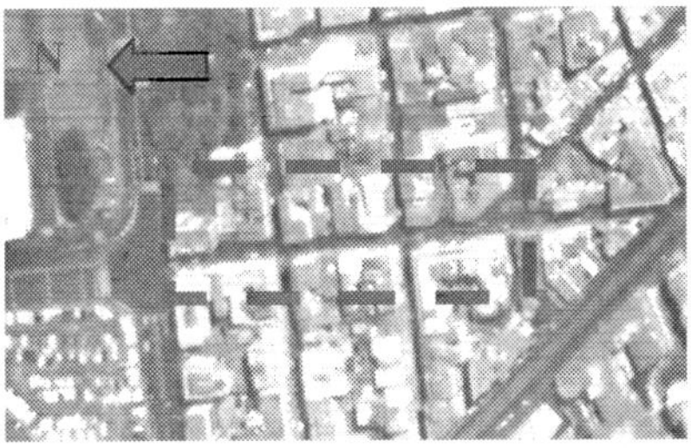

Figure 1: Area of the case study.

The canyon is modelled without trees and parked cars. Undisturbed wind velocity and wind directions were obtained by an anemometer located on the rooftop at a height of 25 m above ground level. Measured wind conditions were used as a parameter input for numerical predictions. A portable anemometer Testo 452, located on 1th floor, was used to compare predicted and measured values.

The wind and turbulence measurements were averaged over a period of 15 minutes.

Carbon monoxide measurements were taken in the street using CO monitor based on the non-dispersive infrared method (NDIR). The carbon monoxide, in street canyons, is primarily produced by petrol engines and it is practically inert on short timescales. It is usually considered as a gas tracer of the dispersion of traffic-related pollutants.

2.2 Traffic flow measurement

The measurements recorded were 15-min average of 1-min sampled.

Traffic data was registered using a video camera. A video camera was mounted at the roadside near the north entrance of the street to record the flow of traffic during the rush hours (7:30–8:30, 10:00–11:00, 13:00–14:00 and 18:00–19:00). The tape was then replayed on the laboratory television to count the number of each vehicles. Vehicles were classified as vehicles (V), heavy vehicles (HV) and motorcycles (MC).

3 Model description

3.1 Governing equation

The numerical model used in this study is Airpak which is a three-dimensional, steady, k–ε turbulence model applied to investigate the high Reynolds number skimming flow field of an urban canyon. The choice comes from the capability of Airpak to deal with complex shaped walls and other boundary conditions using flexible fine-scale grids. Airpak is a CFD modelling based on the numerical solution of the governing fluid flow and dispersion equations which are derived from basic conservation and transport principles: the mass conservation (continuity) equation, the three momentum conservation (Navier–Stokes) equations in x; y; z; the transport equation for pollutant concentration. The equations of state (obtained through the thermodynamic equilibrium assumption) and the Newtonian model of viscous stresses are also enlisted to close the system numerically, while initial and boundary conditions have to be specified to characterize external wind flows by the wind direction, wind speed, its height above the ground, and description of the local terrain.

The code solves the Navier-Stokes equations for transport of mass, momentum, species, and energy when it calculates laminar flow with heat transfer. Additional transport equations are solved when the flow is turbulent or when radioactive heat transfer is included.

The equation for conservation of mass, or continuity equation, can be written as:

$$\frac{\partial \rho}{\partial t} + \frac{\partial \rho}{\partial x_i}(\rho \cdot u_i) = 0 \tag{1}$$

Transport of momentum in the ith direction in an inertial (non-accelerating) reference frame is described by

$$\frac{\partial}{\partial t}(\rho \cdot u_i) + \frac{\partial}{\partial x_j}(\rho \cdot u_i \cdot u_j) = \frac{\partial p}{\partial x_i} + \frac{\partial \tau_{ij}}{\partial x_j} + \rho \cdot g_i + F_i \tag{2}$$

where p is the static pressure, τ_{ij} is the stress tensor, and ρg_i is the gravitational body force in the i direction. F_i contains other source terms that may arise from resistances, sources, etc. The energy conservation equation for a fluid region written in terms of sensible enthalpy h is:

$$\frac{\partial}{\partial t}(\rho \cdot h) + \frac{\partial}{\partial x_i}(\rho \cdot u_i \cdot h) = \frac{\partial}{\partial x_i}(k + k_t) \cdot \frac{\partial T}{\partial x_i} + S_h \tag{3}$$

where k is the molecular conductivity, k_t is the conductivity due to turbulent transport and the source term S_h includes any volumetric heat sources you have defined. The species transport equations takes the following general form

$$\frac{\partial}{\partial t}(\rho \cdot m_{i'}) + \frac{\partial}{\partial x_i}(\rho \cdot u_i \cdot m_{i'}) = -\frac{\partial}{\partial x_i} J_{i'i} + S_{i'} \tag{4}$$

where Si' is the rate of creation by addition from user-defined sources.

The code used the standard k-ε model based on model transport equations for the turbulent kinetic energy (k) and its dissipation rate (ε). The model transport equation for k is derived from the exact equation, while the model transport equation for ε is obtained using physical reasoning and bears little resemblance to its mathematically exact counterpart. The turbulent kinetic energy, *k*, and its rate of dissipation, ε, are obtained from the following transport equations:

$$\rho \frac{Dk}{Dt} = \frac{\partial}{\partial x_i}\left[\left(\mu + \frac{\mu_t}{\sigma_k}\right)\frac{\partial k}{\partial x_i}\right] + G_k + G_b - \rho\varepsilon \tag{5}$$

$$\rho\frac{D\varepsilon}{Dt}=\frac{\partial}{\partial x_i}\left[\left(\mu+\frac{\mu_t}{\sigma_\varepsilon}\right)\frac{\partial\varepsilon}{\partial x_i}\right]+C_{1\varepsilon}\cdot\frac{\varepsilon}{k}\cdot\left(G_k+C_{3\varepsilon}G_b\right)-C_{2\varepsilon}\rho\frac{\varepsilon^2}{k} \quad (6)$$

In these equations, G_k represents the generation of turbulent kinetic energy due to the mean velocity gradients, G_b is the generation of turbulent kinetic energy due to buoyancy,$C_{1\varepsilon}$,$C_{2\varepsilon}$, and,$C_{3\varepsilon}$ are constants. σ_k and σ_ε are the turbulent Prandtl numbers for k and ε respectively The "eddy" or turbulent viscosity, μ_t, is computed by combining k and ε.

$$\mu_t=\rho C_\mu\frac{k^2}{\varepsilon} \quad (7)$$

The model constants have the following default values $C_{1\varepsilon}$ =1,44, $C_{2\varepsilon}$=1.92; C_μ=0.09 σ_ε=1.3.and σ_k=1.0.The degree to which ε is affected by the buoyancy is determined by the constant $C_{3\varepsilon}$

$$C_{3\varepsilon}=\tanh\left|\frac{v}{u}\right| \quad (8)$$

where v is the component of the flow velocity parallel to the gravitational vector and u is the component of the flow velocity perpendicular to the gravitational vector. In this way, $C_{3\varepsilon}$ will become 1 for buoyant shear layers for which the main flow direction is aligned with the direction of gravity. For buoyant shear layers that are perpendicular to the gravitational vector, $C_{3\varepsilon}$ will become zero.

3.2 Estimation of pollutant emission rate by traffic

The rate of emission q_{ik} by traffic in the k_{th} lane, of species i, is given by:

$$q_{ik}(t)=\frac{EF_{ik}(t)\cdot N_k(t)}{A_k\cdot 1000} \quad (9)$$

where EF_{ik} is the emission factor of pollutant i, and N_k is the average traffic flow rate (or number of vehicles per unit time); the subscript k refers to the kth lane. N_k is determined from measurements (see figure 2), so q_{ik} can be evaluated once EF_{ik} is known utilizing COPERT III methodology [8, 9].

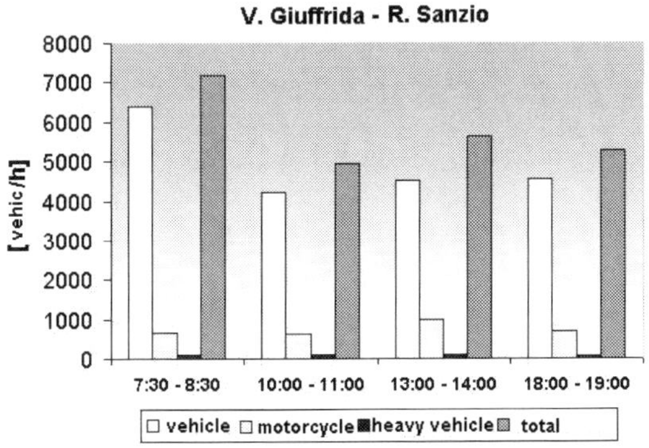

Figure 2: Traffic flow.

A_k (=h_k w_k) in the above equation is the cross-sectional area of the kth line source, where h_k and w_k are the height and bandwidth of the line source. The typical values of h_l=1.2m and w_l=1.9 m for a passenger car.

Finally the continuous linear source, typical of vehicular emission in stationary condition, was schematized as twenty-eight lined up point sources (0,20 m above the ground) uniformly distributed along the axis of the canyon.

The traffic emissions on the streets outside the street canyon were not considered.

3.3 Model domain and boundary conditions

As shown in Fig. 7 the chosen domain of calculus, in Cartesian coordinates, has the following dimensions: x=700 m, y=100 m and z=700 m

In modelling urban flow and dispersion, smaller grid sizes would be better located near buildings to solve flow and dispersion fields, but away from buildings larger grid sizes are allowed. To make the CFD model efficient for a given computing resource, a non-uniform grid system is implemented in the model.

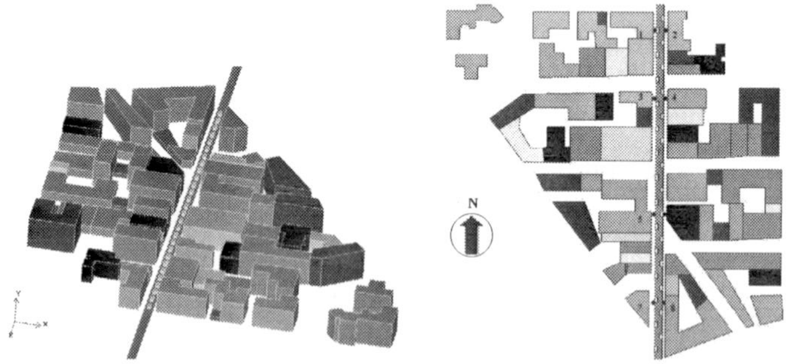

Figure 3: Buildings in 3D Domain (on the right) and plan of the case studied (on the left).

The grid of calculus becomes thicker and thicker inside the urban canyon and close to the ground level. The maximum dimension of the grid is x=25 m, y=10 m and z=25 m on the Atmospheric Boundary Layer (ABL), while the minimum dimension of the mesh is x=y=z=0,01m on the exhausts pipe. Accurate simulation of ABL flow in the computational domain is imperative to obtain accurate and reliable predictions of the related atmospheric processes.

The external wind flows was characterized by the wind direction, wind speed, its height above the ground, and a description of the local terrain using the following equations:

$$U(h) = \begin{cases} U_{met}\left(\dfrac{d_{met}}{H_{met}}\right)^{a_{met}} \cdot \left(\dfrac{h}{d}\right)^{a} \rightarrow h < d \\ U_{met}\left(\dfrac{d_{met}}{H_{met}}\right)^{a_{met}} \rightarrow h \geq d \end{cases} \qquad (10)$$

where U_{met} is the wind speed measured from a nearby meteorological station, H_{met} is the anemometer height, a_{met} is the terrain factor for the meteorological station, d_{met} is the boundary layer thickness at the meteorological station, a is the terrain factor for desired location, d is the boundary layer thickness at desired location.

Moving vehicles intensify both micro and large scale mixing processes in the environment, by inducing turbulence and enhancing advection, by entraining masses of air in the direction of the vehicle motion [11]. So traffic is taken into account in the modelling with a moving layer below the object "car" that is a block used to simulate the vehicle.

Different wind speed scenarios were taken into account (Table 1).

Table 1: Meteorological input data.

Simulation number	1	2	3	4	5	6	7
Wind direction	East	East	East	S-E	S-E	S-E	South
Wind velocity (m/s)	1,0	2,5	4,0	1,0	2,5	4,0	4,0

4 Results and discussion

4.1 Measurements

The CO concentration was measured by CO monitor based on the non-dispersive infrared method (NDIR) with a detection range of 0– 50 ppm and with detection limit of 0.5 ppm.

Fifteen-min average of 1 minute samples (reported at the end of each time period) of CO concentrations were measured in n.8 different points within the street at the height of 3.00 m, allowing an investigation of the spatial variability of the CO concentrations along a street canyon.

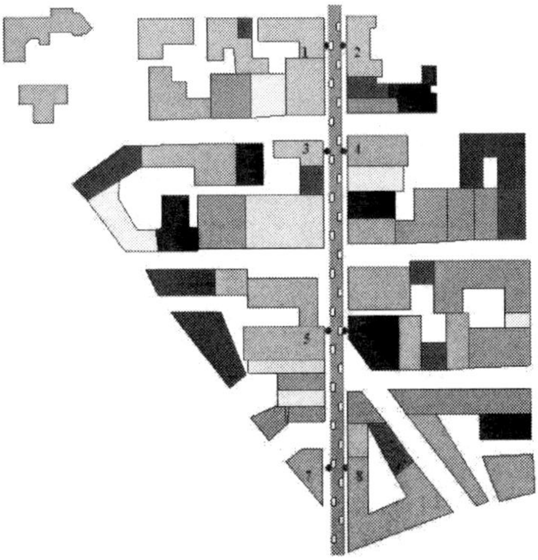

Figure 4: Points of measurement.

The accuracy of the CO monitor sensor was verified by a calibration with a gas containing 5 ppm of CO.

The turbulent mixing of the in-canyon volume of air, implies that any two measurements of a pollutant conducted at close proximity to each other in the street, can differ significantly.

The variables measured, in each point, were: CO maximum, minimum, average; wind speed, maximum, minimum, average; relative humidity, maximum, minimum; temperature, maximum, minimum, average.

In point 5 and 6, CO concentration was measured at 3 m, 6 m, and 9 m above ground.

The distribution of the CO concentration is not homogeneous in the street; the pollutant concentration is higher at the lower levels of the street than at the higher level. Moreover, in the lower part of the street we can observe a higher concentration on the leeward side than on the windward side of the street.

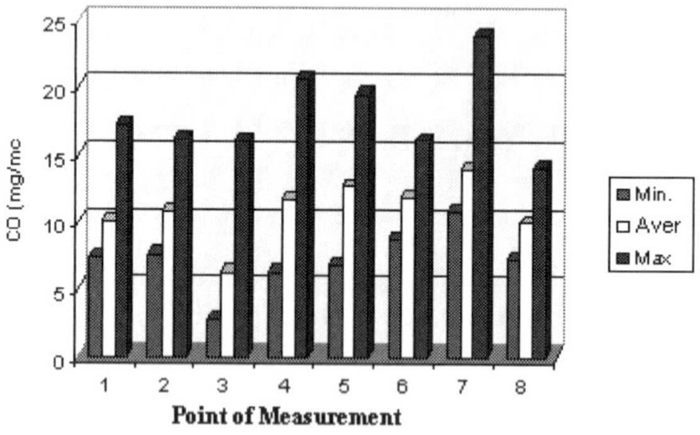

Figure 5: CO concentration of 12 January 2005 from 7:30 to 8:30.

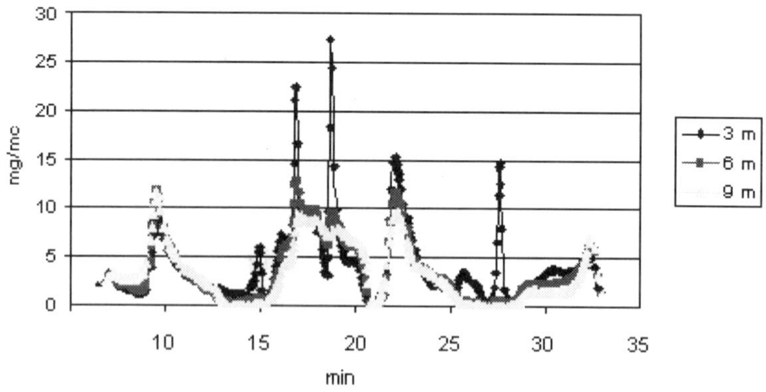

Figure 6: Evolution of CO concentrations in the windward side (point 6) during the 12 January 2005 from 08:00 to 8:30. The mean direction for the period was 82° from the street axis and velocity 2.5 m/s.

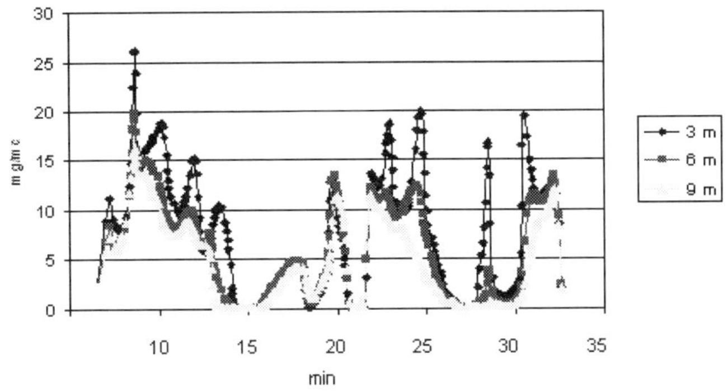

Figure 7: Evolution of CO concentrations in the leeward side (point 5) during the 12 January 2005 from 08:00 to 8:30. The mean direction for the period was 82° from the street axis and velocity 2.5 m/s.

4.2 Simulated velocity field and concentration

In the case of direction of wind perpendicular to the canyon axis, the authors have noticed, according to the theoretical studies:

- The airflow in the canyon can be seen as a secondary circulation feature. If the wind speed out of the canyon is below 2,5 m/sec, the coupling between the upper and secondary flow is lost and the relation between the speed of the wind above the roof and within the canyon is characterized by a considerable scattering.
- Regarding the direction of the vortex, it has to be expected that, as the vortex is driven by a downward transfer of momentum across the roof – level shear zone, the orthogonal flow to the canyon axis create a vortex with the air near the ground flowing opposite to the wind direction outside the canyon.

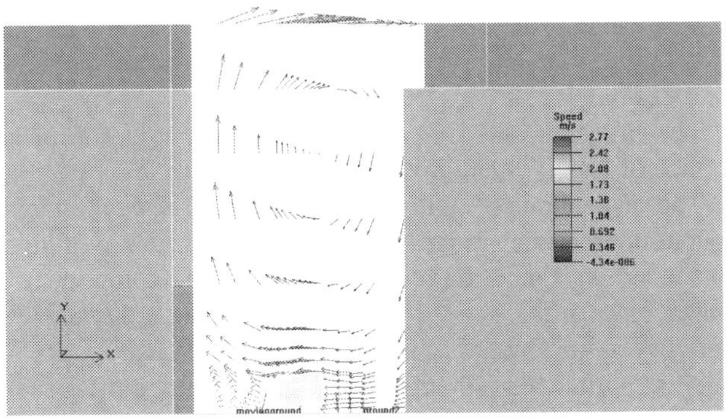

Figure 8: Flow pattern – Pz=60 m - Nz=1 – v=1,0 m/sec ⊥ Canyon.

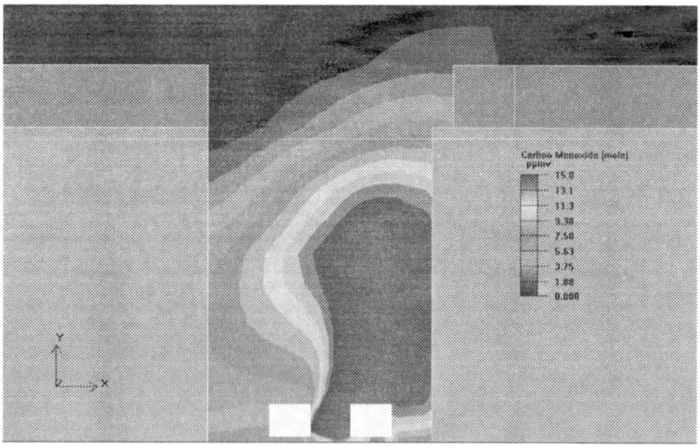

Figure 9: Exemplum of CO iso-concentration Pz=60 m – Nz=1 – v=1,0 m/sec ⊥ canyon.

- The authors have found that high wind speeds (>2,5 m/sec) produce transverse and longitudinal stable whirls between buildings façades.
- The transverse stable whirls are originated by the impact of vertical component of the wind on the windward building façades while the longitudinal stable whirls are originated both by the buildings with different heights, flanking the street, and by transversal streets converging in the canyon, which create vortex circulation.
- In this case the effects of the finite canyon length play an important role on the airflow distribution in the canyon. As a matter of fact the Authors have found intermittent vortices in the buildings corner. These vortices are responsible for the mechanism of advection from the building corner to the mid block creating a convergence zone in the mid block region of the canyon.
- Vertical velocities in the centre of the vortex of the canyon have been measured to be close to zero.

Moreover, the authors have noticed:

- An evident variation of the direction of velocity near the intersection with transversal streets that allow the inlet of air flow coming from zones outside the canyon.
- The biggest value of pollutants concentration forecasted close to the face of leeward buildings near the jets positions.
- The dispersions of pollutant gases inside the canyon very small both in longitudinal and in cross directions. As far as longitudinal direction is concerned, this is due to a very negligible component of wind vector, whereas, in the cross direction, the airflow motion forces the pollution between the facades of the building on both side.
- A very little component of wind velocity along the axis of canyon.
- An accumulation of polluting concentrations on leeward side of the building caused by buildings geometry, vortex circulation, transversal street and courtyards that supports the formation of depression zones.

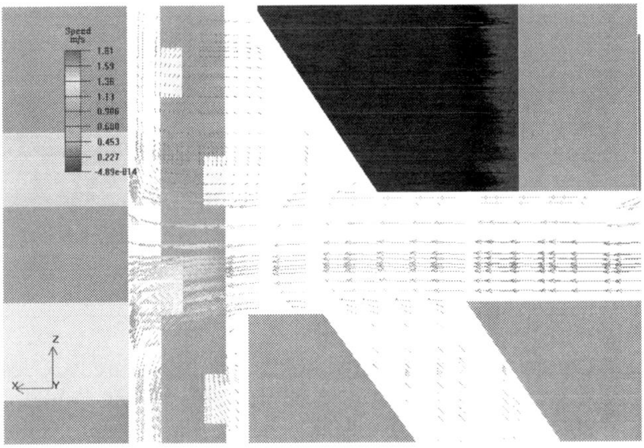

Figure 10: Exemplum of flow pattern - Py=3,00 m – Ny=1 – v=2,5 m/sec.

- Values of concentrations between 510 and 50 mg/m^3 of CO.
- The concentration of CO for leeward façades is twice as high as for the windward façades.

4.3 Comparison between 3D simulations and measurements

Both the above plots reveal that the predicted values generally follow the hourly trend of the measurements, related mainly to variations in traffic flow rate, and the pollutant concentrations on the leeward sides are higher than those on the windward sides. Whatever, the differences of average values of CO concentrations between simulations and measurements are 10–15%. However, the predictions are quite close to the field measurements, although there are some differences the differences may follow from unsteadiness and large fluctuations, which, related to either traffic or approaching wind in actual environments, were not considered in the quasi-steady-state model.

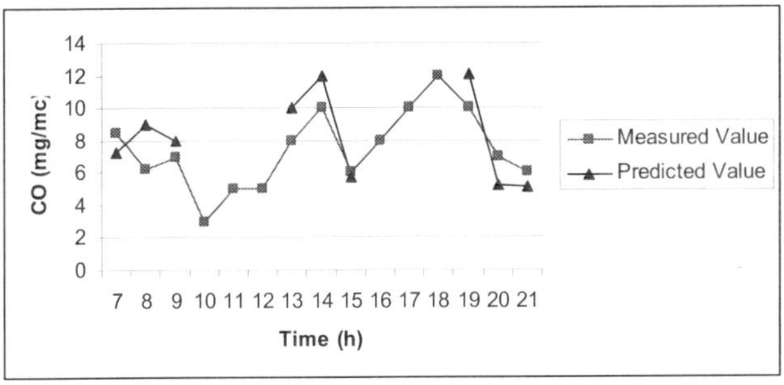

Figure 11: Comparisons between simulated and measured CO concentrations at a sampling height of 3.00 m in the windward side on 20 January 2005 from 07:00 to 21:00- the mean direction for the period was 85° from the street axis and velocity 2.1 m/s.

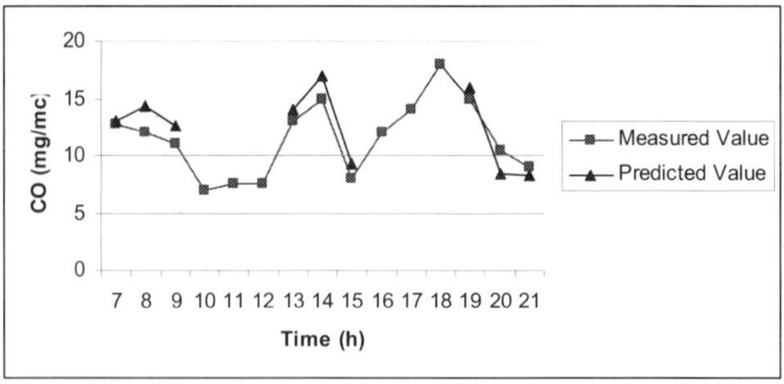

Figure 12: Comparisons between simulated and measured CO concentrations at a sampling height of 3.00 m in the leeward side on 20 January 2005 from 07:00 to 21:00 – the mean direction for the period was 85° from the street axis and velocity 2.1 m/s.

5 Conclusions

The study shows that the model predictions of the spatial variability of any traffic-related pollutant in a canyon with adjoining side streets should be used to understand the complex flow regimes and turbulence structures that develop under certain background wind directions. This has special relevance to predicting the relative exposure of people to pollutants within street canyons.

However the experience obtained by this study shows that comparison of predicted and experimental results is difficult due to very unsteady wind conditions during field measurements. Experimental measurement are a necessary completion as a validation tool for predictions that has a very serious limitation due to unstable meteorological conditions that mostly prevail in a real situation.

References

[1] Croxford, B., Penn, A., Hillier, B., 1996. Spatial distribution of urban pollution: civilising urban traffic. Science of the Total Environment 189/190, 3–9.

[2] Sotiris Vardoulakis, Bernard E.A. Fisher, Koulis Pericleous, Norbert Gonzales-Flesca., 2003. Modelling air quality in street canyons: a review, Atmospheric Environment Vol 37, 155-182

[3] Nakamura, Y., Oke, T.R., 1988. Wind, temperature and stability conditions in an east–west oriented urban canyon. Atmospheric Environment 22, 2691–2700

[4] Oke, T.R., 1988. Street design and urban canopy lay and climate. Energy and Buildings 11, 103–113

[5] Hoydysh, W.G., Dabberdt, W.F., 1994. Concentration fields at urban intersections: fluid modelling studies. Atmospheric Environment 28 (11), 1849–1860.

[6] Kastner-Klein, P., Plate, E.J., 1999. Wind-tunnel study of concentration fields in street canyons. Atmospheric Environment 33, 3973–3979.

[7] Zannetti, P., 1990. Air Pollution Modeling: Theories, Computational Methods, and Available Software. Computational Mechanics Publications, Southampton, UK

[8] Ntziachristos, L., Samaras, Z., 1997. COPERT II: Computer Programme to Calculate Emissions from Road Transport—User's Manual. European Topic Centre on Air Emissions, EEA, Copenhagen, Denmark.

[9] Ntziachristos, L., Samaras, Z., 2000. COPERT III: Computer Programme to Calculate Emissions from Road Transport—Methodology and Emission Factors. European Topic Centre on Air Emissions, EEA, Copenhagen, Denmark

[10] AIRPAK 2.1 User Guide Inc. 2001

[11] Xian-Xiang Li, Chun-Ho Liu, Dennis Y.C. Leung, K.M. Lam "Recent progress in CFD modelling of wind field and pollutant transport in street canyons" Atmospheric Environment 40 (2006) 5640–5658

Use of CALPUFF and CAMx models in regional air quality planning: Italy case studies

C. Trozzi, S. Villa & E. Piscitello
Techne Consulting srl, Rome, Italy

Abstract

Applications of air pollution dispersion models CALPUFF and CAMx have been performed over regional Italy territories in order to evaluate transport, wet and dry deposition of coarse and fine particulates and formation of secondary fine particulates with special attention to nitrates and sulphates.

Two different geographical domains were chosen, Trento and Florence provinces, with horizontal cells the size of 1km x 1km, and meteorological variables were considered from meteorological models and local measurements as well; emissions were taken from high resolution bottom-up emission inventories and the time scale was spanning an entire year on hourly basis.

Results show how the more accurate chemical model contained in CAMx performs better in secondary particulates formation, in particular regarding the sulphates, while CALPUFF seems slighlty better in predicting dispersion paths. Even if both models application suffers from lack of background concentrations of particulate matter, the CALPUFF model produces more accurate hourly concentration values in the single cell, making this model a better choice between the two in total PM evaluation for this particular case study. Correlation between CAMx and CALPUFF predicted data is also strong.

1 Introduction

The best dispersion model choice in a regional air pollution study is often a very debated topic, since every model has his strong and weak spots over a wide variety of characteristics needed by this kind of study.

We choose to compare Calpuff [1] and CAMx [2] models over two very different geographic domains in terms of geomorphology: the Trento province

WIT Transactions on Ecology and the Environment, Vol 123, © 2009 WIT Press
www.witpress.com, ISSN 1743-3541 (on-line)
doi:10.2495/AIR090021

presents a typical alpine or subalpine scenario, with a lot of canyons and high peaks, thus most of human activities, and so pollution, carried out in valleys; the Florence province, on the other hand, presents wider flat territories, gentle hills and sub mountainous terrain only in the northeastern corner, thus with a more spread pollution production in terms of geography.

2 Characteristics of simulations

For both studies and both models, we used a 1 Km wide cells grid covering the entire province, surface weather data from regional weather stations networks, upper air data from application of MM5 [3] and CALMET [4] weather forecast models and emission data from high resolution bottom-up managed by APEX system [5] over a time period of several years. Spatial and temporal subdivision of emissions data was the same for both models and was carried out by appropriate preprocessors [5]. Both models were run over a time period of a year on an hourly basis.

3 Models calibration

Calibration of models were performed over three different kinds of sampling stations; we choose an urban traffic, an urban background and a rural background sampling station for every province; in this way we can see how models perform on those kinds of territories, the results (in monthly mean for January, but results are similar for every month) show clearly (Table 1) a lack of background particulates concentration that could be filled e.g. with accounting of transboundary particulates fluxes from a continental scale model [6].

Since we did not have hourly data of background particulates, the statistical indexes like fractional bias and normalized gross error, calculated on hourly basis, are not shown for PM_{10}. Statistical indexes are calculated for every model in respect to measured data from sampling stations.

In Table 2 computed statistical indexes [7] are reported for selected pollutants and networks. Calibration results show that CAMx model in Florence province performs better than CALPUFF as long as we move from traffic to rural station. This behaviour is opposite in Trento province, especially in terms of fractional bias. This kind of behaviour can be ascribed to the different nature of the models, the lagrangian CALPUFF seems to better reconstruct paths of pollutants in canyons and to overestimate concentrations in urban zones in some cases.

4 Results for particulates

Both models can treat secondary particulate formation and transport, though CALPUFF chemical internal module MESOPUFF II is less complex than CBIV

Table 1: Calibration of models: comparison of measured and computed average concentrations (μg/m³) for selected sampling station.

Province	Station	Measure	NO_2 Calpuff	CAMx
	Urban traffic	101.85	72.25	42.94
Florence	Urban background	42.50	65.17	35.58
	Rural background	19.41	32.29	18.92
	Urban traffic	58.47	28.23	28.18
Trento	Urban background	51.79	29.40	17.95
	Rural background	2.23	2.62	5.54
			SO_2	
	Urban traffic	4.88	6.92	4.76
Florence	Urban background	3.48	5.02	2.53
	Rural background	N/A	2.56	1.00
	Urban traffic	5.65	4.77	6.97
Trento	Urban background	N/A	4.34	3.54
	Rural background	N/A	0.67	0.96
			PM_{10}	
	Urban traffic	42.53	12.09	6.64
Florence	Urban background	29.43	10.73	4.47
	Rural background	N/A	6.96	2.16
	Urban traffic	34.31	4.99	2.84
Trento	Urban background	44.45	4.36	1.69
	Rural background	N/A	0.88	0.32

chemical model contained in CAMx. The former has only capabilities to produce SO_4 and NO_3 aggregates while the latter can treat in addition organic aerosol and elemental carbon as well.

The following graphs will show comparison between the two models in terms of hourly concentrations calculated over the urban traffic stations (Figures 1 and 2), urban background stations (Figures 3 and 4), rural station (Figures 5 and 6) respectively for PM_{10} and secondary particulates expressed as sum of all subspecies produced and treated by the models.

Correlation between the two hourly series produced by the models for every kind of cell is shown in Table 3.

From these results, it is clear that, given the same methodology in terms of meteorology and emissions, the two models perform very different when two geographical domains are of so different nature. While correlation decreases in both cases as long as we move from traffic to rural stations, that quantity shows poor values, below 60%, for Trento province. In addition, concentrations of particulates given by CALPUFF are always higher than those from CAMx in terms of monthly means; thus, regardless of available background concentrations of particulates, the CALPUFF model shows the best performance in total PM determination.

Table 2: Calibration of models: Statistical indexes in selected sampling station.

Province	Station	Fractional Bias		Normalized gross error	
		NO_2		NO_2	
		CALPUFF	CAMx	CALPUFF	CAMx
Florence	Urban traffic	-0.34	-0.81	0.70	0.59
	Urban background	0.42	-0.18	1.14	0.46
	Rural background	0.50	-0.03	1.35	0.71
Trento	Urban traffic	-0.70	-0.70	0.62	0.58
	Urban background	-0.55	-0.97	0.64	0.75
	Rural background	0.16	0.85	1.35	1.65
		SO_2		SO_2	
Florence	Urban traffic	0.35	-0.02	1.11	0.77
	Urban background	0.36	-0.32	1.39	0.84
	Rural background	N/A	N/A	N/A	N/A
Trento	Urban traffic	-0.17	0.21	0.72	0.83
	Urban background	N/A	N/A	N/A	N/A
	Rural background	N/A	N/A	N/A	N/A

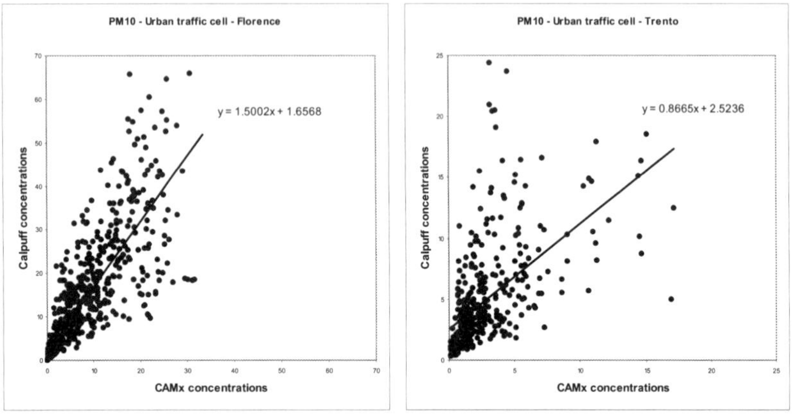

Figure 1: Models comparison: hourly PM_{10} concentrations – urban traffic cell.

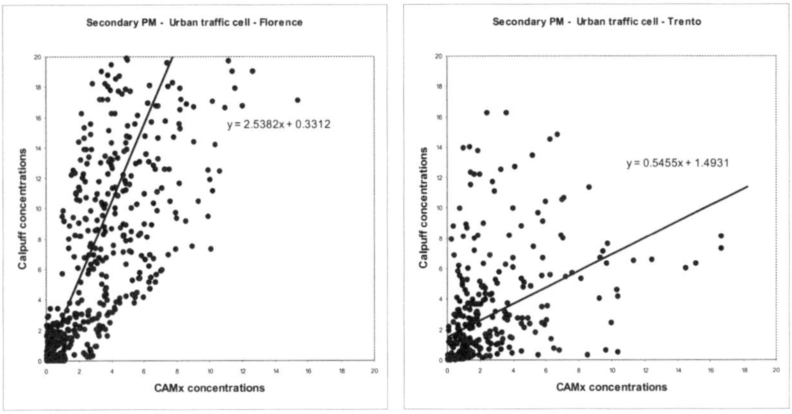

Figure 2: Models comparison: hourly secondary particulates concentrations – urban traffic cell.

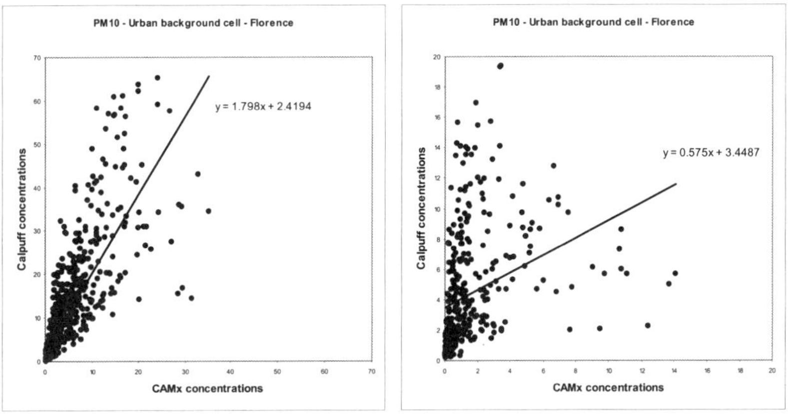

Figure 3: Models comparison: hourly PM_{10} concentrations – urban background cell.

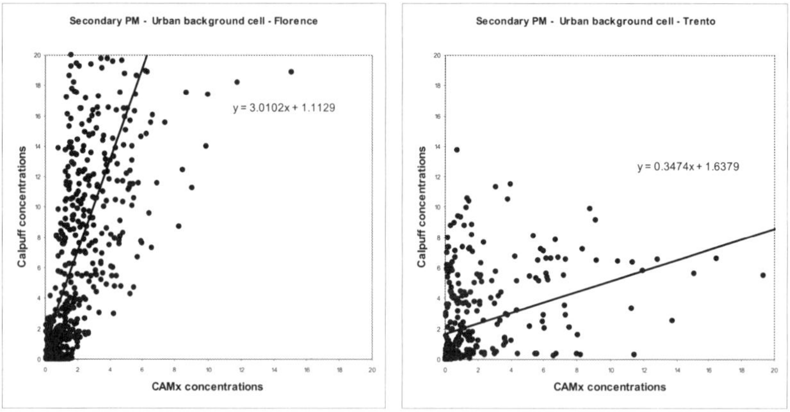

Figure 4: Models comparison: hourly secondary particulates concentrations – urban background cell.

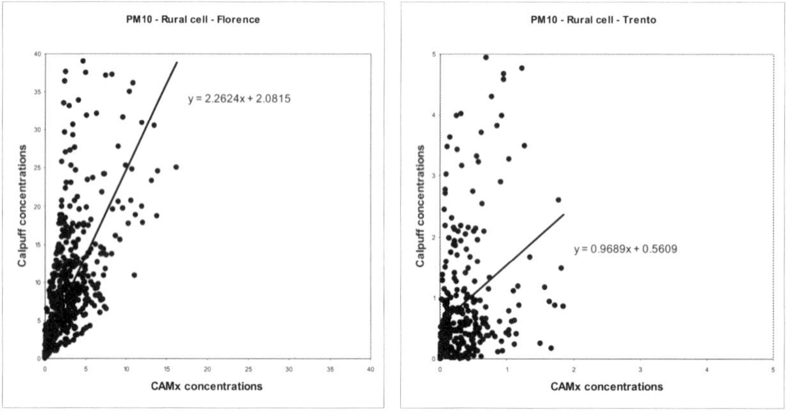

Figure 5: Models comparison: hourly PM_{10} concentrations – rural cell.

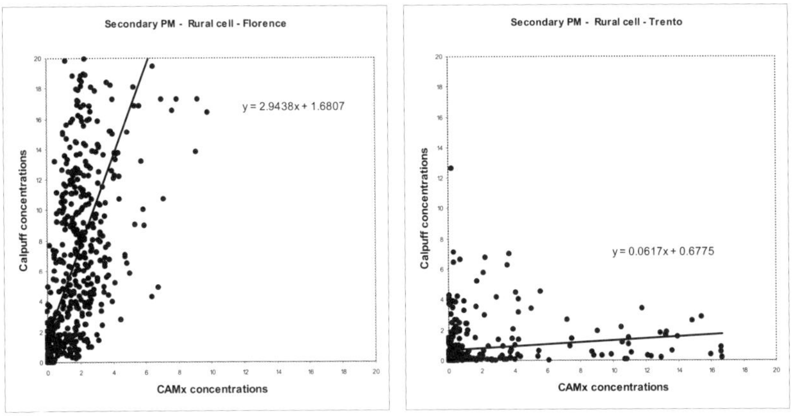

Figure 6: Models comparison: hourly secondary particulates concentrations – rural cell.

Table 3: Models comparison: correlation of hourly PM_{10} and secondary particulates concentrations.

		PM_{10}	Sec. PM
	Urban traffic	0.83	0.78
Florence	Urban background	0.78	0.73
	Rural background	0.71	0.64
	Urban traffic	0.57	0.43
Trento	Urban background	0.33	0.39
	Rural background	0.27	0.14

4.1 Nitrates and sulphates percentages in secondary PM

Trento province presented a higher share of sulphur dioxides emissions than Florence province; this resulted in a different composition of secondary PM for both CALPUFF and CAMx models, that showed higher percentages of sulphates for Trento province as listed in Table 4.

From Table 5, it is clear that CAMx model responds better to emission variation of sulphur dioxide, changing the percentages of sulphates in secondary PM accordingly; where CAMx shows even a variation in percentage of 63 points, like the case of rural stations, CALPUFF only vary a little.

CAMx seems on the other hand to underestimate nitrates, but since we didn't have measured data for secondary PM, it's hard to say what model is nearest to the real values. Surely the most complex chemical mechanism of CAMx can assure a wider variety of chemical species involved and so the possibility to go deeper in determination of secondary particulates formation and transport.

Table 4: Composition of secondary PM.

		Urban Traffic	Urban Background	Rural Background
			CAMx	
Florence	PEC	11.89%	11.80%	12.63%
	PNO_3	2.07%	1.76%	2.47%
	POA	62.69%	63.91%	58.67%
	PSO_4	23.35%	22.54%	26.22%
Trento	PEC	6.93%	4.29%	1.46%
	PNO_3	0.26%	0.14%	0.15%
	POA	38.05%	23.42%	8.45%
	PSO_4	54.76%	72.14%	89.94%
			CALPUFF	
Florence	PEC	N/A	N/A	N/A
	PNO_3	62.12%	62.12%	59.00%
	POA	N/A	N/A	N/A
	PSO_4	37.88%	37.88%	41.00%
Trento	PEC	N/A	N/A	N/A
	PNO_3	55.64%	50.31%	51.50%
	POA	N/A	N/A	N/A
	PSO_4	44.36%	49.69%	49.50%

PEC: Elemental Carbon Aerosol , PNO3: Nitrates Aerosol, POA: Organic Aerosol, PSO4: Sulphates Aerosol.

5 Conclusions

Comparison of the two models not only shows natural differences in pollutants concentrations due to the different approach in physics or chemical modelling, but even due to the kind of geographical domain of application. While a domain mainly composed by plains or gentle hills, like the Florence domain, keeps acceptable values of correlation between CALPUFF and CAMx calculated concentrations, a mountainous domain, with peaks and deep valleys like the Trento's one, does not.

In addition, while correlation decreases in both cases as long as we move from urban to rural cells, analysis of monthly means and statistical indexes shows that CAMx model in Florence province performs better than CALPUFF as long as we move from traffic to rural station. This behaviour is opposite in Trento province, especially in terms of fractional bias.

CAMx model responds better to variation in sulphur emissions in secondary sulphate particulates formation and transport, though the lack of measured data does not allow one to say more in that sense.

Further application of both models on very different geographical domains, accounting of background particles from a continental scale model, and sampling station network capable of measuring secondary particulates will surely contribute to go deeper in this kind of study, especially for particulates.

References

[1] Scire J.S., Yamartino R.J., Strimaitis D.G.: A user's guide for the CALPUFF dispersion model, TRC Companies
[2] Yarwood G. et al.: CAMx user's guide ver. 4.3, Environ Corporation
[3] Dudhia J., Gill D., Manning K.: PSU/NCAR Mesoscale Modeling System Tutorial Class Notes and User's Guide MM5 Modeling System Version 3
[4] Scire J.S., Insley E.M., Yamartino R.J., and Fernau M.E.: A User's Guide for the CALMET Meteorological Model, TRC Companies
[5] C. Trozzi, E. Piscitello, S. Giammarino, R. Vaccaro, Advanced Pollutant Emissions Computer System (APEX 4.0), 9th Int. Conf. Model. Monit. Manag. Envir. Probl. – Envirosoft 2002 – 6-8, May 2002 – Bergen (N), Organized by WIT
[6] EMEP Co-operative Programme for Monitoring and Evaluation of the Long-Range Transmission of Air Pollutants in Europe, Transboundary particulate matter in Europe, Status report 2006, Joint, CCC & MSC-W, Report 2006
[7] Zannetti P. Air pollution modeling. Theories, Computational Methods and Available Software Computational Mechanics Publ.: Southampton & Boston, 1990
[8] Koo, B., A.S. Ansari, and S.N. Pandis. 2003. Integrated Approaches to Modeling the Organic and Inorganic Atmospheric Aerosol Components. *Atmos. Environ., 37, 4757-4768*

Numerical simulation of particle air dispersion around the landfill

N. Samec, M. Hriberšek & J. Ravnik
Faculty of Mechanical Engineering, University of Maribor, Slovenia

Abstract

Paper presents numerical simulation of particles dispersion, lifting from a gypsum landfill under different weather conditions. Simulations consist of two parts: simulation of a long term impact of the particles on the surrounding area, performed by implementation of the Gauss dispersion model based computer code, and second, a CFD based simulation for assessing the flow and mass concentration fields in the vicinity of the landfill for several pre-selected flow cases. The GIS based topography is used for creating computational domain, and the ISC3 software for Gauss dispersion and the CFX5 software for CFD model are implemented. The results of both computational approaches are presented and compared. In the conclusions, a relation of the simulation results with existing environmental pollution levels is made.
Keywords: gypsum particles, gypsum landfill, Gauss dispersion model, computational fluid dynamics, wind induced particle transport.

1 Introduction

Dust clouds are one of the important environmental risks, especially, when they originate from the artificially made landfills of hazardous material. Since particles, forming dust, are generally very small, they are easily lifted from the landfill and transported into the surroundings. The precondition for lift off must be an adequate velocity field in the vicinity of the landfill surface. In the long term, the most important source of strong velocity field near the surface are winds, therefore position of a landfill must always be weighted by the wind rose, valid for the area under consideration.

In the case of the present study, the landfill under consideration was an existing wet landfill of gypsum. Gypsum landfills are frequently presenting

WIT Transactions on Ecology and the Environment, Vol 123, © 2009 WIT Press
www.witpress.com, ISSN 1743-3541 (on-line)
doi:10.2495/AIR090031

environmental risks, especially when gypsum is product of gas cleaning technology [1]. A wet landfill does not present any important environmental risk in terms of dust clouds; however, it does present a technical problem due to low landfill space efficiency. In order to increase the overall mass of deposited gypsum in the landfill, techniques of dry deposition have to be implemented.

Since the impact of the reconstruction on the environment can be experimentally verified only after the reconstruction is finished, a modelling approach to prediction of environmental hazards has to be taken [2]. In order to minimize the impact of the dry gypsum landfill on its surroundings, an extensive experimental and numerical study was carried out. The experimental part was used for determination of mass fluxes from the model landfill surface and for characterisation of basic particle properties. The experimental findings were used in prescribing boundary and initial conditions for numerical simulations of dust propagation.

The main goal of investigations, presented in the paper, was to develop numerical models, which would be able to predict the impact of the gypsum landfill on its environment in terms of solid particles dispersion into the atmosphere. Additionally, during the process, several monitoring stations will be set up, what will later allow a comparison of computed results with results under realistic environmental conditions, and the position of the stations will be determined based on the results of numerical simulations.

2 Particle characteristics

The data on structure and size distribution of gypsum particles is one of the most important physical parameters, needed in setting up the correct physical model for numerical simulation. The data was determined for different compositions of the wrung out and build in gypsum. The main cases were the summer composition, consisting of 1 to 1 ratio of both, the winter composition, consisting of fresh gypsum only, and the deposited (wet) gypsum only.

In all cases the test samples of gypsum were prepared by simulating the wringing out the gypsum, as will be performed on the landfill. The simulation was carried out on a laboratory vacuum filter device, able to produce samples with liquid content < 30%. The samples were then analysed using the Mastersizer 2000 (Mavern Instruments Ltd.) measuring device for the size distribution of particles, and additionally investigated under optical and electron microscope for the shape characterisation of particles.

The average density of particles, dried at 45°C and 0,9% wet, for four different samples, was 2.39 g/cm^3, which is close to the nominal density of gypsum, i.e. density of $CaSO_4$ x $2H_2O$ =2.317 g/cm^3.

The shape of the dried particles is presented in Fig.1, with a close-up of a typical sample. It can be concluded, that the general shape is a cylindrical and the ratio of the length to diameter is approximately 10. This data was later used in selecting an adequate empirical correlation for drag coefficient for a particle.

In order to account for the effect of particles of different sizes, the size spectrum was divided into nine size classes, presented in Table 1.

Figure 1: Dry gypsum particles under the electron microscope.

Table 1: Size distribution of gypsum particles after the mechanical loading experiment.

Particle equivalent diameter [um]	Mass fraction [%]
5	38.7
15	16.5
25	7.5
35	6.6
45	3.8
60	6.0
85	5.8
150	9.6
320	5.5

3 Numerical models

Numerical models form the core of modern engineering simulation tools. They range from complex models, incorporating governing physical phenomena of fluid flow and mass transfer on differential level, to simplified models, based on lumped parameter approach and with simplified solutions for fluid flow and mass transfer. In the case of modelling of dust clouds and deposition of particles complex models produce detailed spatial results, but can become computationally too demanding for simulating large time spans. The latter is the domain of simplified models, which however lack of spatial resolution.

In the present work we performed numerical simulation by using both types of numerical models, specifically:

1. Gauss dispersion model, incorporated in the numerical code ISC-ISCST3 [3–5],
2. Computational Fluid Dynamics model (CFD), in our case in the form of the numerical code ANSYS-CFX [6].

There exist several other models, which were already successfully applied in environmental modelling [7, 8], especially Lagrange models, incorporating lumped parameter approach with control volume movement according to predetermined flow field [9].

Regardless of the numerical model used, the results of numerical simulations always depend strongly on input data, in our case:

1. Particle characteristics: size (equivalent diameter), size distribution, specific weight of wet and dry particles, characteristic shape and corresponding coefficient of dynamic drag.
2. Mass flux of particles, entering the flow domain as a consequence of wind interaction with landfill surface.
3. Direction and magnitude of winds in the surroundings of the landfill, mostly on several years average basis.

4 Computational domain, boundary and initial conditions

4.1 The Gaussian plume model

The modelling was done using the ISCST3 model, developed by the EPA [3, 4]. The model encompassed 100 square kilometres abound the gypsum landfill. Northeastern corner was defined in Gauss-Krueger coordinate system at (5531500m, 5126000m) and the Southwestern corner at (5521092m, 5115633m). Thus the height of the model is 10367m and the width 10407m.

The model requires data on the landfill as well. We chose:

- Size of the open part of the landfill: 25 × 10 m,
- Landfill volume: 250 m^3
- Landfill surface: 250 m^2
- Particles are lifted from the bottom of the landfill.
- Coefficients describing rinsing of gypsum in rain and snow were both set to 0.0001 hr/mm s.

The meteorological data was gathered from the stations for the period between 1.5.2003 and 19.11.2005. The first is located at the landfill measuring the temperature, precipitation and wind direction and velocity. The second is located in the nearby town of Celje, where data on relative humidity, cloud cover, cloud height, sun radiation and air pressure was gathered. The data was written in the SAMSON (Solar and Meteorological Surface Observational Network) format and a RAMMET [5] meteorological pre-processor was used to prepare data for our model. RAMMET requires also additional terrain data, given in Table 2. The AERMIX model [5] was used to estimate the mixing heights.

Table 2: Terrain properties for meteorological data.

Data	Unit	Value
Anemometer height	M	2
Minimum Monin-Obukhov length	m	2
Surface roughness length (measurement site)	m	0.1
Surface roughness length (application site)	m	0.1
Noon-Time Albedo		0.18
Bowen ratio		0.8
Anthropogenic Heat Flux	W/m^2	43
Fraction of net radiation absorbed at the ground		0.15

4.2 The CFD model

The area under consideration encompassed 10.5 km times 10.5 km with the landfill located in the centre. Digital model of heights was used to model the terrain. The upper boundary of the domain was flat, set 2612 m above the lower point in the terrain. The side walls of the domain were vertical. Based on the geometrical model a computational grid was set up having 434,000 elements. The gird is condensed around the landfill and shown in Figure 2. Figure 3 shows the boundary conditions. Inflow of air was prescribed on two vertical walls and an outflow on the other two.

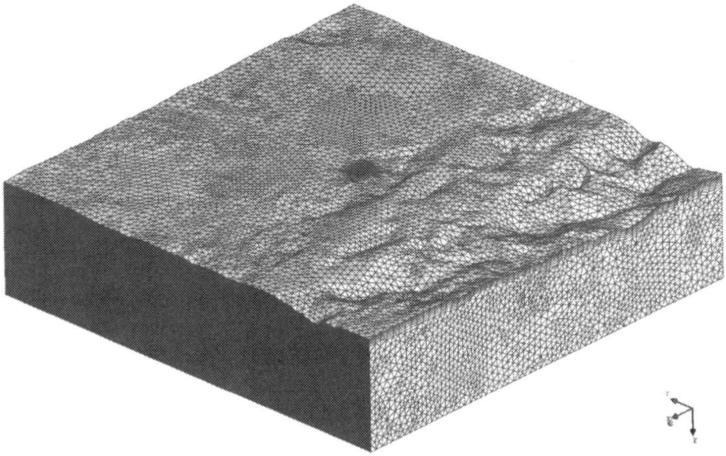

Figure 2: The grid used in the CFD model (view from the bottom).

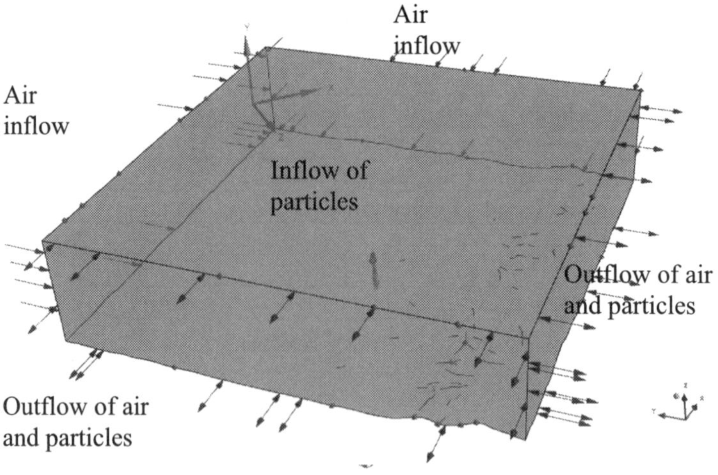

Figure 3: The boundary conditions of the CFD model.

4.3 Determination of inflow particle mass flux

An experiment was set up to determine the flux of particles leaving the landfill. We prepared a flat surface of dry gypsum. A heavy roller drove over the surface a few times to crumble the gypsum. An artificial wind of 1.0m/s – 1.2m/s was directed over the surface. Other conditions were constant (temperature, humidity). The results showed that only particles with sizes lower than 50 microns were lifted. Also, we noticed that the emission index rises substantially when gypsum is crumbled by heavy machinery. The emission index of installed gypsum is more than an order of magnitude smaller. The worst case scenario thus happens in dry conditions on the open part of the landfill, when heavy machinery crumbles the gypsum surface.

5 Results of numerical simulations

5.1 Gaussian plume model

The Gaussian plume model ISCST3 was run with meteorological data between May 2003 and November 2005. Gypsum with density 2210kg/m^3 was chosen. Emission index of installed gypsum 0.0021 g/s/m^2 and of mixed gypsum 0.0026 g/s/m^2 was chosen. The gypsum diameter/mass fractions distribution used in the model is listed in Table 1.

Figure 4 shows isolines of constant gypsum concentration in air, by showing the second worst daily averaged value in each individual node in the period May 2003 – November 2005. We observe that the concentration decreases almost in concentric circles around the landfill. The predominant South-Eastern wind direction is evident from the isolines of low concentration, i.e. 10-15 μg/m^3.

Figure 4: Isolines of concentration of gypsum in air, [$\mu g/m^3$], second highest value, 24 hour averages, density 2210 kg/m^3, emission flux: 0.0021 $g/s/m^2$.

Figure 5: Isolines of concentration of gypsum in air, [$\mu g/m^3$], second highest value, 24 hour averages, density 2210 kg/m^3, emission flux: 0.0026 $g/s/m^2$.

Figure 5 shows the results of simulation of mixed gypsum, where the measures emission flux was 0.0026 g/s/m^2. The results are almost identical to those obtained for installed gypsum.

We may conclude, that in the worst case scenario, we may expect 1 day in two and a half years when 750m away from the centre of the landfill the gypsum concentration in air will reach about 40 μg/m^3 and about 0.15g/m^2 of gypsum will be deposited on this day.

5.2 CFD model

We analysed dry non-stick gypsum with density of 2210 kg/m^3. We analysed three dominant wind directions;. For the 45° North-Eastern wind, Figures 6–8 show isosurfaces of instantaneous concentration of gypsum with 4.5 μm diameter for different values of concentrations.

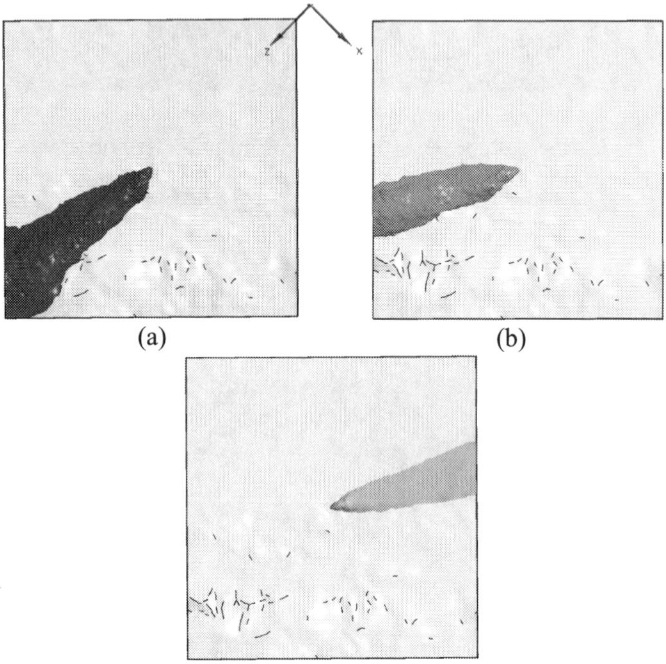

(a) (b)

(c)

Figure 6: Isosurface of constant concentration 1.0×10^{-5} μg/m^3, for particles with diameter of 4.5 μm. Density of gypsum 2210 kg/m^3, wind direction 45° NE, 67.5° ENE and 247.5° WSW, (a) to (c).

It is evident from the results that the hills in the South greatly affect the concentration and deposition fields. Although their presence was included in the Gaussian plume model as well, the CFD model shows their influence better. The high concentration of gypsum 4.5 μm diameter particles, 1.0×10^{-3} μg/m^3, is shown in Figure 7. It reveals that only a narrow area directly in the direction of

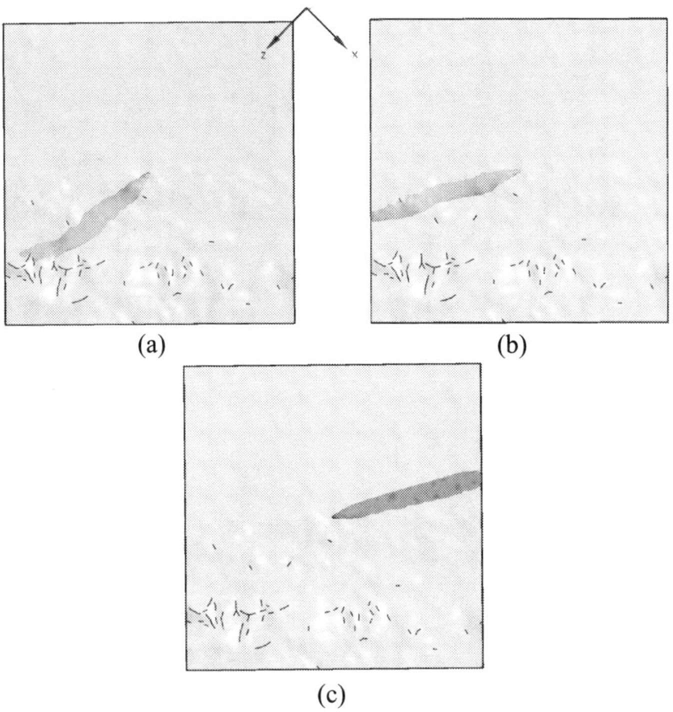

Figure 7: Isosurface of constant concentration 1.0×10^{-3} μg/m^3 for particles with diameter of 4.5 μm. Density of gypsum 2210 kg/m^3, wind direction 45° NE, 67.5° ENE and 247.5° WSW, (a) to (c).

the wind will have high concentration of gypsum in the air. Diffusion in directions perpendicular to the dominant wind direction is relatively small, which can be seen in the plots of isosurfaces of small concentration in Figure 6.

Maximal deposition rates for the computed test cases were:

- Wind 45° NE: $6{,}386 \times 10^{-3}$ μg/m^{2}s,
- Wind 67.5° ENE: $7{,}55 \times 10 \times 10^{-3}$ μg/m^{2}s,
- Wind 247.5° WSW: $7{,}099 \times 10^{-3}$ μg/m^{2}s.

which qualitatively corresponds well with the results of Gaussian model.

A quantitative comparison is directly not possible, as the CFD model gives instantaneous values of concentration fields [10], and their integration in time would not be comparable with results of Gaussian models, which implement varying meteorological conditions.

The simulation confirmed our assumption that the wind direction is the main factor influencing the transport and distribution of gypsum particles. Concentrations of gypsum in the air were found to be relatively small. On average it was found to be approximately 0.12 μg/m^3, while the minimal concentrations were 2.3×10^{-6} μg/m^3. The highest levels were found in the

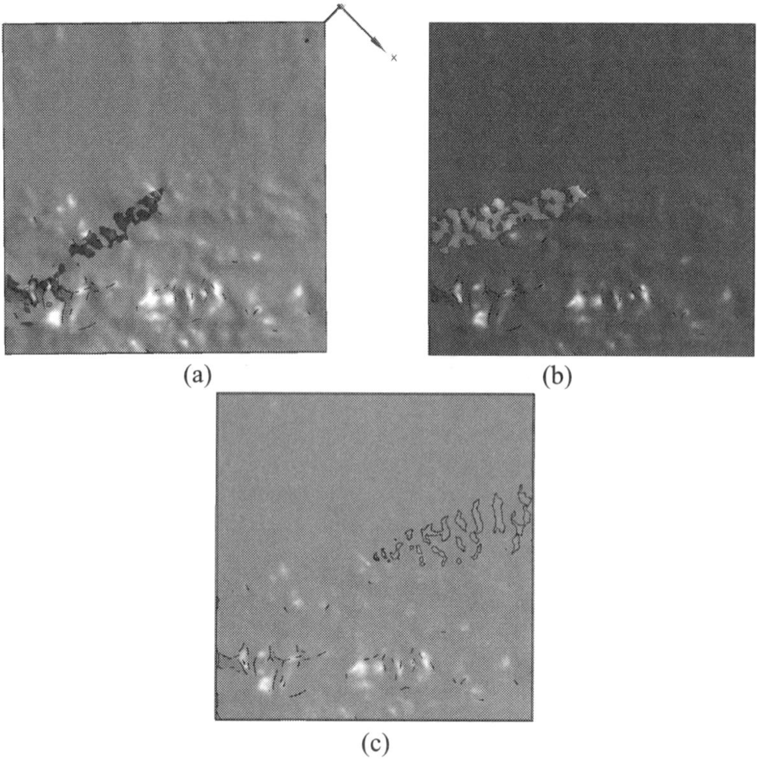

Figure 8: Deposition of gypsum with diameter of 4,5 μm. Wind direction 45° NE, 67.5° ENE and 247.5° WSW, (a) to (c).

immediate surrounding of the landfill. They diminished quickly as we move away from the landfill. Particles of smaller diameter showed better mobility and were found the farthest away from the landfill.

6 Conclusions

In the process of transformation of a wet gypsum landfill into a dry one environmental hazards of the process have to be evaluated in order to adapt the transformation procedures. As one of the hazards is dusting of dry particles due to strong winds, the goal of the presented research work was to find, how dangerous this transformation process would be and how it will affect the surroundings of the landfill. As the meteorological influences are the most important in this case, computational study was performed.

In general, the Gaussian models do give a strong indication on the average concentrations of the particles in the air and their deposition, however, they do lack of spatial resolution. In this sense, the combination with CFD based models can give a deeper insight into the flow field and corresponding particle concentration fields in the directions, in which the critical contamination of the environment is to be expected. The computational results will serve as a basis for long term monitoring of spreading and deposition of particles in the surroundings of the gypsum landfill.

Acknowledgements

The authors thank Mr Kolarič and Mrs Podgoršek-Kovač for their work in the field of CFD modelling and characterisation of particle properties.

References

[1] Plaza, C., Xu, Q., Townsend, T., Bitton, G., Booth, M. Evaluation of alternative landfill cover soils for attenuating hydrogen sulfide from construction and demolition (C&D) debris landfills. Journal of Environmental Management, Volume 84, Issue 3, Pages 314-322, 2007.
[2] Nazaroff, W., W., Alvarez-Cohen, L.: Environmental engineering science. John Wiley and Sons, 2001.
[3] U. S. Environmental Protection Agency: User's guide for the Industrial Source Complex (ISC3) Dispersion Models – Volume 1 (Revised). EPA-454/B-95-003a. Office of Air Quality Planning and Standards, Research Triangle Park, NC, 1995.
[4] U. S. Environmental Protection Agency: User's guide for the Industrial Source Complex (ISC3) Dispersion Models – Volume II – Description of model algorithms. EPA-454/B-95-003b. Office of Air Quality Planning and Standards, Research Triangle Park, NC, 1995.
[5] ISC-AERMOD View, Interface for the U.S. EPA ISC and AERMOD Models, Lakes Environmental, 2006.
[6] Ansys-CFX 5.7, AEA Technology, 2004.
[7] N.S. Holmes and L. Morawska: A review of dispersion modelling and its application to the dispersion of particles: An overview of different dispersion models available. Atmospheric Environment, Volume 40, Issue 30, pp. 5902-5928, September 2006,
[8] VDI 3783 Bl. 12 Entwurf. Umweltmeteorologie. Physikalische Modellierung von Stroemungs- uns Ausbreitungsvorgaengen in der atmosphaerischen Grenzschicht. Berlin: Beuth Verlag, 1999.
[9] Steven R. Hanna, Olav R. Hansen, Seshu Dharmavaram: FLACS CFD air quality model performance evaluation with Kit Fox, MUST, Prairie Grass, and EMU observations. Atmospheric Environment, Volume 38, Issue 28, pp. 4675-4687, September 2004.
[10] Kolarič, D.: Numerical simulation of spreading of solid particles from two dimensional source (in slovene). Master of science thesis, Faculty of Mechanical Engineering, Maribor, Slovenia, 2007.

Selecting a fast air quality calculator for an optimization meta-model

L. Aleluia Reis[1], D. S. Zachary[1], U. Leopold[1] & B. Peters[2]
[1] Resource Centre for Environmental Technologies, Public Research Centre Henri Tudor, Luxembourg
[2] Faculty of Science, Technology and Communication (FSTC), University of Luxembourg, Luxembourg

Abstract

Air pollution models have been developed over the last few decades, ranging from large detailed models, involving complex physical-chemical phenomena, to less detailed models. Air pollution models can also be grouped according to their scale. The air quality model, AYLTP, to be presented in this paper, aims at a spatial grid specificity that falls outside of the typical air quality scales approach. This model requires a spatial domain of approximately 100 km $\times$ 100 km, a spatial grid spacing of approximately 100-500 m, a time step of 10 minutes and a temporal domain of 24 hours. Moreover AYLTP requires a fast core calculator, as it will be incorporated on the Luxembourg Energy and Air Quality meta-model (LEAQ), which is built in an optimization framework. This paper aims at selecting the most suitable code to serve as a core calculator to be incorporated in AYLTP. A set of criteria was established to carry out an analysis of different open source air quality models suitable for the LEAQ meta-model. The selection of the models was based on a space-time graph. For each model, areas of influence were determined, based on the assumption that for a fixed CPU time, the grid spacing increases with the spatial domain size. Two models, AUSTAL2000 and METRAS, fit the required criteria. The choice between these two models was based according to the model's flexibility in terms of resolution and CPU performance. In this paper we briefly review the LEAQ project and discuss the criteria used to find a suitable core calculator. AUSTAL2000 is the model that better suits the criteria due to its simpler characteristics and faster transport calculator.
Keywords: meta-model, air quality model, scales, model comparison.

WIT Transactions on Ecology and the Environment, Vol 123, © 2009 WIT Press
www.witpress.com, ISSN 1743-3541 (on-line)
doi:10.2495/AIR090041

1 Introduction

Air quality is directly related to emissions, air transport and pollutants chemistry. The complexity of the phenomena influencing pollutants concentrations call for the use of modeling tools, termed air pollution models.

This paper presents the ideal attempt to improve the performance of the AYLTP prototype (TAPOM-Lite) [1], by adding the effects of terrain, turbulence and improving the influence of the meteorological factors. The goal of this paper is to find an air quality model, to serve as a core calculator which in turn would be embedded in the AsYmptotic Level Transport Pollution Model (AYLTP).

The Luxembourg Energy and Air quality meta-model (LEAQ) consists of two models, an energy model, GEOECU (Geo-Spatial Energy Optimization CalcUlator), and an Air quality model, AYLTP. The two models are coupled by an optimization routine called OBOE (Oracle Based Optimization Engine) which uses an Analytic Center Cutting point Method (ACCPM) [2]. The energy model, is by itself an energy optimization model which calculates the lowest cost energy arrangement with emission and energy constraints, e.g. demand, operational, technological and seasonal, etc. ACCPM is used to determine an optimal solution for the meta-model. The energy model passes the total cost (objective) to ACCPM, via the procedure called the Oracle. ACCPM uses the objective as well as directional information, termed subgradients, to guide the method to an optimal solution. Subgradients are defined as the sensitivity of change of total cost per change in employed technological device [3]. External constraints, breaches of air quality over maximum allowable levels, are also used as directional information by ACCPM. For each iteration of the optimization routine, the objective and subgradients are then used to calculate new levels of maximum primary emissions levels. The entire process is repeated until a lowest cost energy solution is determined by GEOECU and no air quality breaches are found. Finally a lowest cost energy solution with air quality constraints is achieved.

By their nature, both models are distinct, therefore they have different temporal and spatial scales. GEOECU produces emission data for a 30-40 year time period with a few years time step, e.g. 5 year time steps. This emission value will be distributed over periods of 24 hours for a typical and worst day of a season. The emission distribution will be done according to land use and scheduled corresponding to each economic sector daily emissions profile. Economic sectors are defined as groups of economic activities that share a common feature in terms of the spatio-temporal distribution of their total emissions for example transportation, industry, residential, and commercial.

In the atmosphere, phenomena occur at various scales, although for practical purposes, there is a need to develop specific-scale models. A specific-scale approach allows approximations and parametrization of the different phenomena at different scales [4]. For example, urban scale models include the effects of topography and land-sea-breezes. Smaller scale phenomena such as turbulence, are naturally treated by local scale models. The AYLTP model incorporates large scale characteristics including topographic effects and small scale characteristics such as turbulence. These characteristics, added to the fact that the model must run fast, make the selection of the core calculator challenging. The analysis of the suitability of the most appropriated model is a separate project itself. Therefore this paper presents a simple procedure to guide the AYLTP project in the right direction.

2 AYLTP design

The development of AYLTP rises from the need of an air quality model with specific requirements. AYLTP is connected to an optimization routine that iterates several times until it finds the optimal solution. Potentially, AYLTP model needs to be run approximately 1000 times (30-50 iterations times the number of subgradients). Practically, the air quality model must be run, using no more then a few minutes.

2.1 AYLPT requirements

Generally, air quality models involving detailed chemical reactions are CPU expensive [5]. In order to meet this demand, AYLTP is designed to calculate only the slow ozone reactions, i.e. asymptotic ozone levels. Accordingly, the fast photochemical equations are omitted. Particularly, this will be done using the core calculator, that will calculate ozone using the tabulated asymptotic ozone level for each time step. The 24 hour air quality result is based on the average primary emissions from each five year technoeconomic period of the energy model. Approximations are therefore acceptable. As the GEOECU model predicts for such a long term the whole meta-model has inherently large uncertainties associated with long time scales. Consequently, a highly accurately, CPU costly air quality model would not be required in this application. Accordingly, emissions of the most problematic pollutants will be included, such as NO_x, VOC, SO_2, CO, CO_2, PM_{10} and $PM_{2.5}$, and photochemistry for ozone. Furthermore, AYLTP will treat transport and diffusion of pollutants, turbulence and meteorology and dry deposition.

Despite the need for a simple air quality calculator, this model will include the most significant meteorological factors. The improvement of the meteorological package is important for photochemical reactions. Hence this package will include wind direction and speed, solar irradiation, humidity and temperature.

Air pollution is intimately related with meteorology. Wind can force the movement of the pollutants and affect their mixing ratios by accelerating or slowing the chemical reactions between pollutants. Radiation influences the photochemical processes that generate ozone. Atmospheric stability, is important in the dispersion of the pollutants and influences the chemical reactions. The above relations will be addressed in the model. Turbulence, is also important in fluid flow thus, this model attempts to incorporate turbulence in a very simple way.

The model is currently being constructed for Luxembourg, but it will be applicable to any other city. Luxembourg is a country of small dimensions with irregular terrain. Terrain irregularities play an important role in air flow phenomena. In order to have a good understanding of pollutant transport over Luxembourg, a grid domain of 50 × 80 km with a 100-500 m resolution will be applied. The border regions of the neighbouring countries need to be included in the simulation.

Terrain features will be incorporated due to its importance considering 3D wind fields. The SRTM (Shuttle Radar Topographic Mission) 90m Digital Elevation Data is available on a global scale, with a average resolution of approximately 90 meters [6]. The availability of such a detailed topographic information, is one of the reasons for the choice of the target grid spacing (100-500 m). A summary of the AYLTP requirements is shown in Table 1. The list of requirements are those which would be applicable to a general city/region.

Table 1: AYLTP requirements and inputs.

AYLTP Requirements	
Spatial Domain	100 km × 100 km
Horizontal resolution	100 m to 500 m
Vertical resolution	20 layers
Temporal Domain	24 hour
Temporal resolution	10 min
N^o of cells	20 000 000 cells
CPU time	few minutes, no more than 30 min
AYLTP Inputs	
Meteorology	Wind speed and direction, solar irradiation, humidity, temperature, atmospheric stability
Terrain	Terrain elevation profile
Land use	Urban, agriculture, industrial and transport
Emissions	Sectoral emissions calculated by GEOECU

2.2 Inputs and outputs

As inputs, the model requires emission values, terrain and meteorological information. GEOECU will output emission values for each of the pollutants considered. Land use maps will be used to distribute these emission values over space. Emission values will be distributed according to economic sectors. The emissions will be scheduled accordingly to the daily profiles of each sector. Even tough the time step strongly depends on the input data available, a 10 minute time step is targeted, because is the time required to track the slow photochemical reactions. Table 1 summarizes the AYLTP inputs.

The model outputs 3D hourly concentration maps for each pollutant for 24 hours period. As the meteorological input represents a typical meteorological day and the daily profile scheduling represents a typical emissions day, the output will represent the 'typical' air quality levels related in a certain energy scenario arrangement. Atypical days with poor air quality will also be simulated. Air quality values, averaged over a threshold 60 ppb (AOT60), will be spatially calculated and air quality breaches will be addressed. Furthermore, 3D wind fields will be plotted as well.

3 Core calculator selection

Air pollution modeling is a growing research domain, its applications have been used to support environmental management [4]. There exists a wide range of air pollution models. A review on open source air pollution models was carried out in order to choose the most appropriate code as a basis for AYLTP. The review was based on the list provided by the Model Documentation System [7] developed at the Aristotle University of Thessaloniki and on the COST 728/732 Model Inventory [8].

The model selection was based on a space-time graph. All the graphs presented in this paper were built using R [9]. First, the spatial and temporal scales as well as the resolution were evaluated. Air pollution models can be classified according to their scale focus. Figure 1 shows a comparison between the local, urban and mesoscale models and the AYLTP required scales. AUSTAL2000 was also included in Figure 1 because it was the final choice core calculator. Figure 1 shows that AYLPT does not completely match any of the scale classification models. The spatial domain of AYLTP falls in the range of urban to mesoscale models, whereas the resolution is typical of a local scale model. All these specificities called for a different selection approach of a core calculator.

The analysis of the criteria was made graphically (Figure 2). The graph was constructed using the range of spatial domain and resolution found for each model. Only the open source models were included in this selection process. The models in which the information about the spatial and temporal scale was not available are not shown. In this analysis it is assumed that the grid spacing increases linearly with the spatial domain increases when keeping the CPU time constant. Thus instead of a range box, a triangle is used to convey this relationship. The triangles show that the smallest grid spacing available for each model is in fact not applicable for all domain sizes, if one imposes the constant CPU constraint. In practice, imposing a CPU time constraint, the combinations of grid spacing and spatial domain available lie on the shaded area above the triangle's hypotenuse.

Taking the FARM model as an example, symbolized by the black point in Figure 2. Assuming a spatial domain of 10 000 km, the grid spacing applicable, in practice, would be the range from the horizontal line that crosses the black point up to the top of the shaded area. The same type of analysis can be carried out

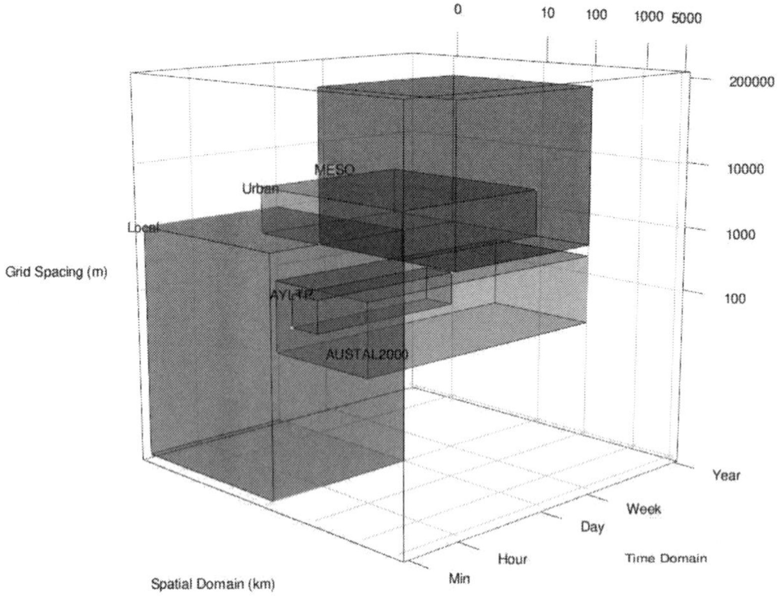

Figure 1: Schematic representation of air pollution models' scales. Red represents local scale models, green represents urban scale, blue the mesoscale. Orange stands for the AYLTP and gray describes AUSTAL2000.

for grid spacing, i.e. for a certain desired grid spacing, the maximum domain size that can be applied lies on the point where the horizontal line crosses the triangle hypotenuse.

4 Results and comparison

One observes that AUSTAL2000, and METRAS are the models that overlap the AYLTP box. Likewise one may observe that the model AURORA also overlaps the AYLPT range. However the CPU time found for AURORA, for a simple grid ($60\times60\times35$) for a month calculation, on a Intel Xeon 2GHz, is on the order of 70 hours [7]. As a result of the selection process, two models, AUSTAL2000 and METRAS, were found to be the best suited to serve as a core calculator (Figure 3). Both models overlap the AYLTP range, although none of them can, for a fixed CPU time, run with the largest domain and the highest resolution. Therefore, an extended analysis on these two models was carried out. A time step and time domain graph was built. Figure 3 shows the time criteria for the two models and its relation with AYLTP time requirements. Neither METRAS nor AUSTAL2000 overlap AYLTP range, though AUSTAL2000 touches the left upper limit of the AYLTP box. This means that the smallest time step allowed by AUSTAL2000 is

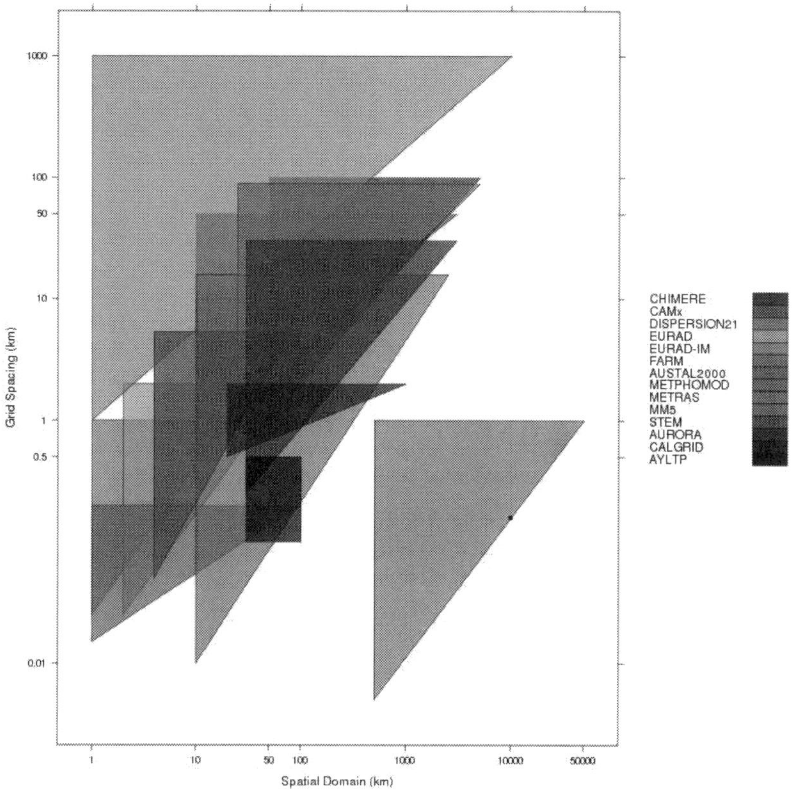

Figure 2: Compilation of models' spatial applications. The black point represents an example of how for a certain domain, the grid spacing applicable lies on the shaded area about the horizontal line.

the highest allowed by AYLTP. Thus, given a fixed CPU time AYLTP could be run with AUSTAL2000 smallest time step, but only for a time domain of one hour. Then METRAS model's time characteristics are quite different from the AYLTP requirements. The METRAS model's smallest temporal domain is equivalent to the largest of AYLTP and the time step falls under AYLTP requirements. Both models fit the spatial prerequisites, the main difference between them is their fluid motion approach. AUSTAL2000 is a Lagrangian particle model while METRAS is an Eulerian model.

The METRAS model calculates atmospheric flows, mesoscale effects, transport of pollutants and deposition of species. It contains turbulence, it can handle a complex terrain and chemistry. Eulerian models use a 3-dimensional computational grid. For each grid cell the mass balance of incoming and outgoing fluxes of the pollutants is calculated, solving the advection diffusion equation.

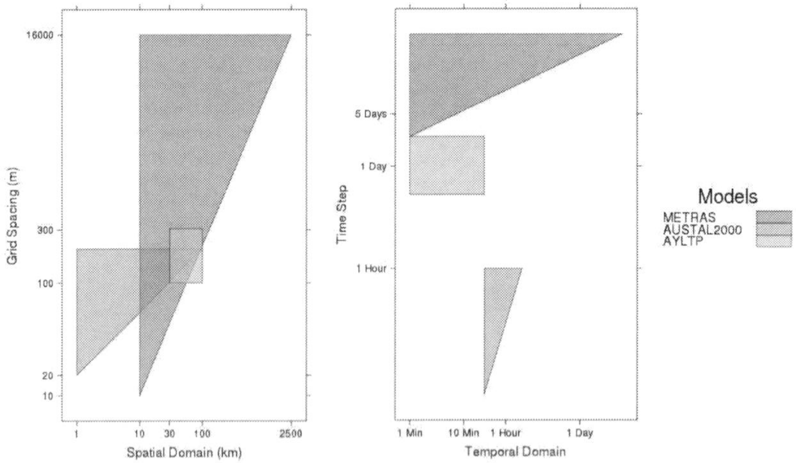

Figure 3: Spacial (left figure) and temporal (right figure) comparison of METRAS and AUSTAL2000 models with AYLTP.

AUSTAL2000 is a Lagrangian particle model the official reference model of the German Regulation on air quality control [10]. It simulates the trajectories of tracer particles immediately instead of investigating the fluxes. This approach offers in general more flexibility and precision in modelling the physical processes involved [11]. It simulates transport by the mean wind field, dispersion in the atmosphere, sedimentation of heavy aerosols, ground deposition and chemical conversion of NO to NO_2. The effect of turbulence on the particles is simulated by a random walk model [11].

Comparing the two models, regarding the AYLTP targeted processes, one observes that METRAS model is more complete, including all the aimed features. Thus, it uses rather complex chemical and dry deposition mechanisms. Nevertheless, ozone chemistry needs to be implemented in AUSTAL2000. The issue arising from this analysis is to decide between the simplification of the chemistry module of METRAS or the implementation of a simplified chemistry package for AUSTAL2000.

As mentioned above CPU time is a key factor for this project. Flexibility is another important issue, as the LEAQ meta-model is meant to be applicable to any city. Hence, a flexible grid spacing is desirable. This point is important when one takes into account the availability of the different quality input information for each city, and the city's dimensions and terrain particularities.

AUSTAL2000 uses a faster methodology to calculate pollutant's transport, whereas reliable numerical schemes, used in Eulerian models, are CPU expensive [12]. Concerning chemistry, AUSTAL2000 only yields a very simple NO to NO_2 conversion. Likewise, Lagrangian dispersion modeling is not based on the advection diffusion equation, but simulates the trajectories of a sample of particles.

This approach is simpler and CPU inexpensive [11]. The particle approach yields more flexibility, because for a fixed grid spacing and spatial domain, it still allows the adjustment of the number of particles. This adjustment enables a compromise between statistical uncertainty and CPU time, tuning the number of particles.

In this sense, AUSTAL2000 better serves the purpose of this work. Its approach is faster and the model structure involves less parameters, thus is more readily adaptable. Nevertheless, it has some disadvantages, mainly because the Lagrangian particle approach is less flexible when dealing with chemistry.

4.1 A trade-off between accuracy and calculation time

This meta-modeling approach has inherently large uncertainties associated with it, which are propagated along the meta-model. Another parallel project will be carried out with the focus on uncertainty propagation through LEAQ [13]. As explained above, AYLTP is an air quality model embedded in a optimization framework. Despite the attempt to include the most important factors influencing air quality, a compromise between CPU time and accuracy had to be performed. Hence, phenomena are treated on a simple level, and more complex physical and chemical interactions are ignored. Typical meteorological scenarios are assumed as being representative of a season. The air quality model is dependent on the energy model, which calculates the energy scenarios for a five year interval. Accordingly, certain assumptions and simplifications can be done.

4.2 Core calculator adaptations to AYLTP

The AUSTAL2000 calculator has some of the requirements already implemented including: turbulence, dry deposition and the inclusion of the species SO_2, NO_x, $PM_{2.5}$ and PM_{10}. Full inclusion into the AYLPT model will require a simple chemistry module, involving the relation of NO_x and VOC in ozone formation. This will be done in a simple way, using an asymptotic level of photochemistry. The number of sources allowed is limited in AUSTAL2000, for input format reasons, thus modifications are needed to make this parameter flexible. The model also needs to incorporate the following species: VOC, O_3 and CO. The pollutant CO_2 is a output of GEOECU model, but it will not be incorporated in AYLPT, as it is its quantitative emission value that is important for decision processes. The meteorology already accounted in AUSTAL2000 includes the wind direction and speed and the atmospheric stability. Therefore, the effect of solar irradiation, temperature and humidity, will still need to be implemented.

5 Conclusions

The increasing concerns with air pollution and the strict EU legislation has triggered the development of a wide number of air quality models. A variety of models are currently available and deal with different scales and parametrization levels according to their scope of application. The LEAQ project requires an efficient air quality model. The AYLTP prototype, embedded in LEAQ, is now under development and a well suitable air quality core calculator is being selected. The spatial domain of AYLTP falls between urban and mesoscale. However, the resolution is typical of a local scale model. A set of criteria has been established to help select the core calculator. The criteria included spatial and temporal domain, resolution and CPU time. Two open source models were found to be suitable for AYLTP specificities. The choice was made based on less expensive CPU demands and spatio-temporal characteristics. The Lagrangian approach, AUSTAL2000, was chosen because it tends to have lower CPU demand and offers a larger flexibility regarding calculation time and resolution scale. The incorporation of AUSTAL2000 in AYLTP will require adaptations. The adaptations will include a fast ozone calculator module, adaptation of the time step, introduction of the species VOC, O_3 and CO, and improvement of the meteorological package. The meteorological package will include wind speed and direction, solar radiation, humidity and temperature. The selection of the most suitable model is a time demanding task, the procedure used in this paper is a simple approach to guide the project in the direction of the LEAQ requirements.

Acknowledgements

We acknowledge FNR - Fonds National de la Recherche Luxembourg for the grant founding, AFR - Aide la Formation-Recherche, under the grant identifier PHD-08-004. We also would like to thank the LEAQ research team. We acknowledge, as well, Professor Dimitris Melas from the Aristotle University of Thessaloniki and Dr. Ulf Janicke from Janicke Consulting, for the useful comments.

References

[1] Haurie, A., Zachary, D.S. & Sivergina I., A Reduced-Order Photo-chemical Air Quality Model, 2003.

[2] Carlson, D., Haurie, A., Vial, J.P. & Zachary, D., Large-scale convex optimization method for air quality policy assessment. *Automatica*, **40(3)**, pp. 385–395, March 2004.

[3] Drouet, L., Beltran, C., Edwards, N., Haurie, A., Vial, J.P. & Zachary, D., An oracle method to couple climate and economic dynamics. *The Coupling of Climate and Economic Dynamics Essays on Integrated Assessment Series: Advances in Global Change Research Coupling Climate and Economic Dynamics*, eds. A. Haurie & L. Viguier, Kluwer, volume 34, 2005.

[4] Moussiopoulos, N., *Air Quality in Cities*. Springer, 2003.

[5] Grell, G., Duddhia, J. & Stauffer, D., Ncar mesoscale model(MM5). Technical Report NCAR/TN-398+STR, NCAR, 1994.

[6] CGIAR - Consortium for Spatial Information (CGIAR-CSI), *SRTM 90m Digital Elevation Data, VERSION 4*. http://srtm.csi.cgiar.org/, 2008.

[7] Moussiopoulos, N., MDS - Model Documentation System.

http://pandora.meng.auth.gr/mds/, 2009.

[8] COST 728/732, *COST 728/732 - Model Inventory*. http://www.mi.uni-hamburg.de/Summary-Tables.510.0.html, 2007.

[9] R Development Core Team, *R: A Language and Environment for Statistical Computing*. R Foundation for Statistical Computing, Vienna, Austria, 2009. ISBN 3-900051-07-0.

[10] Consulting, J., TA Luft / AUSTAL2000, 2009.

[11] Janicke, L., *Lagrangian dispersion modelling*. 235, 37-41, Particulate Matter in and from Agriculture, 2002.

[12] Mathur, R., Hanna, A., Odmanm O., Coats, C., Alapaty, K., Trayanov, A., Xiu, A., Jang, C., Fine, S., Byun, D., Schere, K., Dennis, R., Novak, J., Pierce, T., Young, J. & Gipson, G., The Multiscale Air Quality Simulation Platform (MAQSIP): Model Formulation and Process Considerations. Technical report, arolina Environmental Program, University of North Carolina at Chapel Hill, 2004.

[13] Zachary, D. & Leopold, U., The unLEAQ project, UNcertainty analysis for the Luxembourg Energy Air Quality project, a Luxembourg national research (FNR) proposal. Technical report, Public Research Centre Henri Tudor, 2009.

The role of meteorological factors on year-to-year variability of nitrogen and sulphur deposition in the UK

M. Matejko[1], A. Dore[2], C. Dore[3], M. Błaś[1], M. Kryza[1], R. Smith[2] & D. Fowler[2]
[1]*Department of Meteorology and Climatology, University of Wrocław, Poland*
[2]*Centre for Ecology and Hydrology, UK*
[3]*AETHER Ltd., UK*

Abstract

FRAME is a statistical Lagrangian model, which describes the main atmospheric processes (emission, diffusion, chemistry and deposition) taking place in a column of air. The model is used to calculate maps of dry and wet deposition for sulphur and nitrogen. Historical emissions data are used in the model to calculate changes in deposition of sulphur and oxidised and reduced nitrogen for the UK at a 5 km x 5 km resolution for the years 1990-2005. Emissions of SO_2, NO_x and NH_3 in the UK have fallen by 77%, 47% and 18% during this period. FRAME calculated reductions in wet deposition to the UK of 56%, 17% and 16% for SO_x, NO_y and NH_x respectively. Inter-annual variation in meteorology was found to have a significant influence on pollutant transport and the national wet and dry deposition budget. This occurred due to differing wind direction frequency as well as annual precipitation. When using year with specific wind conditions, wet deposition can even change by more than 20%. It was also observed that wind conditions have a greater influence on deposition budget than precipitation data. Modelled trends in nitrogen and sulphur wet deposition have been compared with measurements from the national acid deposition monitoring network during this period. A more comprehensive monitoring network has been used to verify model results for deposition of SO_4^{2-}, NO^{3-}, and NH^{4+} for the year 2005.

Keywords: emissions reduction, atmospheric circulation, long-range transport, pollutant deposition, UK.

WIT Transactions on Ecology and the Environment, Vol 123, © 2009 WIT Press
www.witpress.com, ISSN 1743-3541 (on-line)
doi:10.2495/AIR090051

1 Introduction

Deposition rate does not necessarily decrease with the distance from emission sites and meteorological factors are crucial for controlling the spatial and temporal distribution of both concentration and deposition of pollutants. Atmospheric resistance times of SO_2 and NO_x are even several days and pollutants can be transported about a thousand of kilometers [20]. In many countries deposition to a large extend comes from outside sources and it is necessary to assess the changes on a large scale, considering the trans-boundary fluxes [10]. The effective pollutants removal is caused by chemical liquid-phase transformations in clouds [12, 21] and water flux from the atmosphere to the ground especially in the form of precipitation. It could affects soils and freshwater, particularly in areas where annual precipitation is high [19].

Analysis of meteorological conditions showed that inter-annual variation of circulation patterns and precipitation play a significant role in causing year-to-year fluctuations. Hence it is an important component in long range transport modelling, pollutant deposition researches and policy analysis [3, 11]. In the UK Lagrangian trajectory models such as HARM [18], TRACK [17] and FRAME (the Fine Resolution Atmospheric Multi-pollutants Exchange model) [22] have been developed to assess acid deposition to sensitive areas. Air pollution predictions are usually constructed by keeping the climate constant and only changing the emission [15]. A first pilot study of the impact of regional climate change on deposition of sulphur and nitrogen in Europe was presented by Langer and Bergstrom [14]. Variability of meteorological factors from year to year is important to understand non-linearity in the emission-deposition relationship. They also complicate the process of assessing the effects of emission reduction strategies [10].

In this paper, we use FRAME model to consider the important contribution of meteorological data to deposition budget of SO_x, NO_y and NH_x in the UK. The accuracy of the model is verified by a detailed comparison with measurements from the UK national monitoring networks for ion concentration in precipitation for the year 2005 and also UK budget comparison for the period 1990-2005.

2 FRAME model

2.1 Model description

The atmospheric transport model, FRAME (Fine Resolution Atmospheric Multi-pollutant Exchange) is used to assess the long-term annual mean sulphur and nitrogen deposition over UK. Detailed description of the FRAME model is provided by Singles *et al.* [22], Fournier *et al.* [9] and Dore *et al.* [7]. FRAME is a Lagrangian model which describes the main atmospheric processes taking place in a column of air moving along straight-line trajectories following specified wind directions. The model consists of 33 vertical layers of varying thickness ranging from 1 m at the surface and increasing to 100 m at the top of the domain [1]. Vertical mixing is described using K-theory eddy diffusivity, and solved with a Finite Volume Method [24]. Dry deposition is calculated by determining a vegetation velocity (V_d) to each chemical species derived from a dry deposition model [23]. This model derives maps of deposition velocity taking into account surface properties and geographical and altitudinal variation of windspeed. Wet deposition is calculated with scavenging coefficient and a constant drizzle approach, using precipitation rates calculated from a map of average annual precipitation. The amount of material removed in a time period (Δt) is given by

$$\Delta c_i = c_i(1-e^{-\lambda i \Delta t})$$

Δc_i – decrease in concentration of species i due to removal by precipitation,
λ_i – scavenging coefficient.

The wet deposition flux to the surface is the sum of wet removal from all volume elements aloft, assuming that scavenged material comes down as precipitation. There is no difference between in-cloud and below-cloud processes and an averaged value of scavenging ratio (Δ_i) is used in the model. To produce scavenging coefficient λ_i, Δ_i is combined with the precipitation rate and the depth of the mixing layer ΔH_{mix}:

$$\lambda_i=(\Delta_i P)/H_{mix}$$

An increased washout rate is assumed over hill areas due to the seeder-feeder effect. It is assumed that the washout rate for the orographic component of rainfall is twice that used for the non-orographic components [5]. As air columns move along its trajectory , chemical interactions between NH_x, SO_x and NO_y take place. The parameterisation combine descriptions of both dry chemistry and aqueous phase chemistry.

The FRAME domain covers the UK and the Republic of Ireland with a grid resolution of 5 km x 5 km and grid dimension of 172 x 244. Input pollutant concentrations at the boundary of the model domain are calculated with FRAME-Europe – a similar model which runs on the EMEP grid at 50 km x 50 km resolution. Trajectories are advected with different starting angles at a 1-degree resolution, using directionally dependent wind speed and wind frequency roses. To create wind speed rose for FRAME radiosonde data are used [6], but wind frequency rose is based on the Jenkinson objective classification).

2.2 Emissions input data

The FRAME model uses a database of SO_2, NO_x, NH_3 emissions with a 5 km x 5 km grid-square resolution as input. Emissions of SO_2 and NO_x are taken directly from the National Atmospheric Emissions Inventory (NAEI, www.naei.org.uk) for the UK. Emissions of ammonia are estimated for each grid square using the AENEID model (Atmospheric Emissions for National Environmental Impacts Determination) that combines data on farm animal numbers, with land cover information, as well as fertiliser application, crops and non-agricultural emissions [8]. In order to estimate the temporal trends in deposition to the UK, it is important for input emissions data to be identically formatted. The background and point sources emissions for the year 2002 were taken to be the baseline year. The data for point sources and area emissions were used to scale emissions backwards and forwards in time and generate new emissions file for the years 1990-2001. Emissions from NAEI were used for 2003, 2004, and 2005. Emissions from the Republic of Ireland were scaled backwards in time in a similar manner to the UK emissions. SO_2 and NO_x emissions from international shipping were also included in the domain and were scaled forwards and backwards in time from the baseline year 2000 according to estimates from the NAEI.

For the period 1990-2005, SO_2 emissions are dominated by coal combustions, primarily in Public Electricity and Heat Generation. The emissions have been reduced from 1859 Gg S in 1990 to 344 Gg S in 2005 and the significant reductions have been caused by fuel switching from coal to gas, and the installation of the abatement equipment at power stations. At the same period the reduction in emissions for NO_x amounts from 903 Gg N to 493 Gg N, where the largest reduction has been from Passenger Cars. This is due to the introduction of three-way-catalysts in the late 1990's. Emissions from power generation have also reduced, primarily due to the increased use of gas over coal-fired stations. Emissions of NH_3 are dominated by agricultural activities. The reduction in emissions observed for NH_3 is not as large as that for SO_2 and NO_X. The total NH_3 emission has changed from 315 Gg N in 1990 to 259 Gg N in 2005. This has primarily been caused by a decrease in livestock numbers or improvements to manure management.

2.3 Meteorological data

Precipitation data used in FRAME comes from the Meteorological Office national network (approximately 5000 stations). The data are in the form of annual rainfall fields for the UK and Ireland, with the resolution 5 km x 5 km. For the period 1990-2005 the mean annual precipitation was 1130 mm yr^{-1}. The wettest years were: 1998, 2000 and 2002 (> 1260 mm), while 1996 and 2003 were dry (< 920 mm), relative to the mean for the period (fig. 1). A higher precipitation amount is noticed at the western costal and at higher altitudes, however during the wet year, hilly regions with precipitation above 2000 mm $year^{-1}$ are considerably larger. The considerable enhancement in rainfall in hilly regions can be partially explained with seeder-feeder mechanism [2, 4].

Wind data (frequency and wind speed information) was taken from the objective classification and radiosonde. Wind direction frequency roses are based on objective classification of Lamb-Jenkinson weather types [13, 16]. For each year there is different wind frequency rose. The average (1990-2005) wind rose illustrates that predominant wind directions are from the SW-W (fig. 2(a)). The years 1996 and 2004 were selected to show the difference in wind frequency from E+SE+S directions (most polluted) within the considered period (fig. 2). The wind speed rose was generated by calculating the harmonic mean from the mean radiosonde data set (fig. 2(b)), The highest wind speeds are observed from the southwest and lower from the east.

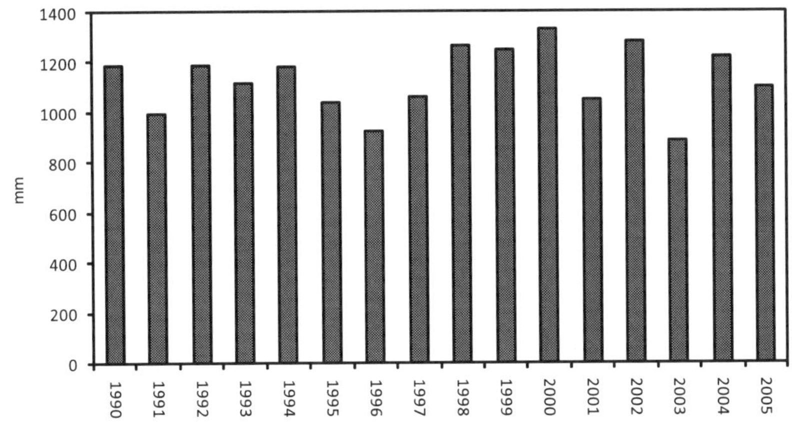

Figure 1: Mean annual UK precipitation 1990-2005.

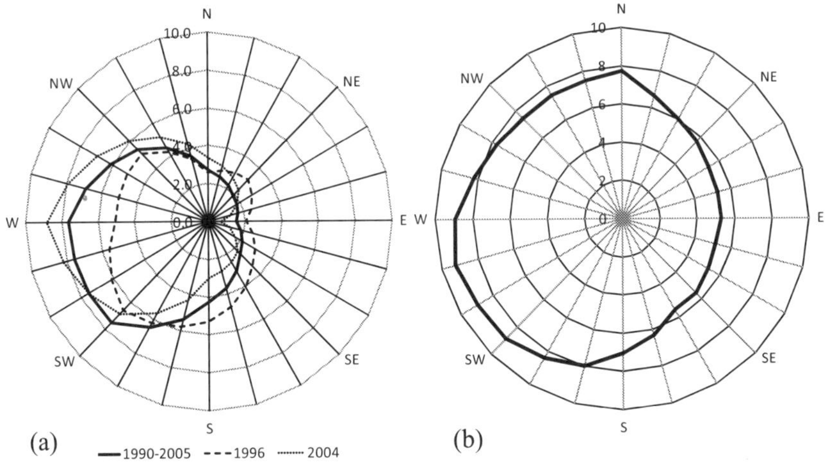

Figure 2: The 1990-2005 average wind frequency rose compare to 1996 and 2004 wind frequency roses (figure (a)), [%]). 1990-2005 harmonic average radiosonde wind speed rose (figure (b)), [m s^{-1}]).

3 Results and discussion

3.1 Comparison of model outputs with measurements

Assessment of the accuracy of FRAME in estimating deposition has been made by comparison modelled data against measurements. Wet deposition data were obtained from the secondary acid precipitation monitoring network (32 sites). Unfortunately, long-term dry deposition is only measured directly at a very few sites in the UK, which means a direct model-measurement comparison dry deposition is not feasible.

The measurements data (NH_4^+, NO_3^-, SO_4^{2-}) were compared with modelled for the year 2005 (fig. 3). A good statistics measures (MB, MAE, R) are evident for all compounds. A satisfactory correlation for wet deposition of sulphate is apparent, with a slope of 1.10 and low intercept of 0.17 for the model-measurement linear regression.

The model performs particularly well against measurements for low deposition and somewhat overestimates higher values (which usually occurs in hilly regions). For NO_3^- and NH_4^+ lower deposition is also represented better but higher values appears both as overestimations and underestimations.

Modelled results were also compared with the country deposition budget (fig. 4). Solid lines, which are quite flat for all compounds show FRAME

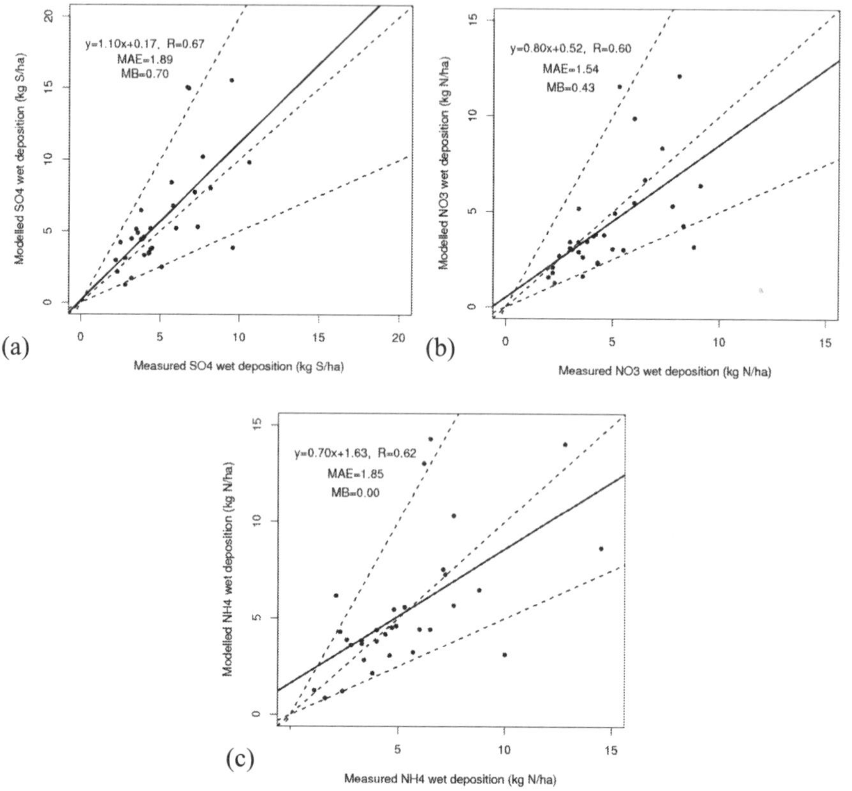

Figure 3: Correlation of modelled wet deposition with measurements from the national monitoring network for 2005: (a) SO_4^{2-}, (b) NO_3^-, (c) NH_4^+. Solid line is the best fit line produced by a regression analysis, dashed lines are for reference: 2:1, 1:2 and 1:1 division.

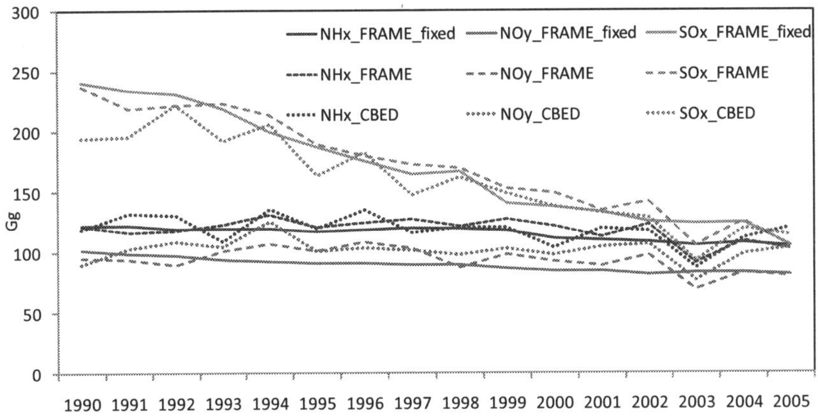

Figure 4: Comparison the country wet deposition budget estimated by FRAME (using specific meteorological conditions or without considering the meteorological influence – "fixed") and CBED (Concentration Based Estimated Deposition).

deposition budget counted without considering the meteorological influence. The FRAME wet deposition budgets for the UK show reductions of 132 Gg S-SO_x yr^{-1}, 15 Gg N-NO_y yr^{-1} and 19 Gg N-NH_x yr^{-1} over the period 1990-2005. There is a very good agreement between modelled and measured lines. FRAME model seems to catch quite well specific meteorological conditions (special peaks in 2002 or 2003).

3.2 Influence of meteorological data on regional budget

To check the influence of meteorological data on deposition budget, FRAME is run using emissions data from the year 2005, and specific wind frequency roses (fig. 2) and annual precipitation (fig. 1) from the period 1990-2005 are selected. First two simulations contain an average (1990-2005) wind frequency rose and precipitation data for the driest (2003) and wettest (2000) year of the analyzed period. For the driest year the annual sum of precipitation is about 22% lower than average and for wettest year by about 18% higher than average. The next two simulations are run using an average (1990-2005) precipitation data and specific wind roses: 1996 and 2004. For 1996 wind rose, there are more easterly directions (NNE-SSW) and for 2004 more westerly and northerly directions (WSW-NE) than average. There is also run simulation with average conditions - wind frequency rose and annual precipitation data averaged for the period.

Fig. 5 shows relative changes in wet and dry deposition results between simulations with specific and average meteorological conditions. Comparing simulations results for wet and dry year, it is characteristic that all species show similar reaction. Simultaneously, meteorological conditions have a greater influence on wet deposition than dry. In 2003 year, with 22% lower annual sum of precipitation corresponds to 13% lower wet deposition. It is clearly seen that greater influence on deposition budget between species have changes of wind conditions than precipitation. When using wind rose from the most polluted directions (1996) wet deposition is higher by about 22%, 14% and 12% for NO_y, SO_x and NH_x, respectively.

The ratio of spatial distribution of wet deposition between simulation with specific wind roses and average simulation are presented in fig. 6. Wet deposition of SO_x, NO_y and NH_x is locally higher than average by about 40% when using wind rose for 1996 (higher contribution of E, SE and S wind sectors). It concerns especially hilly regions on the west and north parts in UK. For the rest areas wet deposition is higher by about 10-15%. There is lower deposition in some areas along the east coast, where major power stations are situated. Using 2004 wind rose (extremely westerly oriented), north parts of UK are seen to have lower deposition by about 10-15% but there are also areas with raised deposition to the south of emission sources in major urban areas of Greater London, Birmingham and Manchester.

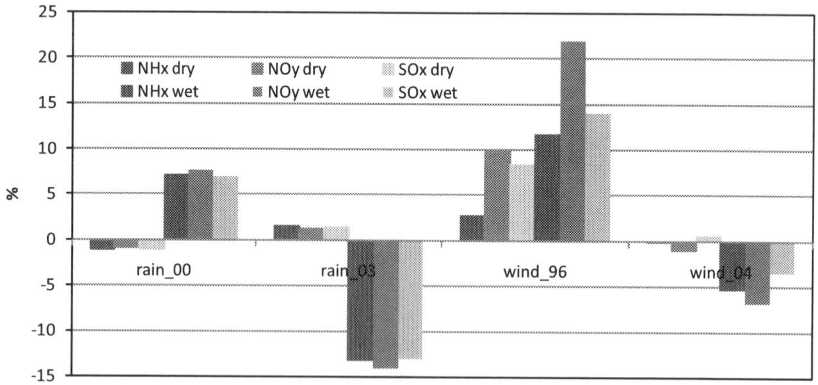

Figure 5: Relative changes in dry and wet deposition of SO_x, NO_y, NH_x between simulations with specific and average meteorological conditions (rain_00 – wettest and rain_03 – driest year; wind_96 – more E directions and wind_04 – more W and NW directions than an average).

4 Conclusions

Inter-annual variability of meteorological factors was found to have a significant influence on pollutant transport and deposition in the UK. Using constant emissions and circulation pattern, but the different rainfall fields, the percentage differences in pollutant deposition between extremes (driest 2003 and 2000 – unusually wet) were almost 20% of wet deposition and 3-4% in case of dry deposition. It was also observed that circulation conditions have a greater influence on deposition budget than precipitation. When using years with specific wind conditions (1996 and 2004), NO_y dry and wet deposition can vary

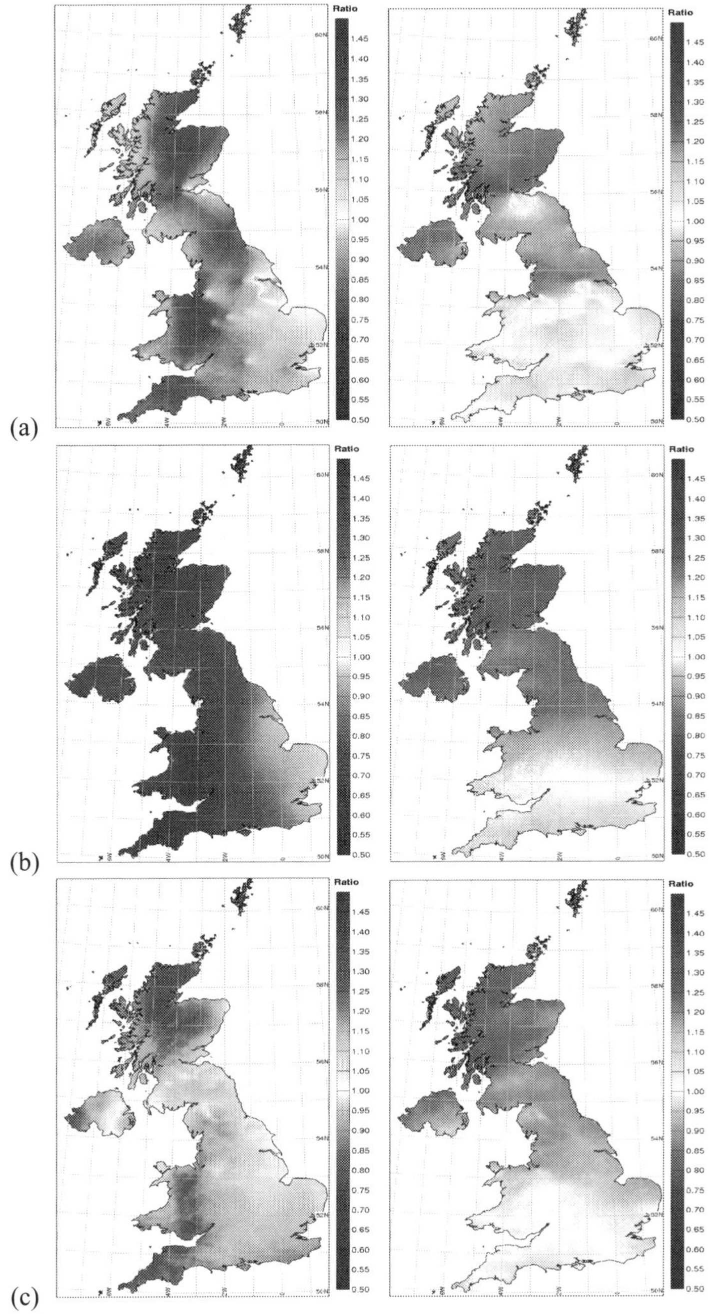

Figure 6: Ratio maps of wet deposition for (a) SO_x, (b) NO_y, (c) NH_x between FRAME simulation for specific wind data (1996 –left, 2004 – right) and average conditions.

by more than 10% and almost 30%, respectively. Such year-to-year variability concerns especially hilly terrain and areas remote from the emission sources. It shows that year-to-year changes in precipitation amount and spatial distribution, together with varying circulation patterns may cover long-term trends in wet and dry deposition due to emission reductions. This is of special importance if the long-term deposition trends are counted in comparison to one selected year, which is considered as a base and then changes are calculated relatively to this year.

References

[1] ApSimon, H. M., Barker B. M., Kayin S., 1994. Modelling studies of the atmospheric release and transport of ammonia – applications of the TERN model to an EMEP site in eastern England in anticyclonic episodes. Atmospheric Environment, 28, 665-678.

[2] Błaś, M., Dore, A.J., Sobik, M., 1999. Distribution or precipitation and wet deposition around an island mountain in southwest Poland. Quarterly Journal of the Royal Meteorological Society 125, 253–270.

[3] Binkowski, F. S., Shankar, U., 1995. The regional particulate model. Model description and preliminary results. Journal of Geophysical Research 100 (D12). 26191-26209

[4] Choularton, T. W., Gay, M. J., Jones, A., Fowler, D., Cape, J. N., Leith, I. D., 1988. The influence of altitude on wet deposition. Comparison between field measurements at Great Dun Fell and the predictions of the seeder-feeder model. Atmospheric Environment 22, 1363-1371.

[5] Dore, A.J., Choularton, T.W. and Fowler, D., 1992. An improved wet deposition map of the United Kingdom incorporating the seeder-feeder effect over mountainous terrain. Atmospheric Environment 26A, 1375-1381.

[6] Dore, A.J., Vieno, M., Fournier, N., Weston, K.J. and Sutton, M.A. 2006. Development of a new wind rose for the British Isles using radiosonde data and application to an atmospheric transport model. Q.J.Roy.Met.Soc. 132, 2769-2784.

[7] Dore, A. J., Vieno, M., Tang, Y. S., Dragosits, U., Dosio, A., Weston, K. J., Sutton, M. A., 2007. Modelling the atmospheric transport and deposition of sulphur and nitrogen over the United Kingdom and assessment of the influence of SO_2 emission from the international shipping, Atmospheric Environment 41, 2355-2367.

[8] Dragosits, U., Sutton, M.A., Place, C.J., Bayley, A., 1998. Modelling the spatial distribution of ammonia emissions in the United Kingdom. Environmental Pollution 102 (S1), 195–203.

[9] Fournier, N., Dore, A. J., Vieno, M., Weston, K. J., Dragosits, U., Sutton, M. A., 2004. Modelling the deposition of atmospheric oxidized nitrogen and sulphur to the UK using a multi-layer long-range transport model, Atmospheric Environment 38, 683-694.

[10] Fowler, D., Smith, R., Müller, J., Cape, J.N., Sutton, M., Erisman, J.W., Fagerli, H., 2007. Long-term trends in sulphur and nitrogen deposition in Europe and the cause of non-linearities. Water, Air, & Soil Pollution, Vol. 7, Numbers 1-3, 41-47.

[11] Hayami, H., Ichikawa, Y., 2001. Sensitivity of long-range transport of sulphur compounds to vertical distribution of sources. Water Air and Soil Pollution130, 283-288.

[12] Hegg, D.A., Hobbs, P.V., 1981. Cloud water chemistry and the production of sulfates in clouds. Atmospheric Environment 15, 1597-1604.

[13] Hulme, M. and Barrow, E. M., 1997. Introducing climate change. pp.1-7 in: Climates of the British Isles: Present, Past and Future M. Hulme and E.M. Barrow, Eds., Routledge, London, pp. 454.

[14] Lagner J., Bergström R., 2001. Impact of climate change on regional air pollution budgets. In: Midgley, P., Reuther, M., Williams, M. (Eds.), Transport and Chemical Transformation in the Troposphere. Springer, Berlin, Heildeberg.

[15] Lagner J., Bergström R., Foltescu V., 2005. Impact of climate change on surface ozone and deposition of sulphur and nitrogen in Europe. Atmospheric Environment 39, 1129-1141.

[16] Lamb, H.H., 1972. British Isles weather types and a register of daily sequence of circulation patterns, 1861-1971. Geophysical Memoir 116, HMSO, London, pp. 85.

[17] Lee, D.S., R.D. Kingdom, Jenkin M., E., Garland J. A., 2000. Modelling the atmospheric oxidised and reduced nitrogen budgets for the UK with a Lagrangian multi-layer long-range transport model. Environmental modelling and assessment 5, 83-104.

[18] Metcalfe, S.E., Whyatt, J.D., Broughton, R., Derwent, R.G., Finnegan, D., Hall, J., Mineter, M., O'Donoghue, M. and Sutton, M.A., 2001. Developing the Hull Acid Rain Model: its validation and implications for policy makers. Journal of Environmental Science and Policy 4, 25-37.

[19] NEGTAP, 2001. Transboundary Air Pollution: Acidification, Eutrophication and Ground Level ozone in the UK. Report of the National Expert Group on Transboundary Air Pollution. DEFRA, London.

[20] O'Neill P., 1998, Chemia środowiska, PWN, Warszawa.

[21] Radojevic, M., Tyler, B.J., Hall, S., Penderghest, N., 1995. Air oxidation of S(IV) in cloud-water samples. Water, Air and Soil Pollution 85, 1985-1990.

[22] Singles, R.J., M.A. Sutton & K.J. Weston, 1998. A multi-layer model to describe the atmospheric transport and deposition of ammonia in Great Britain. Atmospheric Environment 32, 393-399.

[23] Vieno, M., 2005. The use of an Atmospheric Chemistry-Transport Model FRAME over the UK and the development of its numerical and physical schemes. PhD thesis, University of Edinburgh.

ZIMORA – an atmospheric dispersion model

H. R. Zimermann & O. L. L. Moraes
Laboratório de Micrometeorologia,
CRS/INPE – Universidade Federal de Santa Maria, Brazil

Abstract

This paper presents a development and validation of a 3D numerical model for the advection-diffusion equation. Atmospheric flow field generated by mesoscale circulation model is used as input for the wind speed. As the mesoscale model gives information at scale higher than the necessary for description of a plume trajectory a weighted linear average proper interpolation was developed for intermediate these distances. Diffusion coefficients are variables in time and space and are different for lateral and vertical directions. This assumption is important and considers that the turbulence is not isotropic. Numerical scheme is explicit and conservative and has small implicit diffusion in the advection part of the model. Boundary conditions are open at lateral domain and normal in the bottom and at the top of the Planetary Boundary Layer (PBL) height. Output of the model can be viewed at any time step as well as the concentration distributions are showed at horizontal or vertical surfaces. For validation of the model two experiments were carried out near a thermoelectric power plant located in the south of Brazil. Each experiment was forty days longer. Hourly SO_2 concentrations were collected in four receptors around the power plant. During the experiments micrometeorological measurements and tethered balloons were also used in order to describe properly the local atmospheric circulation and PBL characteristics. Various static indices indicate that the model works very well at least for the source and the terrain were it is located, i.e. continuous emission and homogeneous topography.

Keywords: data assimilation, numerical 3D model, atmospheric pollutant dispersion, Zimora, finite differences scheme, K-theory.

WIT Transactions on Ecology and the Environment, Vol 123, © 2009 WIT Press
www.witpress.com, ISSN 1743-3541 (on-line)
doi:10.2495/AIR090061

1 Introduction

One of the most challenging and widespread environmental problems facing the international community today is air pollution. Both the monitoring and modelling of air pollution is essential to provide a picture of the damage humans are doing to the environment, and to enable pollution problems to be discovered and dealt with. The main idea of a model is to describe what may happen to pollutant once it enters the surrounding atmosphere, where it goes and how it is diluted. In general the mathematical description of this process is done by eulerian or lagrangian models.

Understanding the distribution and fluxes of various atmospheric trace constituents requires a proper knowledge of atmospheric transport as well as the relevant physical and chemical transformation and deposition processes for the trace species considered. Numerical modelling currently presents a powerful way of analyzing many problems related to the atmospheric trace constituents. The increasing power of digital computers as well as a steady improvement in the quality and resolution of meteorological data has led to the development and successful application of three-dimensional atmospheric transport models on a range of scales from local to global. In this paper it is presented the development and evaluation of numerical model, Zimora that solves the advection-diffusion equation. The Zimora is and Eulerian based model since it treats conservative equations. It achieves meteorological fields by the use of a dynamical class meteorological pre-processor model (to hereafter MPP) output. In its actual development stage it acquires only wind velocity to numerically solve the prognostic advection-diffusion equation. The dynamical mesoscale numerical model output, used as input into Zimora, for wind field was the Brazilian version of the Regional Atmospheric Modeling System (CSU-RAMS) developed by Colorado State University (USA) and adapted for Brazilian purposes and known by the name BRAMS. Its actual operational status, running as the main weather forecast product from the Atmospheric Modeling Group (GRUMA) facilities in the Santa Maria Federal University, centre of southern state in Brazil, has an output resolution of 3 hours for the time scale and 20 kilometres for horizontal scales and 17 vertical geopotential levels. The proper fields downscaling for input into Zimora will be discussed in details in the 2.1 section.

2 Description of the model

The prognostic equation adopted by this model is based in the physical principles of mass conservation representing the continuity of atmospheric fluid air parcel. This equation, in the general major of the mathematical treatments of diffusion from point sources, is the differential equation that has been used as start point, is a generalization of the classic equation of heat transference in solids, in heat form, by conduction and is essentially one description of mass conservations of the suspended material (Slade [27]).

Consider for example, any scalar concentration quantity C, we express the conservation equation by:

$$\frac{\partial C}{\partial t}+\nabla.(\vec{V}C)=S+D. \tag{1}$$

Here, $\vec{V}$ represents the wind field; S the source/sink term and D represents the molecular diffusion.

Easily expanding the second term on the left side of the eqn. (1), then assuming the atmosphere to be incompressible fluid so that its velocities fields divergent will be null, also neglecting molecular diffusion term and considering that in the realistic atmosphere investigations the velocity fields are tridimensional such that we can express $\vec{V}$ into his components in each direction of Euclidian space, we obtain from equation. (1):

$$\frac{\partial C}{\partial t}+u\frac{\partial C}{\partial x}+v\frac{\partial C}{\partial y}+w\frac{\partial C}{\partial z}=S\,, \tag{2}$$

where u, v and w are the components of the velocity in the longitudinal, lateral and vertical directions respectively. The analysis of the turbulence such that for building turbulent flows (Stull, 1988) and consequently the advection-diffusion equation, it is necessary to separate the movements scales. This can be achieved by representing the instantaneous values of the quantity C in terms of a mean and an eddy fluctuation from the mean, known like Reynolds decomposition (Stull [28]), such that:

$$C=\overline{C}+C'. \tag{3}$$

The instantaneous values of quantities C, u, v and w are then expressed as eqn. (3) and then are substituted into eqn. (2). The terms are expanded, the equations is averaged, following Reynolds average rules (Stull [28]) and we find that:

$$\frac{\partial \overline{C}}{\partial t}+\overline{u}\frac{\partial \overline{C}}{\partial x}+\overline{v}\frac{\partial \overline{C}}{\partial y}+\overline{w}\frac{\partial \overline{C}}{\partial z}+\overline{\frac{\partial u'C'}{\partial x}}+\overline{\frac{\partial v'C'}{\partial y}}+\overline{\frac{\partial w'C'}{\partial z}}=S\,, \tag{4}$$

where bars denote mean and primes turbulent values. The eqn. (4) can be understood as an forecast equation for the concentration C. The over bar in all the terms is associated with the previous process called Reynolds averaging, an applied mathematical method that eliminates small linear terms such those associated with no breaking waves, but retains the nonlinear terms associated with, or affected by turbulence (Wallace 2006). The kinematic fluxes in the last three terms on the left side of the eqn. (4) can be parameterized using a local first-order closure called gradient-transfer or K-theory. In this approximation we assume, analogous to molecular diffusion, that the turbulence causes a net material, released in the atmosphere, direct counter to the local gradient, i.e.:

$$\overline{u'C'}=-\kappa_x\frac{\partial \overline{C}}{\partial x}; \qquad \overline{v'C'}=-\kappa_y\frac{\partial \overline{C}}{\partial y}; \qquad \overline{w'C'}=-\kappa_z\frac{\partial \overline{C}}{\partial z}. \tag{5}$$

The eddy diffusivities, k_x, k_y and k_z in eqn. (5), are used instead of molecular diffusivities and the negative sign stands for movement counter to the local gradient. The set of equations, eqn. (5), are then respectively substituted into last three terms on the left side of the eqn. (4). The terms are expanded using associative properties of the derivative operator and we find that:

$$\frac{\partial \overline{C}}{\partial t}+\overline{u}\frac{\partial \overline{C}}{\partial x}+\overline{v}\frac{\partial \overline{C}}{\partial y}+\overline{w}\frac{\partial \overline{C}}{\partial z}-\frac{\partial \kappa_x}{\partial x}\frac{\partial \overline{C}}{\partial x}-\kappa_x\frac{\partial^2 \overline{C}}{\partial x^2}+ \\ -\frac{\partial \kappa_y}{\partial y}\frac{\partial \overline{C}}{\partial y}-\kappa_y\frac{\partial^2 \overline{C}}{\partial y^2}-\frac{\partial \kappa_z}{\partial z}\frac{\partial \overline{C}}{\partial z}-\kappa_z\frac{\partial^2 \overline{C}}{\partial z^2}=S. \tag{6}$$

The eqn. (6) is the equation of tridimensional advection-diffusion time independent for non isotropic, k_x, k_y and k_z are variable in space and time, turbulence and with source presence. This equation is evaluated using the Lax-Wendroff one-step explicit method, which consists of the approximation for convection equation using Taylor series expansion and second-order, centered-difference, finite difference approximations techniques (Lax-Wendroff, 1960). The time derivative $\partial\overline{C}/\partial t$ is then evaluated and their respective three dimensional fields are projected forward through a short time step δt. The diagnostic equations are then applied to obtain dynamically consistent fields of the other dependent variables at time $t_0+\delta t$. The process is then repeated over succession of time steps to describe the evolution of the C. The time step δt must be short enough to ensure that the fields of the dependent variables are not corrupted by spurious small scale patterns arising from numerical instabilities. The higher the spatial resolution of the model, the shorter the maximum allowable time step and the larger the number calculations are required for each individual time step (Wallace, 2006). The numerical equation to be solved is:

$$C^{n+1}_{i,j,k}=C^{n}_{i,j,k}+S_{i,j,k}+ \\ -\frac{1}{2}\overline{u}\frac{\Delta t}{\Delta x}\left(C^{n}_{i+1,j,k}-C^{n}_{i-1,j,k}\right)-\frac{1}{2}\overline{v}\frac{\Delta t}{\Delta y}\left(C^{n}_{i,j+1,k}-C^{n}_{i,j-1,k}\right)-\frac{1}{2}\overline{w}\frac{\Delta t}{\Delta z}\left(C^{n}_{i,j,k+1}-C^{n}_{i,j,k-1}\right)+ \\ +\frac{1}{4}\frac{\Delta t}{\Delta x}\left(K^{n}_{i+1,j,k}-K^{n}_{i-1,j,k}\right)\left(C^{n}_{i+1,j,k}-C^{n}_{i-1,j,k}\right)+\frac{1}{4}\frac{\Delta t}{\Delta y}\left(K^{n}_{i,j+1,k}-K^{n}_{i,j-1,k}\right)\left(C^{n}_{i,j+1,k}-C^{n}_{i,j-1,k}\right)+ \\ +\frac{1}{4}\frac{\Delta t}{\Delta z}\left(K^{n}_{i,j,k+1}-K^{n}_{i,j,k-1}\right)\left(C^{n}_{i,j,k+1}-C^{n}_{i,j,k-1}\right)+K_x\frac{\Delta t}{\Delta x^2}\left(C^{n}_{i+1,j,k}-2C^{n}_{i,j,k}+C^{n}_{i-1,j,k}\right)+ \\ +K_y\frac{\Delta t}{\Delta y^2}\left(C^{n}_{i,j+1,k}-2C^{n}_{i,j,k}+C^{n}_{i,j-1,k}\right)++K_z\frac{\Delta t}{\Delta z^2}\left(C^{n}_{i,j,k+1}-2C^{n}_{i,j,k}+C^{n}_{i,j,k-1}\right) \tag{7}$$

To avoid this computational instability, we use the Courant Friederichs-Levy (CFL) stability condition limited to values lower than 0.8:

$$\mathrm{CFL}_{adv}=U_i\frac{\delta t}{\delta i}\le 0.8\,,\qquad \mathrm{CFL}_{diff}=K_i\frac{\delta t}{\delta i^2}\le 0.8\,. \tag{8}$$

In eqn. (8) the subscripts i ($i=1,2,3$) represents the components of the wind velocity and diffusivities coefficients in the longitudinal, lateral and vertical directions respectively. The initial source condition for Zimora is a puff release characterized by a Gaussian shape distribution of matter given by (Lyons 1990):

$$\overline{C}(t,x_s,y_s,z_s)=\frac{S(t,x_s,y_s,z_s)}{\sqrt[3]{4\pi t}\sqrt{K_{x_s}K_{y_s}K_{z_s}}}\exp\left(-\frac{1}{4t}\left(\frac{x_s^{\ 2}}{K_{x_s}}+\frac{y_s^{\ 2}}{K_{y_s}}+\frac{z_s^{\ 2}}{K_{z_s}}\right)\right). \tag{9}$$

In eqn. (9) the subscripts s referrers to the coordinates point of the source location. The remaining continuous release of the matter is represented as sequences of this puff by instantaneous releases from the source. Assuming the domain like a box, the boundary conditions are normal in the bottom and at the top of the PBL height:

$$k_z=0\vee z=0;\ z=z_i;\Rightarrow K_z\frac{\partial\overline{C}}{\partial z}=0, \tag{10}$$

and they are open at lateral domain:

$$\frac{\partial\overline{C}(t,x,y,z)}{\partial t}=-\overline{V}\frac{\partial\overline{C}(t-1,x,y,z)}{\partial t}. \tag{11}$$

2.1 Downscaling the wind field

The Zimora seeks to model the atmospheric dispersion with higher spatial resolution than his MPP provides. Hence we developed a multi layer scheme constructed using simple linear weighted average proper to downscale the low spatial resolution of the MPP into any multiple linear spatial scale. To figure out, let's use one dimensional case for simplicity; using two points A and B, representing two known values of any scalar quantity, spaced by a distance D kilometres between them, representing the low resolution. We want to downscale this last to a new high resolution by d kilometre. We evaluate any new point between them using weighted average techniques, attributing to A and B specific weight in each subinterval (D/d) in the new scale between A and B and use weight average. The mathematical expression for weight average is given by:

$$\overline{x}_n=\sum_{i=1}^{m}x_iw_i\Big/\sum_{i=1}^{m}w_i\ , \tag{12}$$

where x_i is the generator entity (A or B values), m is the numbers of points of known values, two in this example (A and B), being averaged and w_i is its weight associated to each interval where we desire evaluate. To determine the weights of A and B, as our example, we need two suppositions: a) At the point of view of A, its weight should contribute the most at his own point and in the B point should be null; b) At the point of view of B the rule is the same, weight of B is maximum in his own point and null in the A point. After that, we can guess mathematical expressions that agrees these suppositions and plays transference function role to the weights of A and B, such mathematically does not alters the values of A and B and allows interpolation between them. Generalizing for 4 known points U_A,U_B,U_C and U_D, equally spaced, like figure (1), we found these expressions:

$$H_{i,j}^{U_A}=\frac{1}{\delta x}*(p-\max(i,j)), \tag{13}$$

$$H_{i,j}^{U_D} = 1 - \left(\frac{1}{\delta x} * (p - \min(i,j)) \right), \tag{14}$$

$$H_{i,j}^{U_B} = H_{p+1-i,j}^{U_D}, \tag{15}$$

$$H_{i,j}^{U_C} = H_{j,i}^{U_B}. \tag{16}$$

In the set of equations (13-16) the i, j represents the grid node in the new scale which we are building, δx is the spatial scale factor that we are reducing from the initial grid, $p = \delta x + 1$ and H represents the weight function respective for its superscript known point in each new grid node. So, the set of equations (13-16) plays the role of W_i into a generalized two-dimensional form of the eqn. (12). The Zimora model assimilates the wind and turbulent diffusivities coefficients fields, with resolution of 20 km, from the output of the MPP; after that it increase those resolutions to 5 km. The fig. 1 shows the Zimora's non-staggered grid where the four black nodes represents the MPP output and the grey nodes represents the new resolution with unknown values to be interpolated for that field.

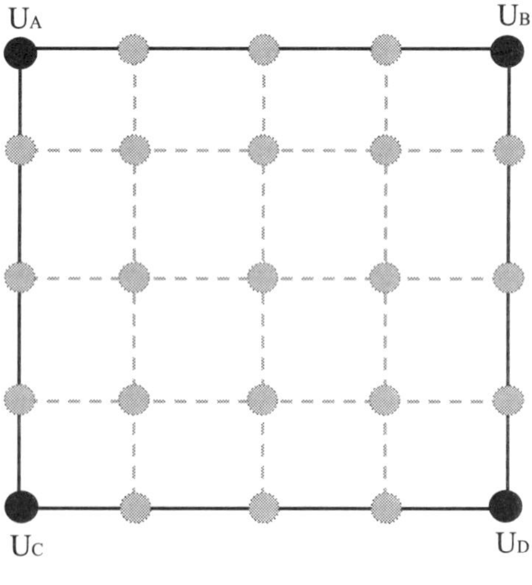

Figure 1: Example of non-staggered grid with four known points (black) and unknown points (grey).

Using the figure 1, example of data input scheme for Zimora model from mesoscale forecast model for the wind velocity, we can demonstrate a practical example. Let us assume that the velocities in the corners are (U_A=3.629, U_B=3.630, U_C=4.025 and U_D=3.994) for wind field in the 20 km spatial resolution from MPP output. In order to evaluate Zamora's' grey unknown points with spatial resolution of 5 km non-staggered grid, we first find the spatial resolution factor $\delta x = 20\,km/5\,km = 4$. Then, applying this into the set of eqn. (13-16) and letting the i, j vary for $p = 1$ to $p = \delta x + 1$, we get $p-1$ elements of a matrix $\widetilde{H}_{pxp}$. Calling each of these layers and using into a generalized two dimensional form of the equation (12):

$$Z_{ij} = \frac{\sum_{k=1}^{m} U_k H_{ij}^k}{\sum_{k=1}^{m} H_{ij}^k}, \tag{17}$$

here, the index k represents each one of known points U_A,U_B,U_C and U_D, and evaluating for i, j ranging from $p = 1$ to $p = \delta x + 1$, we get a $\widetilde{Z}_{5x5}$ matrix which represents the Zimora new nodes values as shown in table 1.

Table 1: Values of the final $\widetilde{Z}_{5x5}$ downscaled wind field in the new resolution non-staggered grid by interpolation using weight layers.

3.629	3.629	3.630	3.630	3.630
3.728	3.756	3.756	3.757	3.721
3.827	3.822	3.820	3.817	3.812
3.926	3.888	3.883	3.878	3.903
4.025	4.017	4.010	4.002	3.994

This same procedure is applied, repeated, until the last MPP output grid points, always in groups of four nodes. Generalizations are the same for vertical, changing only the spatial resolutions both in MPP output and Zamora's' grid points, such in the final process we construct volumetric cells of 5km x 5km x 100m.

3 The boundary layer parameterizations

In order to evaluate the diffusion coefficients that are in numerical equation, we have used the boundary layer parameterization proposed by Moraes [22]. This parameterization is based on Hanna [4] assumption that the mixing efficiency of the air is completely determined by the properties of the vertical velocity spectrum, and that the parameters, vertical dispersion parameter σ_w, and k_m are sufficient to describe the vertical velocity spectrum. Then it follows from dimensional reasoning that the vertical eddy diffusivity coefficient K_z may be written as

$$K_z = c_1 \sigma_w k_m^{-1} \tag{18}$$

where c_1 is a constant equal to 0.09.

From the measurements done in the surface layer and using local scalings assumptions we obtained for the vertical turbulent diffusion coefficient in the whole boundary layer the following expressions for stable and convective atmospheric stability regimes

$$K_z = c_1 z u_* (1 - z/h)^{\alpha_1/2} \frac{1 + (z/L)(1 - z/h)^{\alpha_2 - 1.5\alpha_1}}{0.27 + 0.50(z/L)(1 - z/h)^{\alpha_2 - 1.5\alpha_1}}, \quad (19)$$

$$K_z = 0.5 w_* z (1 - z/z_i), \quad (20)$$

where z is the height above ground, z_i is the boundary layer height, u_* is the friction velocity, α_1 and α_2 are constants determined experimentally (Moraes *et. al.* 1991; Moraes [22]), L is the local Monin-Obukhov length and w_* is the vertical convective scale.

The lateral diffusion coefficient is estimated as $0.1 w_* z_i$ in unstable conditions and as $2K_{Mz}$ in neutral-stable conditions, where K_{Mz} is the maximum of K_z.

4 Verification of Zimora against experimental data results

In order to test and validate the model a micrometeorological experiment was conducted in the 2007 spring season near a thermoelectric power plant located in the Candiota town, south of Brazil near Uruguay border. This power plant can be considered an ideal Laboratory since it is the major source of some pollutants in a large area and it is located in a region of smooth topography. The power plant has a 150-m height stack and the rate of emission of e.g. SO_2 is continuously monitored. The objective of the campaign was to collect a comprehensive, meteorological diffusion and dispersion database to be used to evaluate the dispersion numerical model. The campaign site (31.40S, 53.70W, 250 m above sea level) has previously been used for a number of experiments related to research on atmospheric boundary layer (for details and results from the micrometeorological point of view see Moraes [22]). It is located in a region known as the South America (SA) Pampa. The Pampa is the part of SA that covers all Uruguay territory, the south of Brazil and northwest of Argentina

Meteorological data was collected over a 5-week time resulting in a full set of daytime period and the database comprises measurements of wind speed, wind direction, turbulence, temperature, humidity, pressure, solar radiation, the boundary layer structure and upper-air-soundings. The data records were broken into 1-hr segments and those collected at 10 Hz were used to compute micrometeorological parameters (u_*, w_*, L) and data collected at 1 Hz were used to estimate wind speed, wind direction and air temperature. Tethered balloon was launched each two hours and gives information mainly of the evolution of the PBL height. Ground level concentrations, used in this paper, were measured in one position located 6.8 km to the southwest of the source. The concentration data set consists of three hours values of SO_2.

Figure 2 and 3 below as well as table 2 presents the model performance.

Figure 2 is an example of the SO_2 plume evolution for one day of simulation. Each panel has the hourly mean concentration for some specific hours of the day. The selected day choose for this presentation was characterized by a cold front

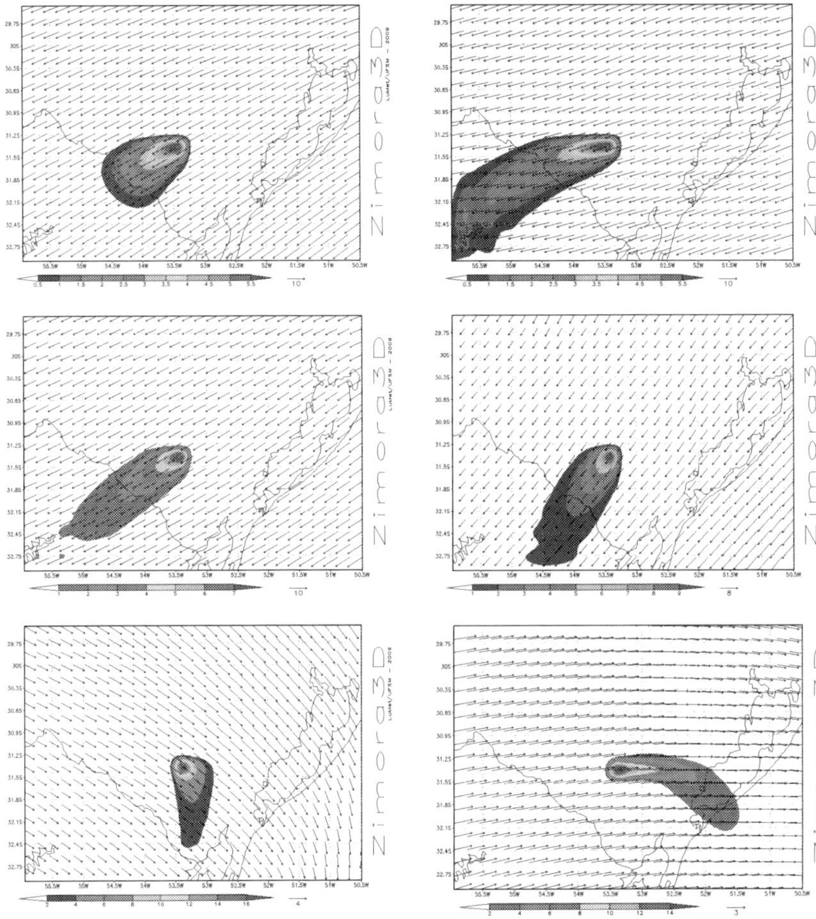

Figure 2: From left to right and top to bottom, 1, 4, 8, 12, 16 and 20 hours SO_2 plume time evolution simulated for the day 27/09/2008. The concentrations in the bar scale are $\mu g / m^3$ and wind speed scale is *m/s* which magnitude scale is shown by the arrow over number on the bottom corner of each graph.

that passed in the region and we can see that the model represents very well the change of the wind direction due this cold front.

To avoid some misunderstanding, due to the gray scale shaded allowed by data visualization software, as far from the center point the shaded contour refers for the leftmost concentrations values in the bottom scale.

Figure 3 is the comparison between model prediction and observations of SO_2 ground concentration. Table 2 presents some statistical indices, defined as the normalized mean square error (nmse), correlation coefficient (r), factor of 2 (fa2), fractional bias (fb), and fractional standard deviation (fd). It should be stressed that from the observed concentration values no background value has been subtracted.

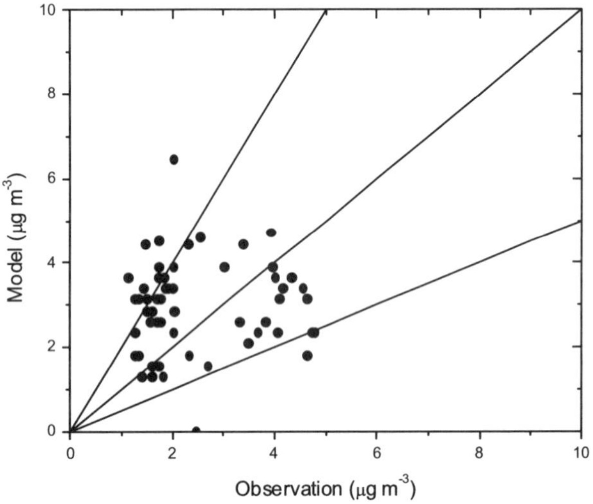

Figure 3: Model versus observed SO_2 data comparison scatter plot.

Table 2: Statistical index evaluation of the model prediction.

nmse	r	fa2	fb	fs
0.14	0.78	0.92	-0.12	-0,221

5 Conclusion

The numerical scheme adopted sounds to agree very well for simulating the advection-diffusion equation. As a result, the simulated time concentrations plume evolution, for all the used datasets, presented a good agreement with behavior of the geophysical wind field acquired from the RAMS and downscaled by Zimora model.

The comparison, between observed versus modeled data by Zimora, indicates that for the cases when low observed concentrations happens, the model overestimates those concentrations. The cases when observed concentrations data are higher, the model sub estimates those concentrations. Therefore, in the general, the model satisfactorily describes the observed data.

The interface, between numerical regional meteorological model and the Zimora numerical dispersion model works fine, showing ableness synergy between these two systems.

For future enhancements, it can be implemented into Zimora model, a lot of new relevant features, like automation between source emission monitor, enhance the parameterizations and development of graphical interfaces to uses it as air quality monitor and virtual research lab.

Acknowledgements

This work was supported by the Companhia de Geração Térmica de Energia Elétrica (CGTEE) and by the Brasilian Agency for the Development of Science and Technology (CNPq).

References

[1] Barnes, L. S.: 1964,A Technique for maximizing Details in Numerical Weather Map Analysis, Journal of Applied Meteorology, Vol. 3, 396-409.

[2] Bendat, J. S., Piersol, A. G.:1990, Random Data - Analysis And Measurement Procedures, 2nd Ed., Wiley, New York.

[3] Cressman, G. P.:1959,An Operational Objective Analysis System, Monthly Weather Review. 87, 367-374.

[4] Hanna, S. R.:1968, A method of estimating vertical eddy transport in the planetary boundary layer using characteristics of the vertical velocity spectrum., Journal of Atmospheric Sciences, Boston, Vol. 25,1026.

[5] Haugen, D. A.:1978,Effects os sampling rates and averaging periods on meteorological measurements., In. Proc. Fourth Symposium on Meteorological Observations and Instrumentation, April 1978, Denver, CO, American Meteorological Society, Boston, MA, 15-18.

[6] Hibbard, W. L., Wylie, D. P.:1985,An Efficient Method of Interpolating Observations to Uniformly Spaced Grids., Conf. Interactive Information and Processing systems (IIPS) for Meteorology, Oceanography and Hydrology.

[7] Hinze, J. O.:1975, Turbulence., McGraw-Hill.

[8] Jesus, S.M.:1999/00, Estimação Espectral e Aplicações,Curso: Engenharia de Sistemas e Computação,Faculdade de Ciências e Tecnologia, University of Algarve.

[9] Hoffman, J. D.. 1993. Numerical Methods for Engineers and Scientists. Mechanical Engineering Series, McGraw- Hill International Editions.

[10] Kaimal, J. C and Finnigan, J. J.:1994, Atmospheric Boundary Layer Flows: Their Structure and Measurements, Oxford Press. 289pp.

[11] Kovalets, I. V., Tsiouti, V., Andropoulos, S., Bartzis, J. G.:2008, Improvement of Source and Wind Field Input of Atmospheric Dispersion Model by Assimilation of Concentration Measurements: Method and Applications in Idealized Settings, Applied Mathematical Modelling, In Press.

[12] Lamb R. G. (1978). A numerical simulation of dispersion from an elevated point source in the convective boundary layer. Atmospheric Environment 12, 12971304

[13] Lax, P.D., Weak solutions of nonlinear hyperbolic equations and the numerical computation, Comm. Pure Appl. Math.7 (1954), 159-193.

[14] Lax, P.D., Hyperbolic systems of conservation laws II, Comm. Pure Appl. Math., 10 (1957), 537-566.
[15] Lax, P. D., and Wendroff, B. (1960). Systems of conservation laws. Commun. Pure Appl. Math. 13, 217-237.
[16] Lin, C. C.:1953, On Taylor's Hypothesis and the Acceleration terms in the Navier-Stokes Equation., Quarterly Applied Math., 10:295
[17] Lin, C. C; Reid, W.H.:1963, Handbuc der Physic., volume VIII/2, chapter Turbulent Flow. Theoretical Aspects., Springer-Verlag, 438-523.
[18] Lisieur, M.: 1997,Turbulence In Fluids, 3rd Ed., Kluwer Academic Publishers, Dordrecht.
[19] Lumley, J. L., Panofsky, H. A.: 1964,The Structure Of Atmospheric Turbulence, John Wiley & Sons, London, 232pp.
[20] Lyons, T. J. and Scott, W. D.: 1990, Principles of Air Pollution Meteorology, Belhaven Press, 22pp.
[21] Moraes O. L. L., R. C. M. Alves, R. Silva, A. C. Siqueira, P. D. Borges, T. Tirabassi, U. Rizza. 2001 25th NATO/CCMS International Technical Meeting on Air Pollution Modeling and its Application, Louvain-la-Neuve 15-19, Belgium, p. 359.
[22] _____.:2000. Turbulence Characteristics In The Surface Boundary Layer Over The South American Pampa. Boundary-Layer Meteorology 96: 317-335.
[23] _____., Ferro, M., Alves, R.C.M., and Tirabassi, T 1998.Estimating eddy diffusivities coefficients from spectra of turbulence. Air Pollution, v. VI, p. 57-65.
[24] _____., Degrazia, G.A., and Tirabassi, T. 1998 Using the Prandtl-Kolmogorov relationship and spectral modeling to derive an expression for the eddy diffusivity coefficient for the stable boundary layer. IL Nouvo Cimento, v. 20D, n. 6, p. 791-798.
[25] Panofsky, H. A.:1949,Objective Weather Map Analysis, Journal of Meteorology, No. 6, Vol. 6, 386-392.
[26] Seaman, L. N.:2000, Meteorological Modeling for Air-Quality Assessments, Atmospheric Environment, N. 34, 2231-2259.
[27] Slade, D. H.:1968, Meteorology and Atomic Energy-1968, USAEC, TID-24190, 445pp.
[28] Stull, R. B.:1988, An Introduction to Boundary Layer Meteorology, Kluwer Academic, 637pp

The application of GIS to air quality analysis in Enna City (Italy)

F. Patania, A. Gagliano, F. Nocera & A. Galesi
Energy and Environment Division of D.I.I.M.,
Engineering Faculty of University of Catania, Italy

Abstract

Air quality models play a very important role in formulating air pollution control and management strategies by providing information about better and more efficient air quality planning.

This paper describes an urban air quality modelling system for evaluating the environmental effects of transport related air pollution. A preliminary evaluation of the model has been performed for one site in Enna City using data from an extensive monitoring and measuring campaign. Then the authors utilize a geographic information system (GIS), which integrates a vehicle emission model, pollutant dispersion model and related databases to estimate the emissions and spatial distribution of traffic pollutants. The model can not only analyze the current pollution situations, but also can predict the emissions influenced by exhaust coming from changes in specific traffic conditions or management policies.
Keywords: CALINE 4, GIS, air pollution.

1 Introduction

Air quality models have a very important role in developing air pollution control and management strategies. Line source models are often used to simulate the dispersion of pollutants near the highways, the roads, roundabouts, bypasses and so on. These dispersion models, mostly Gaussian based, despite several assumptions and limitations, are used by the regulatory agencies to carry out air pollution prediction analysis due to vehicular traffic near the roads as a part of the Environmental Impact Assessment (EIA) procedure.

WIT Transactions on Ecology and the Environment, Vol 123, © 2009 WIT Press
www.witpress.com, ISSN 1743-3541 (on-line)
doi:10.2495/AIR090071

The integration of the geographical information system (GIS) with the existing air pollution modelling approach can help the decision making process of the transportation planners who design and implement control strategies.

Authors used GIS in developing microscale analyses of the old town centre of Enna City. They linked a GIS with CALINE 4 to predict pollutant concentration levels.

The value of GIS (outside of spatial data storage and data visualization) was its ability to compare concentration results to other non-related data.

2 Methodology of research

The authors have developed and applied a model, which integrated several submodels, such as the emission estimation model, dispersion model and relevant spatial and attribute data in a GIS framework. The entire system is applied to estimating emissions from motor vehicles and analyzing the spatial distributions of air pollutants in the old town centre of Enna city. Figure 1 shows the structure of the entire system.

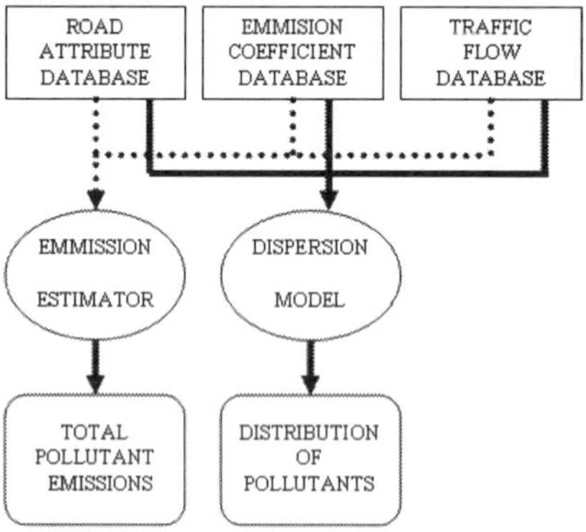

Figure 1: Structure of the GIS system.

The GIS platform used in this research is ArcGIS, a software developed by ESRI, which provides a variety of tools for spatial data exploration, identification of data anomalies, optimum prediction, evaluation of prediction uncertainty, and surface creation.

The research has been carried out in accordance with following steps:

- data acquisition and analysis of traffic flows;
- analysis of meteoclimatic historical data;
- application of COPERT III emission Model
- application of CALINE 4 quality model implemented by Authors
- validation of adopted model through comparison between polluting concentration data forecasted and that ones measured in situ during the campaign of measurements.
- application of GIS Framework

3 Study case

The investigated area is located in the old town centre of Enna city as shown in figure 2. The pollutants selected for study are Carbon Monoxide (CO) and Particular Matter (PM_{10}) because they are the typical gas emitted by vehicular traffic.

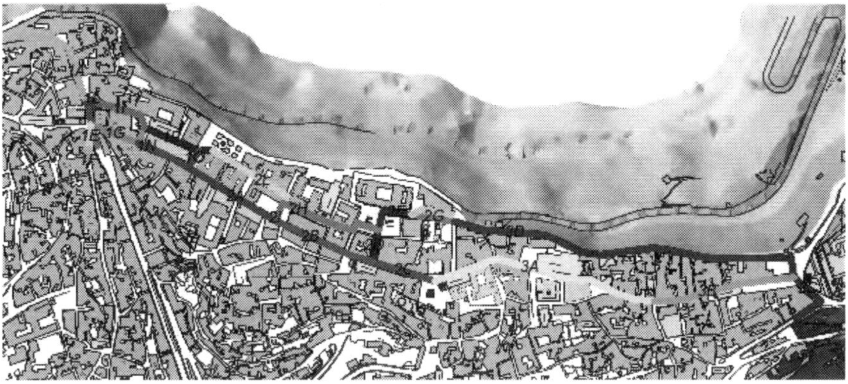

Figure 2: Old town centre of Enna and linked road.

Firstly, the Authors have identified twenty-seven main roads within the survey area and each road was characterized by geometrical features (typology, width and length of carriage way), traffic parameters and georeferenced coordinates.

Geometrically the streets, which presents the roadway less ten meters wide, are continuously bordered in both sides by buildings that are between twelve and fourteen meters high on average.

Basic data on the roads, buildings and other features of Enna City were digitized on a scale of 25.000:1 with AcrGis software. Each segment of the road network was assigned a road code. The road code was the same of link code of CALINE 4.

To monitor the traffic flows people used Manual Classified Counts (MCCs) placed in eight point for a period of fifteen weekdays. People take in account two-wheeled motor vehicles, cars and taxis, buses and coaches, light and heavy vans for each direction of the twenty-seven roads. Figure 3 shows the bar chart of daily average and maximum traffic flows through the twenty-seven investigated roads.

As regard meteoclimatic data, people adopted the historical data series collected by meteo stations of "Enna Ambiente Company" from 1994 to 1996.

To determine the meteoclimatic conditions associated to traffic flow, it was utilized a portable meteoclimatic station (Testo 452) placed at 1,30 m up the ground. It was discovered that wind velocity was always less than 1,5 m/s.

To estimate the amount of polluting gasses produced along the twenty-seven roads, the "Computer Programme to calculate Emission from Road

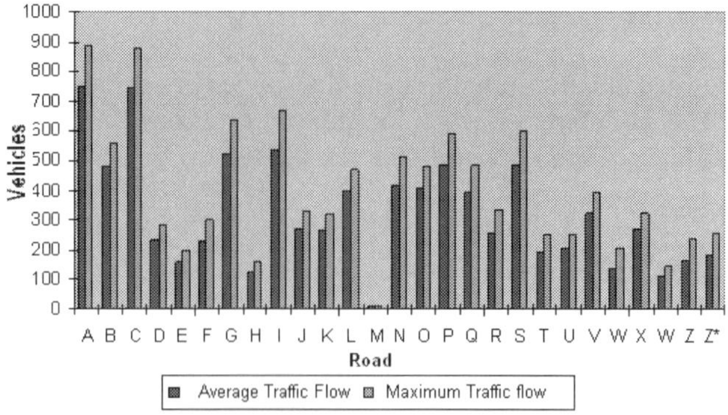

Figure 3: Daily average and maximum traffic flows.

Transport" (COPERT III) was adopted. The programme bases calculus on "emission factor approach" and values the emissions of each category of vehicles by "emission curves speed dependent". The calculus is based on the knowledge of parameters as likes as:

- Composition of circulating vehicles fleet through the investigated areas in function of both kinds of feeding (petrol, diesel oil, GPL) and regulation of reference (conventional or euro). In this case it was used data given by ACI (Automobile Club of Italy),
- The global emissions of circulating fleet.
- The cycle of drive in percentage of kind of run (urban, outside the town, motorway, etc.) the mean velocity for each cycle of drive. The velocity was determined with a campaign of measurements
- Year's run for each class of vehicle (total km/year· vehicle).
- Some other parameters as like as features of fuels, climatic conditions and so on.

COPERT III programme generates as intermediate results both the "emission factor" referred to each pollutant (g/km) for each kind of vehicle and the pollutant weigh (tons) that each kind of vehicle gives out during investigated year. In our case it was necessary to calculate "emission factor" in "main weighted mode" to take in account of real contribution that each category of vehicles gives for each polluting substance of the investigated area.

4 Experimental measurements and equipments

People carried out a measurement campaign of pollutant concentrations to be able to compare data forecasted by CALINE 4 quality model and those coming from measurements. Measures have been done by means of portable analyzers at the same time of the measures of traffic flows.

GRIMM MODEL 107 was utilized to measure the fractions of PM10, PM2,5 and PM 1,0 in the air while Multi Rae Plus PGM 50 was utilized to measure concentrations of various gasses (CO, CO_2).

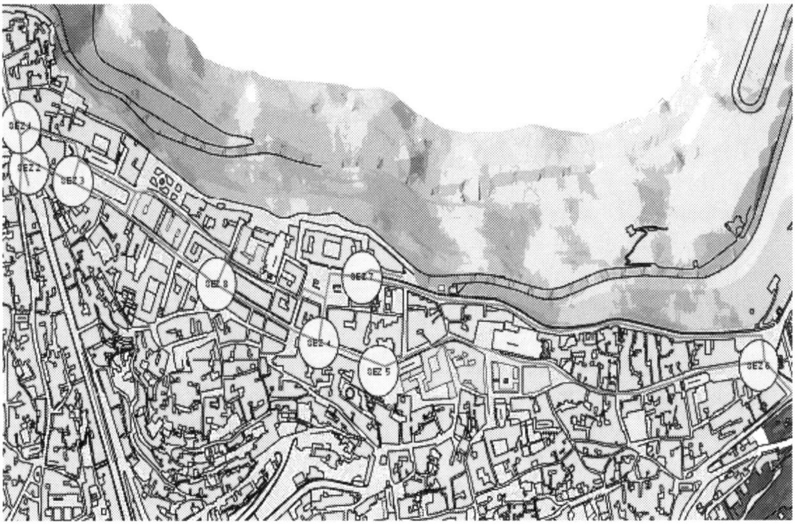

Point of measurement

Figure 4: Points of measurement.

Table 1: Exemplum of tabular datasheet.

Point	Hour	CO	PM 10	PM 2.5	PM 1.0
	8.57		75.80	12.20	9.40
	8.58		55.50	12.00	8.80
	8.59		91.50	17.30	12.20
	9.00		57.10	18.80	16.20
	9.01		25.70	9.00	7.00
	9.02		42.40	9.30	6.80
	9.03		54.70	13.30	8.90
1	9.04	AVERAGE	46.90	13.20	8.30
	9.05		65.90	9.70	6.60
	9.06		30.50	7.40	5.30
	9.07		47.50	8.30	6
	9.08		39.10	9.40	6.50
	9.09		28.20	8.50	5.90
	9.10		39.60	7.60	5.50
	9.11		37.20	8.70	5.80
	15 MIN	1.9	49.2	11.0	7.9

Figure 4 shows the map of the position of each point of measurement in the investigated area.

Both measurements of polluting gases and meteoclimatic conditions have been carried out as follows:

- Four days of measurements for each station: three in working days and one in holiday period
- Four interval of time (Observation Time) during each day of measurements, that is: 06-10h; 12-15h; 18-21h; 18-22h; 22-06h.

- Fifteen minutes of measurements for each Observation Time

Results coming from measurements have been collected in tabular data. Table n.1 shows an exemplum of tabular datasheet.

5 Dispersion model

The modellation of investigated areas has been done by CALINE 4 Caltrans implemented by authors.

CALINE–4 is a fourth generation line source air quality model developed by the California Department of Transportation that predicts CO and PM_{10} impacts near roadways. Its main objective is to assist planners to protect public health from adverse effects of excessive CO and PM_{10} exposure. The model is based on the Gaussian diffusion equation and employs a mixing zone concept to characterize pollutant dispersion over roadways. For given source strength, meteorology, site geometry and site features the model can reliably to predict (1-hour and 8-hours) pollutant concentrations for receptors located within 150 meters of the roadway.

People defined the area of study by specific calculus grid as shown in Figure 4 and in Table 2. The origin of the grid was referred to Gauss – Boaga Coordinates (2456472.819 – 4158201.432)

The roads network was subdivided into 60 stretches. Each of previous related sixty stretches was considered as one linear source of emission being each of them homogeneous as regard:

- Geometrical features: typology, width of carriage way and altitude with reference to local ground.
- Traffic features: yearly amount of running motor vehicles and their typology.

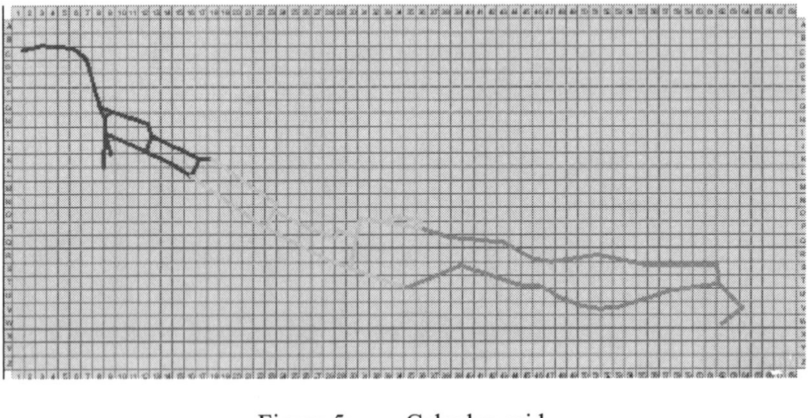

Figure 5: Calculus grid.

With reference to definition of angles of wind, people chose azimuthal ones regarding the North direction. The ruggedness of ground surfaces around the road route has been considered unvarying in all area crossed by single stretch.

The meteorologic variables of atmospheric stability (wind velocity and direction) have been considered unvarying too for each stretch.

Table 2: Calculus grid.

GRID	
Origin: x-axis; y-axis	2456472.819 m; 4158201.432 m
N point x-axis; y-axis	84; 26
Cell Dimension Dx Dy	20 m; 20 m
Mean Altitude	290 m

As regard meteoclimatic data, it was adopted one set of data linked to all "stability class" adopting wind velocity always less than 1,5 m/s as discovered through the campaigns of measurements for investigated area. As regard wind direction, measurements in situ show a predominant wind direction orthogonal to the streets.

Table 3: Exemplum of tabular datasheet used to compare measured and forecasted CO concentration values.

Pollutant CO	Measured		Forecasted (CALINE 4)		ERROR	
N. RECEPTOR	Start Hour	Average Value. PPM	Hour Range	CO PPM	%	ASS PPM
2B-1	08.57	1.9	09-10	1.8	5.26	0.1
6C-2	09.20	1.7	09-10	1.9	11.76	0.2
	18.19	4.9	18-19	4.5	8.16	0.4
7E-1	19.27	4.9	19-20	4.7	4.1	0.2
8G-1	09.36	2	09-10	2.5	25	0.5
	18.39	3.5	18-19	4.7	34.2	1.2
8I-1	09.54	1.8	09-10	1.8	0	0
9I-3	10.20	1.6	10-11	1.7	6.2	0.1
10H-1	17.57	2.2	18-19	3.7	68.7	1.5
11J-1	10.36	1.9	09-10	1.8	5.26	0.1
	19.09	2.3	18-19	4.4	42.0	2.1
12I-1	09.22	1.4	09-10	2.0	42.8	0.6
14K-3	10.53	1.8	10-11	1.9	5.5	0.1
17L-1	08.09	3.4	08-09	2.1	38.2	1.3

The validation of model has been done through comparison between data coming from campaigns of measurements, as previously related, and the ones by application of CALINE 4 model: both data related to the same time of observation.

Table 2 shows an exemplum of tabular datasheet used to compare measured and forecasted values.

The tabular datasheet, also, allows to determine the range of diverging, that is the divergence of the values, between the two series of data (measured and forecasted) so to discover if such a difference can cause the overcoming of limits of concentrations fixed by rules in each single receptor.

6 Results of simulations

Subsequently, a geostatistical interpolation technique, known as universal kriging, was employed to map the concentration of pollutants (Carbon Monoxide, Nitrogen Dioxide, and Particular Matter) over the whole area of study. The general concept is that the prediction of the value $\hat{C}(s)$ of pollutant concentration at any location s (spatial coordinates) is obtained as a weighted average of neighboured data:

$$\hat{C}(s) = \sum_{i=1}^{n} w_i(s)C(s_i) \tag{1}$$

The ultimate goal is to estimate the optimal values of the weights $w_i(s)$, i=1…n. This is accomplished by selecting the weights so that the expected mean square error becomes as small as possible.

So the estimation of the CO, and PM_{10} concentration, using universal kriging, has required the calibration of a variogram model. Figure 5 shows the empirical variogram of PM_{10} daily concentration, figure 6 shows the comparison of measured and predicted daily value of PM_{10} while figure 7 shows the standardized error. Figure 8 displays PM_{10} concentration map using Kriging geostatistical interpolation. Four main hot spot area of maximum concentration can be identified in the PM_{10} pollution map. Table 3 summarize the prediction errors of the universal kriging estimates for PM_{10} concentration

Table 4: Prediction errors of universal kriging for PM_{10} concentration.

Prediction Error	Value
Mean	-0.002684
Root Mean Square	4.625
Average standard Error	4.11
Mean standardized	-0.003464
Root Mean Square Standardized	1.127

The value of mean error (-0.002684) being close to zero, indicates that the predicted values are unbiased. Similar information is provided by the mean standardized prediction error (-0.003464). Also, the average standard error (4,11) is lower than the root-mean square of predicted errors (4.625). This shows that our model slightly under-estimates the variability of PM_{10} concentration. The root-mean square prediction error is a measure of the error that occurs when predicting data from point observations and provides the means for deriving confidence intervals for the predictions. Finally the root mean square standardized (1.127) is very close to 1, and thus corresponds to a very good fit between the point estimates of CALINE 4 and the geostatistical model using universal kriging.

Table 4 summarize the prediction errors of the universal kriging estimates for CO concentration

Table 5: Prediction errors of universal kriging for PM_{10} concentration.

Prediction Error	Value
Mean	0.0000568
Root Mean Square	0.05256
Average standard Error	0.05475
Mean standardized	0.0004092
Root Mean Square Standardized	0.9666

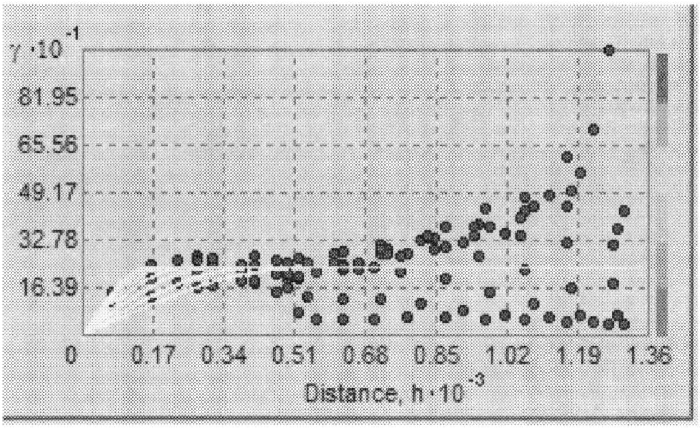

Figure 6: Variogram of PM_{10} daily concentration with average traffic flow.

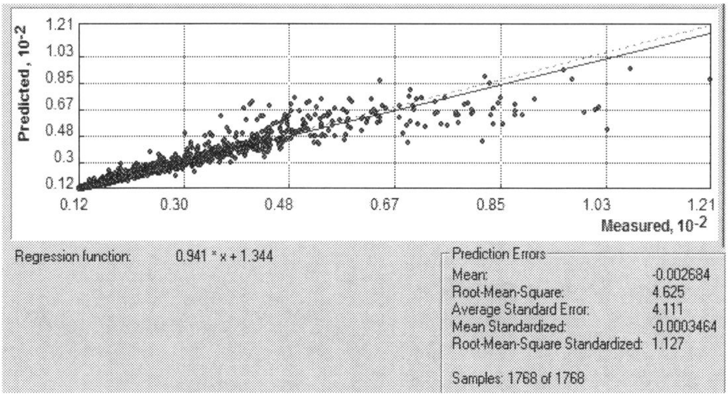

Figure 7: Comparison of measured and predicted daily value of PM_{10} with average traffic flow.

Like so PM_{10} concentration, the value of mean error (0.0000568) and the mean standardized prediction error (0.000409) are very close to zero. On the other way, the average standard error (0.05475) is greater than the root-mean square of predicted errors (0.05256). This shows that our model over-estimates the variability of CO concentration. Finally, the root mean square standardized (0.966) is very close to 1, and thus corresponds to a very good fit between the point estimates of CALINE 4 and the geostatistical model using universal kriging.

Figure 9 displays CO concentration map using Kriging geostatistical interpolation.

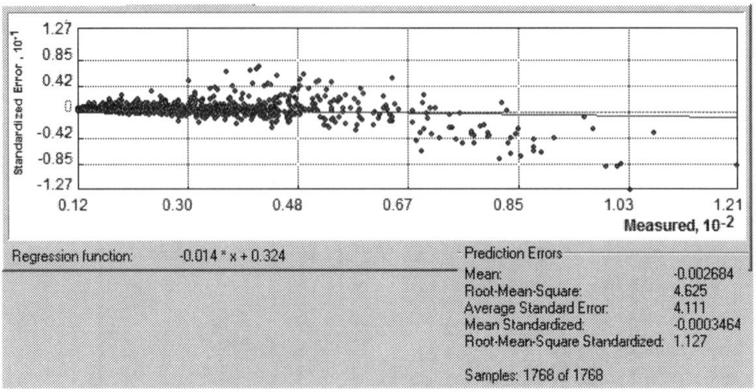

Figure 8: The standardized error of the interpolated map of PM_{10} concentration.

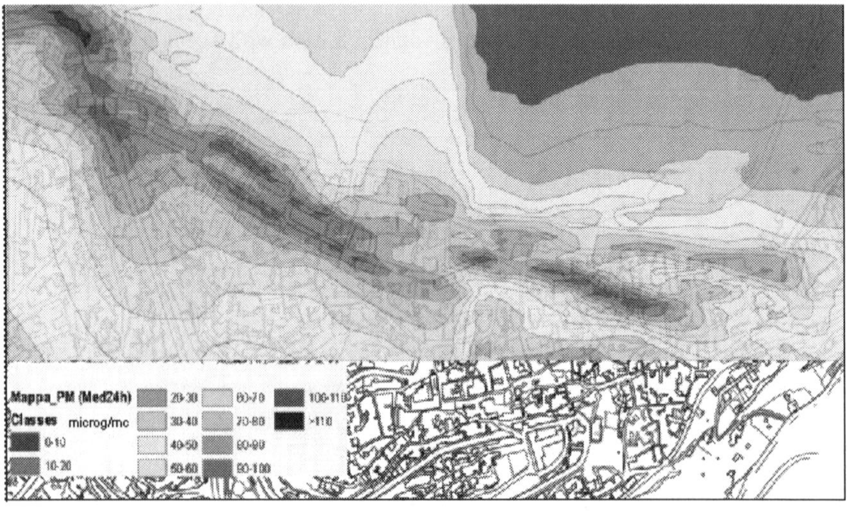

Figure 9: Map of PM_{10} daily concentration.

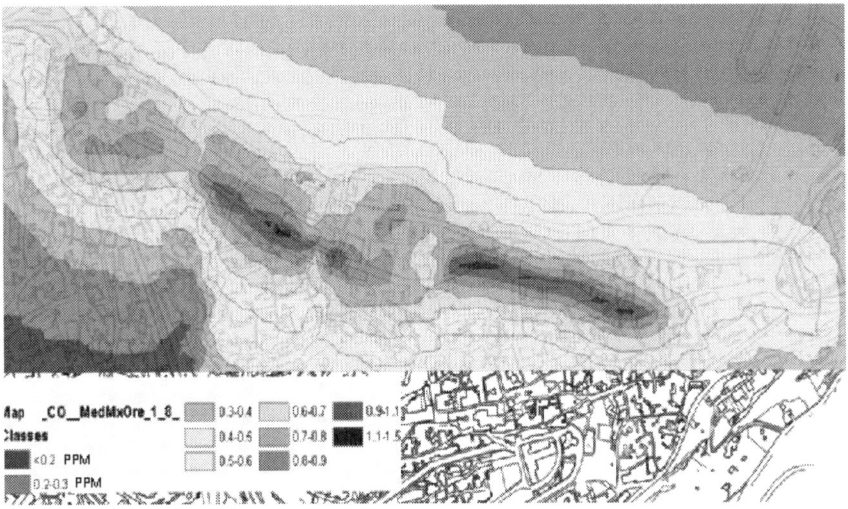

Figure 10: Map of CO concentration (8h).

7 Conclusions

This paper has described a methodology that provides to estimate vehicular air pollution and mapping using the CALINE 4 dispersion model and spatial data analysis. Forecasted traffic volumes and emission factors are used as input to CALINE 4, which estimates CO and PM_{10} concentrations at predefined point locations. Afterwards, universal kriging implemented by AcrGis is used to estimate and map the concentration of the pollutants over the study area. The results of this research show that the visualization and analytical features of GIS provide more information and convenience to users. It also makes the model more efficient and flexible.

The results presented in this paper are preliminary outcomes of ongoing research. The model will be further developed, e.g. expanding its submodels to include things such as transportation models and risk assessment models, with the eventual goal being to develop an Urban Air Quality Decision Support System.

References

[1] USEPA (1998). A GIS based modal model of automobile exhaust emissions. USEPA Report No. EPA-600/-98 –097, Research Triangle Park, NC (USA).

[2] European Environment Agency, "COPERTIII - Methodology and emission factors" Anno 2000

[3] European Environment Agency, "COPERTIII-User's manual" Anno 2000

[4] California Department Of Transportation, " CALINE4 – A dispersion Model for Predicting Air Pollution Concentrations Near Roadways" 1984

[5] California Department Of Transportation, " User's guide for CL4: a user friendly interface for the CALINE4 model for transportation project impact assessment" giugno 1998
[6] Autoritratto ACI – Provincia Enna, 2002
[7] Agenzia Nazionale per la Prevenzione e l'Ambiente, "Le emissioni in atmosfera da trasporto stradale", Serie Stato dell'ambiente, n.12, 2000
[8] ENEA, ERG-SIRE DBT, Dati Temperatura Enna 2000
[9] ENNA AMBIENTE, Monitoraggi Ambientali Periodo 1994–96
[10] Johnston, Ver Hoef, "Using ArcGis Geostatistical Analyst", GIS by ESRI

Section 2
Air management and policy

Effects of road traffic scenarios on human exposure to air pollution

C. Borrego, A. M. Costa, R. Tavares, M. Lopes, J. Valente,
J. H. Amorim, A. I. Miranda, I. Ribeiro & E. Sá
CESAM, Department of Environment and Planning,
University of Aveiro, Portugal

Abstract

Human exposure to air pollution has been identified as a major problem due to its known impact on human health. Particulate matter is a pollutant which rises special concern due to the adverse health effects on sensitive groups of the population, such as asthmatic children. This study is part of the SaudAr research project which main objective was to assess the air quality effects on the health of a population group risk (asthmatic school children) living in an urban area (Viseu). The aim of this paper is to investigate the influence of road traffic emissions on air quality and consequently, on human exposure. For this purpose, the CFD model VADIS integrating an exposure module has been applied over the town of Viseu, for the periods of one week in winter and one week in summer, to four different situations: the reference year (2006) and three future scenarios for the year 2030, BAU, Green and Grey scenario. The differences among the scenarios include changes on the existing land use, the vehicle fleet composition, the mobility, the vehicle technologies and the fuel types. Field campaigns were performed in order to obtain information about vehicle fleet in the town of Viseu and mobility patterns. The quantification of road traffic emissions and the hourly traffic emissions patterns for all scenarios was carried out by the application of the TREM model. The results reveal an increase in PM_{10} emissions, concentrations and exposure in all future scenarios, particularly in winter with an increase around 80% in the BAU and Grey scenarios and only 34% in the Green scenario.

Keywords: traffic emissions, air quality modelling, human exposure, development scenarios.

WIT Transactions on Ecology and the Environment, Vol 123, © 2009 WIT Press
www.witpress.com, ISSN 1743-3541 (on-line)
doi:10.2495/AIR090081

1 Introduction

Despite the fact that in some European urban areas the atmospheric emissions resulting from industry, power production and households contribute substantially to air pollution, the dominant source is the continuing growth in road transport [1]. According to statistics compiled by the European Environment Agency [2], a significant proportion of Europe's urban population still lives in cities where EU air quality limits for the protection of human health are regularly exceeded, stressing the importance of air pollution as one of the major environment-related health threats in Europe. The impacts of air pollution on human health can be expressed in terms of a reduction in average life expectancy and increased premature deaths, hospital admissions, use of medication and days of restricted activity.

Especially the air pollution caused by fine particles represents the highest risk to public health in Europe [1]. On the basis of the emissions in 2000, the EU's CAFE programme estimated a total of 348 000 premature deaths per year due to exposure to anthropogenic $PM_{2.5}$ [3]. In agreement, data from the World Health Organization [4] show that, as a result of human exposure to particulate matter, approximately three million deaths occur globally per year.

In particular the traffic-related air pollution was estimated in 2001 to account, each year, for more than 25 000 new cases of chronic bronchitis in adults, more than 290 000 episodes of bronchitis in children, more than 0.5 million asthma attacks, and more than 16 million person-days of restricted activity [5]. Several studies in adults and children have shown that asthma, among other respiratory health problems, and cardiovascular diseases are aggravated by the exposure to traffic-emitted pollution in urban areas [4, 6–8].

The large fraction of European citizens living in urban areas (about 80%), the continuous growth of road transport, the significant number of air pollutants emitted by road traffic and the close proximity between the emission source and the human receptors are responsible for the contribution of urban air pollution to various health problems and a decreased quality of life. In this context, traffic emission scenarios are privileged tools for evaluating the contribution of road transport to future atmospheric emissions [9–12], while air quality modelling is a useful tool for population exposure studies, especially when exposure monitoring is not applicable and/or air quality monitoring data are not representative for the area of interest [13].

The research on the relations among air pollution, human exposure and road transport, is important in the support to sustainable transport policies, emission control measures, traffic management and urban planning. Therefore, the aim of this study is the assessment of the effects of current and future road traffic emissions on the human exposure to urban air pollution. A number of development scenarios were defined by projecting the expected changes for the year of 2030 in terms of land use, vehicle fleet composition, mobility, and vehicle technologies, in a town in mainland of Portugal characterised by a significant population growth and an expecting economic development.

2 Study approach overview

This study is part of a large project involving researchers from environmental and medical sciences – the "The Health and the Air we breath" (SaudAr) Project [14] – which main objective was to contribute for the sustainable development of a region by preventing air pollution problems and health related diseases in the future due to expected economical growth. The project wide methodological approach is presented in Figure 1. A detailed description of the project is made by Borrego et al. [14].

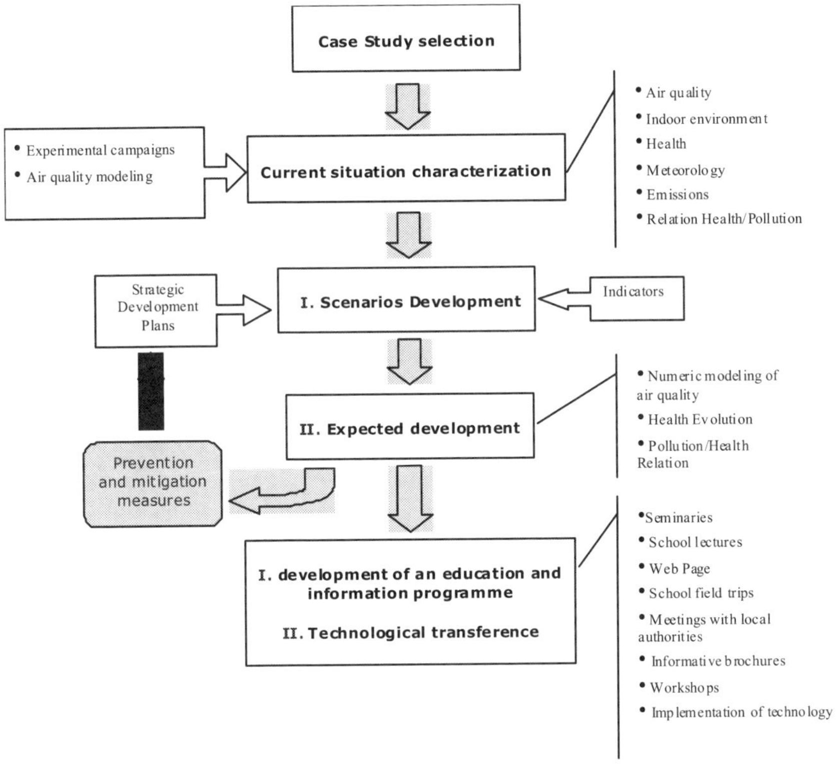

Figure 1: Global scheme of the SaudAr project.

This paper explores the impact on emissions, air quality and individual exposure of different development scenarios for the year 2030. The study analyse the effects related to different mobility patterns connected to three different development scenarios by comparison with the current situation (reference scenario).

The BAU (Business As Usual) scenario keeps the trends in land use and emissions patterns of the last decades. Hence the total emission amount will increase in accordance only to the increase of the total population projected to the year 2030. The other two scenarios, named Green and Grey, try to reflect different options on future development as described forward.

The Green scenario considers a change on land use pattern based on a compact dense occupation in the city centre that is consolidated and rehabilitated. The use of "soft" modes of transport (walking and cycling) becomes an option. More efficient and cleaner public transportation modes are also considered; thus a decreased use of the private cars is expected. The mileage per capita is also decreased due to the compact land use.

Like the Green scenario, the Grey scenario considers a change on land use pattern, which is however based on a more dispersed land use. The city centre is now consolidated, but the new areas of expansion are further away from the centre, have lower densities and are more sprawl with several small household's cores. This means that a higher number of daily trips contribute to an increase of mileage per capita. With a slightly smaller population on the central villages, the implementation of more efficient public transport systems is not so feasible. The "soft" modes, walking and cycling are not also attractive in this kind of land use. Private cars are then the best economical solution for most population displacements, which increased traffic related emissions.

For this work, a town in the central part of Portugal – Viseu - was chosen taking into account its current characteristics in terms of air pollution and urban development. A group of seven children with asthma problems attending the same classroom of the Marzovelos elementary school were selected as study case. These children live, study and have their activities within the selected study domain located in the centre of Viseu (Figure 2). The houses where the selected children live are located in the same figure and are identified by the numbers 1 to 7.

The individual exposure to PM_{10} for the seven children was estimated for the winter and summer periods of the experimental campaigns conducted under the SaudAr project, at 20-27 January 2006 and 19-26 June 2006, and for the three defined scenarios. The individual exposure was estimated using the microenvironment approach [16] which is based on: a) the definition of the daily activity profile of each child for a typical winter and summer school week, allowing the identification of the microenvironments frequented by those children and the time spent in each one; and b) the characterisation of the air quality in those microenvironments.

To estimate the air quality in each microenvironment, as a complement to the experimental campaigns measurements, a cascade of air quality numerical models was applied. The results of a mesoscale model simulation over a large domain were used as initial and boundary condition to the local scale CFD model simulation. The CFD model had as input the road traffic emission estimated for each scenario as presented in §3. The ambient air pollutants concentration calculated with the CFD model were the basis to estimate the indoor concentrations in the different microenvironments using indoor/outdoor relations [16].

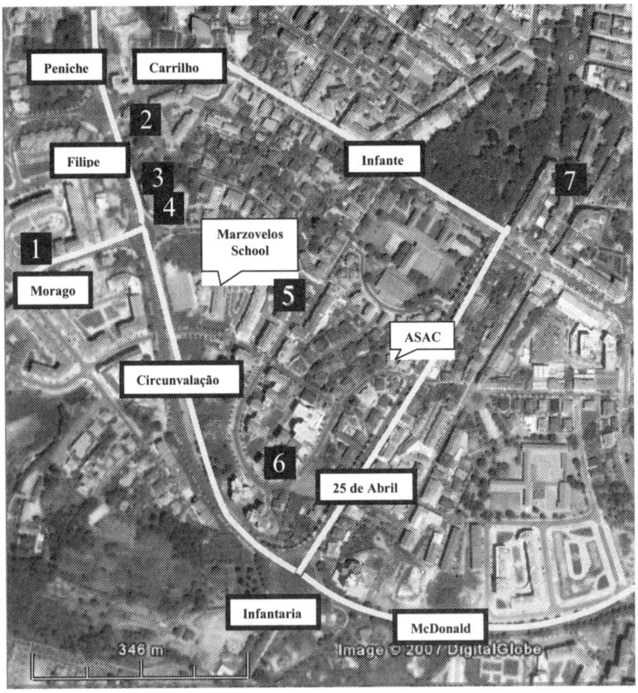

Figure 2: Selected study domain located in the Viseu town centre.

3 Road traffic scenarios

In order to evaluate the effects related to different mobility patterns, three different road traffic emission trends were developed and applied to the study domain (Figure 2), based on currently adopted measures and feasible emissions reduction perspectives. The scenarios were achieved by changing the land use, vehicle fleet composition, mobility patterns and vehicle technologies from the reference situation.

3.1 Scenarios development

The BAU scenario keeps all current trends in vehicle fleet and mobility patterns observed during the experimental campaigns. Vehicle fleet composition trends are based on current and evolution perspectives of Portuguese population and vehicle fleet [17]. This scenario corresponds to a 'does nothing' approach, where the penetration rate of new technologies is in line with the past trends of vehicle turnover. The Green and Grey scenarios try to show different options on future development.

The main change in these scenarios is the introduction of more efficient and cleaner technologies, representing a fast technology renewal, that supposes a complete turnover of the older vehicles replaced by EURO V and EURO VI in

2030. The principal difference between Green and Grey scenarios is the vehicle fleet composition and mobility patterns. Such difference results from the land use considered in each scenario. Grey scenario mobility patterns and fast fleet renewal may result in a decrease of the life time average of vehicle fleet. A slight decrease on public transportation modes and an increase of private vehicles will be the best economical solution for most population displacements.

3.2 Road traffic emission quantification

The quantification of road traffic emissions was performed by the application of the Transport Emission Model for Line Sources (TREM). The TREM model [18] is based on the average speed approach and it applies the emission functions derived from MEET/COST methodology. Different technologies (engine type, model year) and engine capacity are distinguished in TREM to derive emission factors. Furthermore, TREM allows the estimation of several pollutants, including PM_{10}. Recently, the model has been updated based on the EMEP/CORINAIR emission function methodologies [19] to include new vehicle technologies (EURO IV, V and VI) and fuel types (liquefied natural gas, compressed natural gas, hybrids and biodiesel).

For road traffic emissions, this study includes passenger cars (PC), light and heavy duty vehicles (LDV and HDV), heavy duty passenger vehicles (HDPV), mopeds (MP) and motorcycles (MC) parameters. Assuming that the fleet renewal rate in the future will be, at least, similar to the one observed in the recent past and present [17], by 2030 almost all of the existing vehicles will have been replaced by new ones, fulfilling, at least, the EURO V measures.

To predict the vehicle fleet in 2030, up to date information for fleet composition and population were used [17]. The BAU and Green scenarios maintain current vehicle fleet composition and an increase of 8.7% for the total number of vehicles; whereas the Grey comprises an increase of PC and an increase of 10% for the total number of vehicles.

During the experimental campaigns, vehicle fleet and composition data were collected to characterize road traffic hourly profiles in the study domain. Hourly traffic PM_{10} emissions profiles were estimated using the TREM model for typical winter and summer days, according to the scenarios.

In order to evaluate the variation of PM_{10} traffic emissions along the winter and summer periods, total emissions have been estimated for the future scenarios. Table 1 resumes the variation of vehicle fleet and total PM_{10} emissions estimated by TREM for the baseline (reference) and future (BAU, Green and Grey) scenarios.

It is possible to verify that the total number of on-road vehicles during the summer period is significantly higher (+75%) than during the winter period. This results in a significant difference in the PM_{10} emissions estimated by TREM. When compared to the reference scenario values, BAU scenario is characterized as the highest PM_{10} emission values (+24%), followed by the Grey (+10%) and the Green (+9%) scenarios. The slight difference between Green and Grey scenarios results from the differences in the vehicle number and fleet composition as a consequence of the land use change.

Table 1: Total PM_{10} emissions [g.km^{-1}] and total number of vehicles considered in the local scale study domain, for the winter and summer periods.

	Reference		BAU		Green		Grey	
	PM_{10} [g.km^{-1}]	Vehicles	PM_{10} [g.km^{-1}]	Vehicles	PM_{10} [g.km^{-1}]	Vehicles	PM_{10} [g.km^{-1}]	Vehicles
Winter	1820	20092	2266	20611	1985	20611	2024	22728
Summer	3197	35286	3980	36199	3486	36199	3555	39916
Averaged	2508	27689	3123	28405	2735	28405	2790	31322

4 Local scale simulations

The road traffic emissions estimated for the reference and for the future scenarios were used as input data to model the PM_{10} concentrations over the town of Viseu. In this sense, the modelled air quality data allowed the assessment of the short-term individual exposure to PM_{10} for the group of selected children.

4.1 Air quality results

The local scale model VADIS was applied to evaluate the PM_{10} air pollutant concentrations within the urban area of Viseu. VADIS is a CFD model [20] that allows the assessment of short-term local concentrations in urban geometries due to road traffic emissions. Its structure is based on two modules: FLOW and DISPER. The first module, FLOW, is a Reynolds Averaged Navier-Stokes (RANS) prognostic model with a standard k-ε turbulence closure that calculates the wind components, the turbulent viscosity, the pressure, the turbulent kinetic energy and the temperature 3D fields through the finite volume method. The second module, DISPER, applies the Lagrangian approach to the computation of the 3D concentration field of inert pollutants using the wind field estimated by FLOW. The model has been validated with wind tunnel measurements and with data from real case applications, demonstrating a good performance in the calculation of flow and pollutants dispersion around obstacles under variable wind conditions [20].

The study area covers 1300 m x 1300 m x 90 m, with a resolution of 5 m x 5 m x 3 m, and is located in the city centre of Viseu (Figure 2).

The input data to perform the local scale simulation comprises information on the characterisation of the built-up area of the study domain, namely: buildings volumetry and respectively location within the domain; traffic emissions; meteorological and pollutant background concentrations provided by a mesoscale air quality modelling system.

The urban built-up area of the study domain was represented by 142 buildings with different configurations ranging from 9 to 27 m height.

Outdoor 3D fields of flow and PM_{10} concentrations were simulated with VADIS for the reference situation and the future scenarios. Figures 3 presents, as

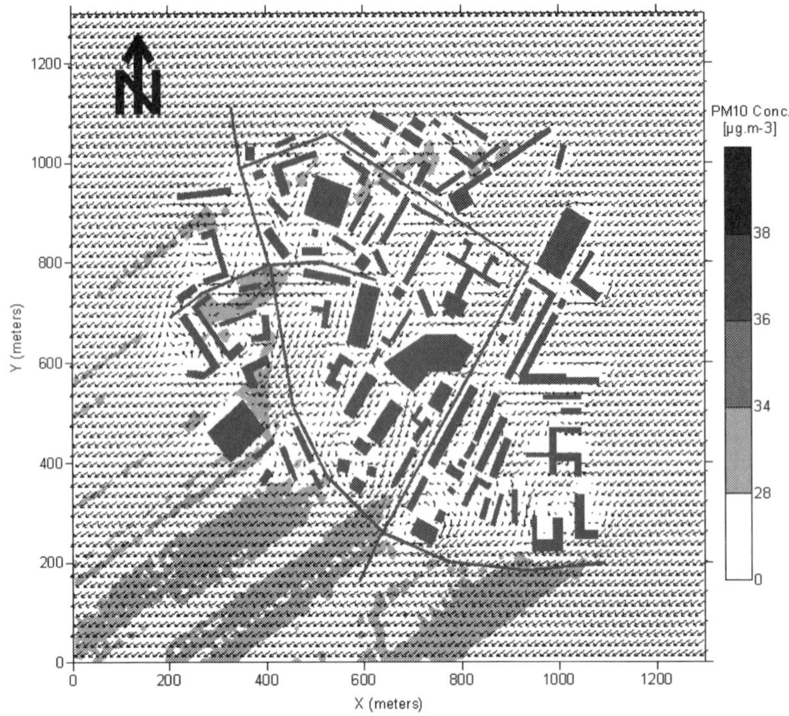

Figure 3: Horizontal view at 4.5 m height of flow and PM_{10} concentration fields for the 20th January 2006 at 4 p.m.

example, the horizontal view at 4.5 m height of flow and concentration fields of PM_{10} over the simulation domain for the 20th January 2006 at 4 p.m., for the reference scenario.

The analysis of the PM_{10} concentration patterns for the winter example shows the transport of the main traffic emissions to southwest, according to the prevailing wind conditions (Figure 3). Hot-spots are observed at the downstream buildings, but with PM_{10} concentration values below the reference daily limit value of 50 $\mu g.m^{-3}$. However, the different temporal resolution of the averages requires some caution in this analysis.

According to the local scale simulations for the reference scenario, the outdoor PM_{10} concentrations nearby the houses of the seven children and the elementary and ASAC (After School Activity Centre) schools are mainly affected by the existent background concentrations determined by the mesoscale modelling system.

The future air quality in the study area was simulated using the road traffic emissions and the PM_{10} background concentrations for the future scenarios (BAU, Grey and Green), maintaining the meteorological conditions used for the reference scenario. Results from the VADIS simulations show an increase of the PM_{10} concentrations for all scenarios for both winter and summer periods with an important influence of the background concentrations. The average PM_{10} concentrations in the domain for the winter period in the three scenarios reveal an increment of 18.0% (Grey), 23.0% (Green) and 16.5% (BAU), relatively to the reference situation. Thus, the Green scenario was identified as the worst case considering the PM_{10} air quality levels, justified by the increase of the population density and the corresponding emissions in a compact city centre [15].

For the summer period, it is observed a higher increase of PM_{10} concentrations for all scenarios regarding the winter period, namely 59.0% (Grey), 58.9% (Green) and 59.9% (BAU). Nevertheless, no significant differences were obtained between the analyzed scenarios.

4.2 Human exposure results

The human exposure was based on the children time-activity pattern and on the concentration levels in the visited microenvironments [16].

Five distinct microenvironments were considered in the study, corresponding to the main locations where the selected children spend their time during the winter and summer campaigns: outdoor, residential (bedroom), school, ASAC and transport.

The time-activity pattern of the seven children was recorded for both winter and summer periods, characterizing their movement within the study domain. For all children, the higher fraction of time is spent indoors, more particularly, at home, followed by school and ASAC. Also, little time is spent outdoors or in transport microenvironments.

The PM_{10} concentrations at the different indoor locations were estimated using indoor/outdoor relations [16].

The average individual PM_{10} exposure in the winter period for the reference and the future scenarios is presented in Figure 4.

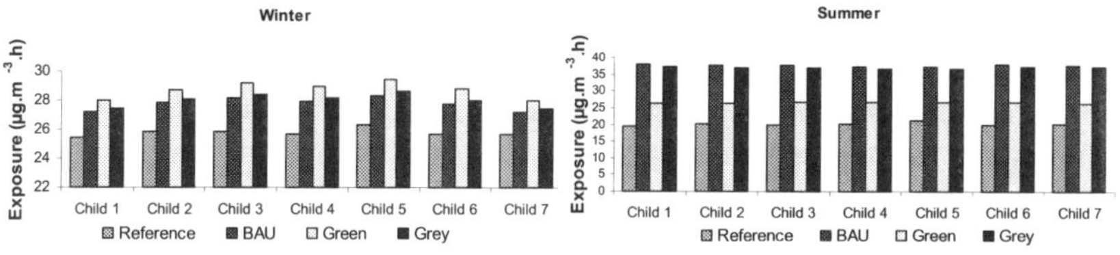

Figure 4: Children individual exposure to PM_{10} for the winter and summer periods.

The analysis of Figure 4 for the winter period allows concluding that the highest and the lowest average exposure values to PM_{10} are found for Children 5 and 1, respectively, for the reference and the future scenarios. Nevertheless, it is important to mention that the PM_{10} concentration that children are exposed to is below the indicative PM_{10} daily limit value of 50 $\mu g.m^{-3}$ imposed by the air quality legislation for the protection of human health. These results are directly related to the spatial pattern of the outdoor PM_{10} concentrations near the different microenvironments that determine the indoor levels and also by the time that each child spent in these indoor spaces. Moreover, the individual exposure for the winter period increases for all scenarios regarding the reference situation, with the highest value found for the Green scenario (+12%), followed by the Grey (+9%) and the BAU (+8%) scenarios. This result is in agreement with the air quality results previously obtained for the three simulated scenarios and for the winter period. These results allow identifying that Children 3, 4 and 5 may have in the future a higher probability to aggravate their respiratory problems.

Figure 4 also presents the average individual exposure to PM_{10} for the group of studied children along the summer period. There is an increase of the individual exposure for all scenarios in regard to the reference situation, with the highest value found for the BAU scenario (+87%), followed by the Grey (+84%) and the Green (+34%) scenarios.

The results of the summer period are higher than the ones found for the winter period, which are in agreement with the obtained air quality results; nevertheless the Green scenario is considered here the best case scenario for the future.

According to the exposure results for the reference and future scenarios, children 1 and 6, and children 4 and 5 are exposed to the highest and the lowest PM_{10} concentrations, respectively. Children 1 and 6 have therefore a highest probability to aggravate their respiratory problems in the future. These results are related to the outdoor PM_{10} levels determined by the VADIS model and used for the estimation of indoor concentrations, meaning that the microenvironments frequented by children 1 and 6 in the summer period are characterized by higher PM_{10} concentrations.

The time-activity patterns of the studied children during the summer period are also a relevant factor on the obtained results.

5 Conclusions

The work presented in this paper is part of a wider project [14], which aimed at the development of scientific elements that can be considered and incorporated into local development plans and in the exercise of the territorial planning, contributing to the sustainable development of a town and therefore, to the improvement of the quality of life of the populations in what concerns air quality and health.

The developed methodology is adequate to the intended study, especially regarding the defined objectives. The developed scenarios show that, despite the introduction of cleaner technologies, it is expected that PM_{10} emissions from the transportation sector grow, leading to a deterioration of the air quality. The BAU scenario is the one that presents the highest emissions, with the Green and Grey scenarios presenting quite similar values. All future scenarios developed point to an air quality degradation, with a mean increase of 20% and 60% of PM_{10} concentration values for the winter and summer periods, respectively. Nevertheless, these results are not linearly translated into exposure values. In winter, results show a small and similar increase in exposure to PM_{10} for all scenarios, while in summer the BAU and Grey scenarios indicate a significant raise of exposure values.

This study reveals the non linearity between emissions, air quality and exposure values, confirming the importance of using exposure, instead of ambient air concentrations, in studies where the relation between air quality and health is investigated.

Acknowledgements

The authors are grateful to Calouste Gulbenkian Foundation for the SaudAr project's financial support. The authors are also grateful for the financial support under the 3rd EU Framework Program and the Portuguese 'Ministério da Ciência, da Tecnologia e do Ensino Superior' for the Ph.D. grant of J. Valente (SFRH/BD/22687/2005) and Richard Tavares (SFRH/BD/22741/2005), and for the PAREXPO project (POCI/AMB/57393/2004).

References

[1] EEA, Europe's environment – The fourth assessment. Office for Official Publications of the European Communities. European Environmental Agency (EEA). Luxembourg. 452 p., 2007 (ISBN: 978-92-9167-932-4).

[2] EEA, EEA Signals 2009 – Key environmental issues facing Europe. Office for Official Publications of the European Communities. European Environmental Agency (EEA). Luxembourg. 40 p., 2009 (ISBN: 978-92-9167-381-0).

[3] EC, The Communication on Thematic Strategy on Air Pollution and the Directive on Ambient Air Quality and Cleaner Air for Europe, Impact Assessment. European Commission (EC), 2005.

[4] WHO, WHO Strategy on Air Quality and Health, Occupational and Environmental Health Protection of the Human Environment. World Health Organization (WHO), Geneva. 12 p., 2001.

[5] Dora, C. and Racioppi, F., Transport, environment and health in Europe: Knowledge of impacts and policy implications. Rome, 2001.

[6] Nicolai T., Carr D., Weiland S.K., Duhme H., von Ehrenstein O., Wagner C. and von Mutius E., Urban traffic and pollutant exposure related to respiratory outcomes and atopy in a large sample of children, European Respiratory Journal 21, 856-963, 2003.

[7] WHO, Health effects of transport-related air pollution. M. Krzyzanowski, B. Kuna-Dibbert and J. Schneider (Eds.). World Health Organization (WHO) Regional Office for Europe. Copenhagen, Denmark. 190 p., 2005 (ISBN: 92-890-1373-7).

[8] WHO, Children's health and the environment in Europe: a baseline assessment. World Health Organization (WHO) Regional Office for Europe. Copenhagen, Denmark. 125 p., 2007 (ISBN: 978-92-890-7297-7).

[9] Reis S., Simpson D., Friedrich R., Jonson J.E., Unger S. and Obermeier A., Road traffic emissions – predictions of future contributions to regional ozone levels in Europe. Atmospheric Environment 34(27). 4701-4710, 2000.

[10] Caserini S., Giugliano M. and Pastorello C., Traffic emission scenarios in Lombardy region in 1998–2015. Science of the Total Environment 389(2-3). 453-465, 2008.

[11] Zamboni G., Capobianco M. and Daminelli Giorgio, Estimation of road vehicle exhaust emissions from 1992 to 2010 and comparison with air quality measurements in Genoa, Italy. Atmospheric Environment 43(5). 1086-1092, 2009.

[12] Gonçalves M., Jiménez-Guerrero P. and Baldasano J.M., Emissions variation in urban areas resulting from the introduction of natural gas vehicles: Application to Barcelona and Madrid Greater Areas (Spain). Science of the Total Environment 407. 3269-3281, 2009.

[13] Borrego C., Tchepel O., Costa A.M., Martins H., Ferreira J. and Miranda A.I., Traffic-related particulate air pollution exposure in urban areas. Atmospheric Environment 40(37). 7205-7214, 2006.

[14] Borrego C., Lopes M., Valente J., Santos J., Nunes T., Martins H., Miranda A.I., Neuparth N. Air Pollution and child respiratory diseases: the Viseu case-study, Portugal. *Air Pollution 2007. 23 – 25 April, Algarve, Portugal. Eds. C.A. Borrego, C.A. Brebbia, WIT Press, Southampton, UK. pp 15 – 24,* 2007.

[15] Ferreira, J.; Martins, H.; Miranda, A.I. e Borrego, C. (2005) - Population Exposure to Atmospheric Pollutants: the influence of Urban Structure. In 1st Environmental Exposure and Health Conference, Atlanta, EUA, 5-7 October 2005 – Environmental Exposure and Health. Eds. M..M. Aral, C.A. Brebbia, M.L. Masila and T. Sinks. WIT Press, pp 13-22.

[16] Borrego C., Lopes M., Valente J., Tchepel O., Miranda A.I., Ferreira J., The role of PM_{10} in air quality and exposure in urban areas. *Air Pollution 2008, 22 – 24 September, Skiathos, Greece. Eds. Eds. C.A. Brebbia, J.W.S. Longhurst. WIT Press, Southampton, UK. pp 511 - 520.*, 2008.

[17] ACAP, Estatísticas do Sector Automóvel – Edição 2008 (cd-rom), ACAP, 2008.

[18] Tchepel, O., Modelo de emissões para apoio à decisão na Gestão da Qualidade do Ar. Dissertação apresentada à Universidade de Aveiro para obtenção do grau de Doutor em Ciências Aplicadas ao Ambiente, 2003.

[19] European Environmental Agency, EMEP/CORINAIR Emission Inventory Guidebook 2007, Technical Report Nº 16/2007, EEA, 2007.

[20] Borrego C., Tchepel O., Costa A., Amorim J., Miranda A.I., Emission and dispersion modelling of Lisbon air quality at local scale. Atmospheric Environment, 37, 5197-5205, 2003.

Exploring barriers to and opportunities for the co-management of air quality and carbon in South West England: a review of progress

S. T. Baldwin[1], M. Everard[2], E. T. Hayes[1], J. W. S. Longhurst[1] & J. R. Merefield[3]
[1]*Faculty of Environment and Technology, University of the West of England, UK*
[2]*Environment Agency, Kings Meadow House, UK*
[3]*Department of Life Long Learning, University of Exeter, UK*

Abstract

Due to the common sources of emissions of both air quality pollutants and greenhouse gases, management measures directed at one category of emissions are likely to positively impact the other. Through the local air quality management (LAQM) process, local authorities are required to monitor and measure specified air pollutants, the sources of which are also common to the primary sources of carbon emissions at a local level. This research tracks the progression of local authority management of carbon emissions and examines the barriers and opportunities for the integration of carbon emissions into the LAQM process. Results are triangulated from three core research methods deployed in South West England: a time series of local authority questionnaire surveys; secondary data analysis of active Air Quality Action Plans; and case study interviews of six local authorities in the region. The research concludes that the absence of statutory targets for carbon emission reductions remains a substantial barrier for local authority carbon management initiatives. However, in order to utilise scarce resources in the most efficient manner, local authorities should draw upon the existing skill set of their Air Quality Officers.

Keywords: air quality, carbon management, co-benefit, synergy, local government.

WIT Transactions on Ecology and the Environment, Vol 123, © 2009 WIT Press
www.witpress.com, ISSN 1743-3541 (on-line)
doi:10.2495/AIR090091

1 Introduction

In 2008, the UK passed legislation which introduced the world's first long-term legally binding framework to tackling the dangers of climate change. The Climate Change Act 2008 commits the UK to greenhouse gas emission reductions through action in the UK and abroad of at least 80% by 2050, and reductions in CO_2 emissions of at least 26% by 2020, against a 1990 baseline [1]. In order to meet these ambitious targets, the UK must take substantive measures to reduce carbon emissions both nationally and locally. To date, the UK has pursued its goals on carbon mitigation mainly at an international level through programmes such as the United Nations Framework Convention on Climate Change [2], the Kyoto Protocol [3], the European Union Emissions Trading Scheme [4] and nationally through the Climate Change Programme [5]. By contrast the UK has been slow to recognise the important role that locally-driven carbon mitigation can play in contributing to its national and international obligations. While the Climate Change Programme advocated early interventions by public sector organisations, both in reducing the carbon intensity of their own operations and by raising awareness in their local communities [5], it did not go so far as to place a legal duty on local government to take action.

1.1 Local authority climate change initiatives

Many local authorities in the UK now recognise the role they can play and have been engaging with voluntary initiatives for tacking climate change and mitigation of carbon emissions. Notable amongst these voluntary initiatives are two schemes. Firstly, the Nottingham Declaration on Climate Change [6] which has been signed by 236 local authorities (as of April 2009) and which commits signatories to three broad aims: acknowledging that climate change is occurring; welcoming and engaging with the government targets; and committing to working at a local level on climate change management. Secondly, the Local Authority Carbon Management Programme [7], through which the Carbon Trust works with 216 participating local authorities to improve energy efficiency and reduce their carbon emissions. In April 2008, the new National Performance Reporting Framework [8] introduced responsibilities for local authorities to achieve year-on-year reductions in carbon emissions from their own operations (National Indicator 185) and reductions in per capita emissions in their administrative area (National Indicator 186). Although the new framework does not explicitly require local authorities to set binding targets for reductions in carbon emissions, it does represent the first consistent local authority-wide approach to monitoring CO_2 emissions. The new carbon monitoring responsibilities will require local authorities to build carbon management policies into the full spectrum of their duties and responsibilities. However, the limited resources available to carry out their new duties means local authorities would be wise to utilise existing skills within their authority, particularly in the comparable skill set developed through the LAQM regime.

There is also growing acknowledgment that air pollution and climate change goals are complementary and should not be pursued in isolation [9]. However, to date, management of the two areas has been largely separate both in Europe and further afield. Current research suggests that considerable ancillary benefits could arise from a more holistic and integrated approach to the management of air pollution and mitigation strategies for carbon emissions [10]. The rationale for these benefits can be divided into four main arguments: the economic benefits of integrated policies [11, 12]; the role of air pollutants as greenhouse gases; the effect of future climatic conditions on air quality [13]; and synergies and trade-offs arising from management techniques [13].

1.2 The Local Air Quality Management process

The Environment Act 1995 required the publication of the first National Air Quality Strategy, the core of which was the setting of a series of health-based standards for eight pollutants [15]. This introduced a statutory requirement for local authorities to periodically review and assess air quality in their area for specified pollutants and to assess current and projected future levels of air quality [16]. These reviews are assessed against a number of national Air Quality Objectives (AQOs) established for seven of the eight priority pollutants. Where an area exceeds, or is likely to exceed, by a stated date, local authorities are required to designate an Air Quality Management Area (AQMA) and develop an Air Quality Action Plan (AQAP) outlining measures they will take to work towards remediating the problem [15]. The application of this process requires many of the same methods, skills and collaborative networks that would be required for an effective carbon management framework at the local level, including: the production of a robust emissions inventory; embedding carbon in the local and regional Air Quality Strategies; collaborative networks between key stakeholders influencing atmospheric emissions; and joint carbon/air quality action plans. Moreover, the Air Quality Strategy for England, Scotland, Wales and Northern Ireland (2007) recognises the importance of considering the impact of air quality measures on carbon emissions by stating that, '...in the development of Air Quality Action Plans and where appropriate, LTPs, local authorities should bear in mind the synergies and trade-offs between air quality and climate change, and the added benefits to the local, regional and global environment of having an integrated approach to tackling both climate change and air quality goals' [17]. This research investigates the opportunities and barriers for local authorities to co-manage carbon emissions at a local level through existing air quality management processes.

2 Methodology

Results presented here are taken from three components of a longitudinal study investigating the progress, barriers and opportunities for an integrated approach to the management of air quality and carbon emissions through the Local Air

Quality Management process in the South West Region of England. Three methods of data collection are presented: two questionnaire surveys (administered in 2008 and 2009); 12 case study interviews in six local authorities; and secondary data analysis of the ten active AQAPs in the region.

The 2008 questionnaire survey was designed to investigate the extent to which traditional air pollutants and carbon emissions are being co-managed at a local governance level in the South West Region of England. The 2009 questionnaire survey was designed to explore the key themes derived from the previous survey and to establish the progression of emerging carbon management strategies within the region. The target group of the questionnaire surveys consisted of all local government bodies in the South West Region of England. Both surveys focused on two key categories of officer in local government: Air Quality Officers; and officers with primary responsibility for carbon management or, where they exist, Climate Change Officers.

The secondary data collection component reviewed the established AQAPs produced by local authorities in response to the declaration of AQMAs. A coding form (survey protocol) was developed to establish the extent to which the action plans considered their impact on climate change. A matrix was constructed to examine specific measures outlined in the AQAPs in more detail. This allowed the identified measures to be categorised as 'synergy', 'trade-off' and 'lose-lose' outcomes regarding the co-management of air quality and climate change.

The case study interviews served to enhance, expand and create depth to research findings. The interviews provided three main purposes: to triangulate data obtained from the questionnaire and secondary data analysis stages of the project; to investigate some of the issues highlighted by the questionnaire surveys; and to gain a clearer understanding of how local authorities are beginning to deal with the climate change issue within the structure and organisation of local government.

3 Research findings

Selected results are presented from: two questionnaire surveys (comprising a total of 22 Air Quality Officers and 12 officers with primary responsibility for climate change from South West England); six case study authorities (comprising 12 interviews distributed evenly between Air Quality Officers and officers with carbon management responsibilities); and secondary data analysis of the nine active AQAPs implemented in the region. Results are presented for the following thematic areas: progress in local government carbon management strategies and integrated management of air pollution and carbon emissions; opportunities for integrated management of air pollution and carbon emissions; barriers to integrated management; and challenges for integrated management.

3.1 Progress in local government carbon management strategies and integrated management of air pollution and carbon emissions

The LAQM process has become well established within local governments in the UK since its inception over a decade ago, driven in part by a strong legislative framework (the Environment Act 1995). The South West Region of England is no exception to this trend, with 100% of the local authorities in the 2009 questionnaire survey employing a minimum of one officer with responsibility for management of air quality within their administrative area. Conversely, local government has not benefited from a comparable legislative intervention for locally-driven carbon management. Despite this lack of a formalised requirement, local government organisations in the UK have shown increasing recognition in recent years of the role they can play in managing and influencing carbon emissions within their jurisdictions. In the 2009 questionnaire survey, 89% (8) of respondents reported employing an officer with primary responsibility for carbon management. The production of a Climate Change Strategy is an important first step in highlighting a local authority's commitment to action and organising the measures by which they will achieve reductions and adapt to the future effects of a changing climate. 67% (6) of respondents had produced a strategy for tackling climate change in their area. This represents a modest increase when compared to the 2008 questionnaire survey response in which 56% (5) of the same survey group reported having developed a strategy.

3.2 Opportunities for integrated management of air pollution and carbon emissions

A number of elements required to implement the LAQM process effectively could directly lend themselves to carbon management at a local level. The multiplicity of air pollution sources (many of which are beyond direct local authority control) often necessitates local authorities to establish strong networks with the organisations, functions and bodies responsible for, or with an influence over, locally-derived emissions. These stakeholders are also likely to be the main sources of locally-generated carbon emissions (i.e. transport, domestic and business). The 2009 questionnaire survey data showed that 65% (13) of respondents have established an external steering group for LAQM whereas only 27% (3) have established an equivalent group for carbon management. The opportunity to utilise established LAQM networks for the purpose of carbon management has been recognised by a number of air quality officers and articulated in case study interviews. For example: *"To make a difference, there has to be contact with emitters at a local level, and air quality officers are best suited for this role as there would already be contacts and expertise regarding air quality, emissions, etc."* (City Council Air Quality Officer).

Air Quality Officers have also developed a great deal of technical expertise through the LAQM process. Many of these skills would also be required for implementing a comprehensive strategy for carbon management such as the production of robust emissions inventories and identification of priority polluters, both necessary for establishing a baseline from which emissions reductions can be measured. Responding Air Quality Officers confirmed this point in the 2008 questionnaire survey by ranking the production of emission inventories as the most transferable aspect of the LAQM process (see Table 1).

Table 1: Mean ranked score of the most transferable aspects of LAQM for the management of carbon emissions.

Components of LAQM process	Mean score
Emissions Inventories	4.16
Action Plan	3.53
Updating and Screening Assessment	3.00
Detailed Assessment	2.26
Further Assessment	2.00
Progress Report	2.00

(scale: 1-6, where 1= not transferable and 6=highly transferable).

The skills developed through a decade of LAQM could be exploited as an efficient and effective use of scarce local authority resources for the purposes of carbon management. Further to this point, some air quality officers expressed concern that not utilising the LAQM regulatory framework, and the existing skills and networks developed through the process, would result in a duplication of effort and funding streams. For example: *"...it would be senseless to invent a duplicate regulatory regime to deal with carbon emissions when a regime with complementary skills already exists. It is particularly important if there likely to be an AQ benefit through the management of carbon."* (District Authority Air Quality Officer).

AQAPs are the vehicle by which local authorities outline the measures they will take to work towards meeting the AQOs. Local authorities must take care to ensure that the measures they select for inclusion in the AQAP do not impact negatively on other environmental concerns. Due to the strong commonality between the sources of air pollution and carbon emissions at a local level, it would be judicious for local authorities to identify measures that will improve air quality while reducing overall emissions of carbon and, where possible, avoid actions that will result in an increase in carbon emissions. The 2009 questionnaire survey showed that 63% (10) of the respondents think that AQAPs should consider their impact on carbon emissions. This recognition of the potential impact of air pollution control measures on carbon emissions was further highlighted by one Air Quality Officer, who stated that: *"Carbon emissions should be considered within the AQAP as it is important that measures that reduce NO_2 and PM_{10}* [the main pollutants of concern in most AQAPs] *also reduce CO_2. A measure that reduces these pollutants but increases CO_2 should be avoided."* (District Authority Air Quality Officer). This positive expression of the importance of considering the wider non-air quality impacts of measures illustrates a shift in attitude in recent years. In the 2008 questionnaire survey of the same group, only 27% (3) of the respondents considered non-air quality issues in the development of their AQAP. Furthermore, the mean score on a Likert scale ranked question showed that consideration of carbon emissions was 'limited or absent' in the development of measures for AQAPs. However, in the 2009 questionnaire survey, 73% (11) of respondents stated that they were 'likely' or 'highly likely' to prioritise actions likely to reduce air pollution and carbon emissions in tandem in future AQAPs.

3.3 Barriers to integrated management of air pollution and carbon emissions

Despite the opportunities the LAQM regime presents for aiding local authority carbon management initiatives, there are a number of barriers that exist to the achievement of a comprehensive integration of the two policy areas. As previously mentioned, there remains a lack of dedicated climate change staff that can co-ordinate a local authorities approach to carbon management. This results in action being fragmented within authorities, and an inconsistency in approach between authorities. While 89% (8) of respondents to the 2009 questionnaire survey reported having an officer with responsibility for carbon management, this role is predominantly an auxiliary function of their primary profession (e.g. forward planning, facilities management, environmental protection or waste management). Case study participants expressed frustration that their duty to fulfil their primary role leaves insufficient time and resources to coordinate and effectively implement carbon abatement measures. For example, one officer stated:*" I'm dealing with climate change but I am still a planner. We haven't got a dedicated officer. I have to do it along side my planning job so one must question the commitment of this authority."* (District Authority Officer with carbon management responsibilities).

There are various possible explanations for the lack of commitment displayed by local authorities in the South West Region of England. During the 2008 questionnaire survey, respondents that had not yet produced a strategy for tackling climate change were asked to rank a compendium of reasons for inaction. The highest scoring reason given was 'not a statutory requirement' followed by 'other issues having higher priority' and 'lack of time'. This suggests that the lack of a strong statutory framework is a primary barrier to local authority action on climate change and carbon management. The need for a statutory requirement to drive action on carbon mitigation was echoed in the case study interviews. *"You can't get away from the need for statutory targets and regulation from central government. Even in a single-tiered government. They* [local authorities] *will only do what they are told to do."* (City Council Officer with carbon management responsibilities).

A statutory requirement could impel management of carbon emissions at a local level and necessitate the allocation of time and resources. Integration of elements of carbon emission management into the LAQM framework could prove an efficient use of these resources: *"A statutory requirement is needed to force local governments to release funding to enable us to do it properly, rather than just trying to do it in our spare time within our existing operational budgets. However, utilising existing skills and networks of the Air Quality Officers could reduce the overall cost of required action."* (District Authority Officer with carbon management responsibilities). This principal explanation, in addition to others deduced by this research, is applicable to other local government organisation across the UK.

Table 2: Number of measures identified in AQAPs that are likely to result in synergistic or trade-off outcomes.

	Synergies	Trade-offs	Total
Number	141	69	210

Table 3: Number of direct/indirect measure in AQAPs that are likely to result in synergistic or trade-off outcomes.

	Synergies	Trade-offs
Direct Measures	47	64
Indirect Measures	94	5

3.4 Challenges for integrated management of air pollution and carbon emissions

One element of the secondary data analysis was to examine the measures identified by AQAPs to assess their likely impact on air pollution and carbon emissions. Part of this process involved the construction of a matrix to examine the likelihood that the specified measures would result in a synergy or trade-off for air quality and carbon emissions. Table 2 shows the aggregate number of measures in South West England that are likely to result in a synergy or trade-off outcome.

The above table shows that 67% (141) of the 210 measures being implemented specifically for air quality in the AQAPs are likely to simultaneously benefit air quality and reduce carbon emissions. The remaining 33% may represent currently missed opportunities. The professed absence of consideration of carbon emissions in the development of AQAPs suggests that most actions taken to improve air quality will be co-beneficial, resulting in complementary reductions in carbon emissions. It would seem likely therefore that, if local authorities were to consider carbon emissions during the development of AQAPs, the percentage of co-beneficial actions could be substantially increased. However, care should be taken to ensure that the pursuit of synergistic measures for air and carbon emissions does not incapacitate the ability of AQAPs to deliver tangible improvements in local air quality. To investigate this issue further, the identified measures were collated into direct measures (measures which local authorities have direct powers implement) and indirect measures (measures that local authorities cannot directly implement but seek to influence indirectly). The results are summarised in Table 3.

It is apparent that, while only 42% (47) of direct measures implemented in AQAPs will result in associated reductions in carbon emissions, 95% (94) of indirect measures result in synergy between the two areas. This would suggest that local authorities should pursue indirect actions in their AQAPs if they wish to deliver simultaneous reductions in carbon emissions. However, in many cases, the indirect measure will be poorly suited to improve the hotspots identified by the LAQM process. Conversely the more physical management (rerouting traffic, bypass, low emission zone etc.) that characterises the direct action may be sufficient to deliver such improvements. This observation was mirrored by one case study authority when discussing the implications of direct and indirect measures: *"They* [indirect measures] *are what I would call 'softer' measures, less dramatic in terms of AQ impact. If the introduction of carbon emissions into AQAPs eliminates the option of direct measures I think we may struggle to bring significant improvements in air quality in the hotspots."* (District Authority Air Quality Officers). Another possible explanation is that indirect measures require negotiation with partner organisations with different priorities and perspectives, amplifying the value of co-creation of more far-sighted and inclusive outcomes by networks of stakeholders.

This also highlights one of the core differences between carbon management and air quality management in the UK. LAQM is a health-based framework focusing on receptor exposure. Thus, AQOs are only exceeded in areas were receptors are likely to be exposed to the offending pollutant(s). This provides scope for the sources of emissions to be isolated and separated from the receptor without the need to reduce overall emissions. Conversely, carbon mitigation is concerned with reduction in total load of emissions. It follows that, while it the LAQM process is focused on managing relatively small 'pockets' of poor air quality, it will be difficult to effectively integrate the management carbon emissions within the existing framework and ancillary benefits for carbon reductions will be limited.

4 Conclusions

Local authorities in South West England appear to be making progress in the area of carbon management. This has predominantly been driven by proactive local authority engagement in voluntary initiatives. The National Performance Reporting Framework is likely to further stimulate growth in the area. However, in the absence of statutory targets for reducing carbon emissions, local governments seem unwilling to provide additional funding and resources to address the issue. In order to utilise scarce resources in the most efficient manner, local authorities should draw upon the existing complementary skills developed by Air Quality Officers through the LAQM regime.

Due to the common sources of emission of both air quality pollutants and greenhouse gases, management measures are likely to impact upon one another. The Air Quality Strategy recognises the importance of these links and is urging local governments to consider their impact on carbon emissions and prioritise synergistic actions where possible. The majority of air quality officers participating in this research project believe that AQAPs should consider their impact on carbon emissions. However, while the LAQM process continues to focus on a 'hotspot' approach, local authorities may find it difficult to remediate areas of poor air quality without impacting negatively on carbon emissions.
Acknowledgment: the authors gratefully acknowledge the support of Great Western Research, the Environment Agency and AAR Environmental Ltd which enabled this research to be undertaken.

References

[1] HM Government, Climate Change Act 2008. The Stationary Office, 2008.
[2] United Nations, United Nations Framework Convention on Climate Change. http://unfccc.int/2860.php
[3] Kyoto Protocol. http://unfccc.int/
[4] European Commission, 2007. EU Action Against Climate Change. EU Emissions Trading: An open system promoting global innovation. European Union Publications Office, 2007.
[5] Defra, Climate Change: the UK Programme. The Stationary Office, 2006.
[6] Energy Savings Trust, Nottingham Declaration on Climate Change. http://www.energysavingtrust.org.uk/housingbuildings/localauthorities/NottinghamDeclaration.
[7] Carbon Trust, Carbon Trust's Local Authority Carbon Management Programme. Available from: http://www.carbontrust.co.uk, 2008.
[8] HM Government and LGA, An introduction to the Local Performance Framework - delivering better outcomes for local people. Wetherby: Communities and Local Government Publications. 2007
[9] Hayes, E.T., Leksmono, N.S., Chatterton, T.J., Symons, J.K., Baldwin, S.T., and Longhurst, J.W.S., Co-management of carbon dioxide and local air quality pollutants: identifying the 'win-win' actions. Proc. Of the 14th Int. Conf. IUAPPA World Congress: Brisbane, 2007.
[10] Vuuren, D., Cofala, J., Eerens, H., Oostenrijk, R., Heyes, C., Klimont, K., den Elzen, M. & Amann, M., Exploring the ancillary benefits of the Kyoto Protocol for air pollution in Europe. *Energy Policy* **34**, pp. 444–460, 2006.
[11] Stern, N., The Economics of Climate Change: The Stern Review, Cambridge University Press: Cambridge, UK, 2007.
[12] European Environment Agency, Air quality and ancillary benefits of climate change policies. Luxembourg: Office for Official Publications of the European Communities, 2006.
[13] Air Quality Expert Group, Air quality and climate change: a UK perspective. Defra: London, 2006.
[14] HM Government, Environment Act 1995. The Stationery Office, 1995.
[15] Longhurst, J.W.S., Beattie, C.I., Chatterton, T.J., Hayes, E.T., Leksmono, N.S., & Woodfield, N.K., Local air quality management as a risk management process: assessing, managing and remediating the risk of exceeding an air quality objective in Great Britain. *Environment International* **32**, pp. 934-947, 2006.
[16] Beattie, C.I. & Longhurst, J.W.S., Joint Working within Local Government: air quality as a case study. *Local Environment* **5**, pp. 401-414, 2000.
[17] Defra, The Air Quality Strategy for England, Scotland, Wales and Northern Ireland. The Stationery Office, 2007.

An urban environmental monitoring and information system

J. F. G. Mendes[1,2], L. T. Silva[1], P. Ribeiro[1] & A. Magalhães[2]
[1]*Department of Civil Engineering, University of Minho, Portugal*
[2]*Innovation Point – S.A., Portugal*

Abstract

Evaluating, monitoring and informing about urban environmental quality has become a main issue, particularly important when considered as a decision-making tool that contributes to more habitable and sustainable cities. Following a tendency observed in other European cities, the city of Braga (Portugal) has decided to create an infrastructure for environmental data acquisition and a web-based platform as a public information system. Some of the innovations introduced in this new platform include the use of mobile instrumented units, the extensive use of simulation software to create long-term pollution (air and noise) maps, and the presentation of the information through a geographical interface developed over Google Maps technology. This paper discusses some of the critical aspects regarding the conceptual design of such an information system, and presents the actual information system developed for Braga, named SmarBRAGA.
Keywords: environmental monitoring, public information, air pollution, noise.

1 Introduction

The growth of the world's population has been followed by the increase of the population living in urban areas, which very often results in additional pressures over space, ecosystems, infrastructures, facilities and the way of life.

Domestic and industrial sources, and mainly motorised traffic, are responsible for pollutant emissions and noise which decisively affect life in today's cities [1]. In this context, evaluating and monitoring the urban environmental quality has become a main issue, particularly important when considered as a decision-making tool that contributes to more liveable and sustainable cities.

WIT Transactions on Ecology and the Environment, Vol 123, © 2009 WIT Press
www.witpress.com, ISSN 1743-3541 (on-line)
doi:10.2495/AIR090101

Noise caused by road traffic is the nuisance most often mentioned by roadside residents. Urban air pollution became one of the main factors of degradation of the quality of life in cities, mainly in roadside areas. This problem tends to worsen due to the unbalanced development of urban spaces and the significant increase of mobility and road traffic. The quantitative evaluation of traffic noise levels and air pollutant concentrations is the basis upon which noise and air pollution control policies stand [2].

Braga is a mid-sized city located in the North of Portugal, where an environmental program was developed leading to its integration in the BragaDigital project funded by EU through the "POS-Conhecimento" program. Within this program, the urban pollution assessment and a public information system were considered a priority.

The project, named SmarBRAGA, obtained public funding resulting on its development by the private company Innovation Point, in close collaboration with the University of Minho and the municipal enterprise AGERE.

2 The conceptual model

The aim of a monitoring and information system is to support the formulation of control strategies and, additionally, to inform the population about the urban environmental situation. In this paper we focus on the information objective.

The critical aspects of such a monitoring and information system are discussed in the following sections and include: i) the quality label-based information; ii) the long-term environmental maps versus the point-measurements; iii) the importance of motor traffic as the main pollution source; iv) the selection of air pollution and noise descriptors; and v) the algorithm which generates environmental quality labels.

2.1 Monitoring and informing by means of quality labels (green to red)

No matter how the environmental measurements are acquired, numerical information needs to be processed and formatted in a way that can be easily understood to the public.

In the case described in this paper, three algorithms were developed and used to convert the measurements taken on a regular basis into a classification scale of quality labels (green to red [left to right]), as represented in Table 1.

Table 1: Urban pollution indicators.

Air Quality, Urban Noise and Heat Index				
Very Good	Good	Moderate	Poor	Very Poor

Ônveli seu.

WIT Press

55780

Proforma

B 13.01.11

BK: 1007650

2.2 Long-term maps versus point-measurements

Point-measurements of noise and air pollution data are very useful inputs to understand and evaluate the environmental quality of an urban area. However, these values are strongly linked with a particular location in a given measurement time, which means that the information is limited and has to be interpreted in that sense. Further extensions or conclusions require other tools, namely the use of simulation software based on mathematical models.

A comprehensive urban environmental information system for the public must include long-term pollution (or quality) maps covering the city and point-measurements. The first reflect an "average" situation of the environmental quality over the city, useful whenever a citizen needs an overview on a particular part of the city, for instance in the case he is selecting where to buy a flat. The last is useful whenever a citizen wants to have an up-to-date measurement of environmental local conditions, for instance in the case he is deciding whether to go for a bike tour.

Long-term horizontal pollution maps are developed through the modelling of the dispersion of air pollution in built-up urban areas and integrate all the parameters which influence the dispersion, such as topography and meteorological conditions. For the noise simulation, the situation is similar, but the mathematical model is obviously a specific one.

As the main source of pollution is typically the traffic, a campaign of traffic counts across the city is required, which should be carried out in at least one typical weekday. Ideally, a traffic count information system should be maintained in order to maintain frequent updates.

2.3 The importance of traffic data

The evaluation of urban environmental quality is closely related with noise and air pollution descriptors, which are intrinsically correlated with the characteristics of traffic. Characterization of road traffic is of extreme importance in order to deliver accurate outputs of long-term air and noise pollution. This must be done considering an all-day (24 hours) period to obtain a complete and reliable idea of the traffic flows on the road network of a city. For this reason, traffic counts must be well-planned, based on the city road hierarchy, and a complete profile of the traffic behaviour during a day for all road categories must be achieved. Furthermore, speed must be evaluated, or at least an average standard speed applied to each road category, as well as the decomposition of traffic in different categories, such as trucks, cars, buses, two-wheelers, among others. The percentage of heavy-vehicles is one of the most important factors for modelling long-term environmental noise and air pollution.

2.4 Selection of air pollution and noise descriptors

Domestic and industrial sources and mainly motorised traffic are responsible for pollutant emissions and noise which decisively affect living in today's cities [3, 4].

The combustion of hydrocarbon fuel in the air generates mainly carbon dioxide (CO_2) and water (H_2O). However, the combustion engines are not totally efficient, which means that the fuel is not totally burned. In this process the product of the combustion is more complex and could be composed by hydrocarbons and other organic compounds, carbon monoxide (CO) and particles (PM) containing carbon and other pollutants [5]. On the other hand, the combustion conditions – high pressures and temperatures – originate partial oxidation of the nitrogen present in the air and in the fuel, forming oxides of nitrogen (NO + NO2) conventionally designated by NOx. Many of the atmospheric pollutants once emitted from the road vehicles react with the air components or react together and form the so-called secondary pollutant (for instance the O3)[6].

The descriptors typically adopted to assess urban air pollution are: Nitrogen Dioxide (NO2, $\mu g/m^3$), Ozone (O3, $\mu g/m^3$), Carbon Monoxide (CO, mg/m^3) and Particulate Matter (PM10, $\mu g/m^3$). For specific purposes others like CO_2 and COV are as well evaluated.

Regarding noise, the classical long-term map descriptor is the A-weighted Equivalent Continuous Sound Level – Leq(A) for three time periods: Lday (day-noise indicator; period between 7h00 am and 8.00 pm); Levening (evening-noise indicator; period between 8h00 pm and 11.00 pm); and Lnight (night-time noise indicator; period between 11h00 pm and 7h00 am). Additionally, an aggregated descriptor is used, the Lden (day-evening-night). These descriptors are the ones adopted and recommended by the European Directives and the national legislation of most of the European countries.

2.5 Quality labels algorithm

Quality labels are generated daily for air quality, urban noise and heat index, based on a multicriteria algorithm which combines normalized values of point-measurements.

The calculation of the air quality label takes into account hourly averages of nitrogen dioxide (NO2), ozone (O3), carbon monoxide (CO) and particulate matter (PM10), for which the quality levels to be combined are presented in Table 2.

Table 2: Air quality and noise.

Air quality				Urban noise	Classification
CO (mg/m^3)	**NO2 ($\mu g/m^3$)**	**O3 ($\mu g/m^3$)**	**PM10 ($\mu g/m^3$)**	**Leq (dBA)**	
] -, 4.9]	] -, 99.9]	] -, 59.9]	] -, 19.9]	] -, 55]	Very Good *(4)*
[5.0, 6.9]	[100.0, 139.9]	[60.0, 119.9]	[20.0, 34.9]	] 55, 62.5]	Good *(3)*
[7.0, 8.4]	[140.0, 199.9]	[120.0, 179.9]	[35.0, 49.9]	] 62.5, 67.5]	Moderate *(2)*
[8.5, 9.9]	[200.0, 399.9]	[180.0, 239.9]	[50.0, 119.9]	] 67.5, 75]	Poor *(1)*
[10.0, + [	[400.0, + [	[240.0, + [	[120.0, + [	] 75, +]	Very Poor *(0)*

Table 3: Heat index (values in ºC).

Month	Very Good	Good	Moderate	Poor	Very Poor
January	] 16, 24]	]12, 16];]24, 28]	] 8, 12];] 28, 32]	] 5, 8];] 32, 35]	] -, 5];] 35, +[
February	] 17, 23]	]14, 17];]23, 26]	]10, 14];]26, 30]	] 7, 10];] 30, 33]	] -, 7];] 33, +[
March	] 15, 25]	]11, 15];]25, 29]	] 6, 11];] 29, 34]	] 2, 6];] 34, 38]	] -, 2];] 38, +[
April	] 20, 30]	]14, 20];]30, 36]	] 9, 14];] 36, 41]	] 4, 9];] 41, 46]	] -, 4];] 46, +[
May	] 20, 30]	]15, 20];]30, 35]	] 9, 15];] 35, 41]	] 4, 9];] 41, 46]	] -, 4];] 46, +[
June	] 21, 29]	]17, 21];]29, 33]	]12, 17];]33, 38]	] 8, 12];] 38, 42]	] -, 8];] 42, +[
July	] 20, 30]	]15, 20];]30, 35]	] 9, 15];] 35, 41]	]4, 9];] 41, 46]	] -, 4];] 46, +[
August	] 20, 30]	]16, 20];]30, 34]	]11, 16];]34, 39]	]7, 11];] 39, 43]	] -, 7];] 43, +[
September	] 20, 30]	]15, 20];]30, 35]	]9, 15];]35, 41]	]4, 9];] 41, 46]	] -, 4];] 46, +[
October	] 15, 25]	]11, 15];]25, 29]	]6, 11];] 29, 34]	] 2, 6];] 34, 38]	] -, 2];] 38, +[
November	] 15, 25]	]10, 15];]25, 30]	]4, 10];] 30, 36]	] -1, 4];] 36, 41]	] -, -1];] 41, +[
December	] 16, 24]	]12, 16];]24, 28]	] 7, 12];] 28, 33]	] 3, 7];] 33, 37]	] -, 3];] 37, +[

For the noise quality level, the classification presented in Table 2 is used.

The heat index depends on the month of the year, assuming quality levels from Very Poor to Very Good according to the classification system summarized in Table 3 (used for the case of Braga).

3 The web-based monitoring and information system for Braga

The evaluation of the urban environmental quality of a city has become a major concern to public authorities and especially for residents. At this point, the architecture of an urban system to monitor the air and noise for the city of Braga (Portugal) named smarBRAGA will be characterized, as well as the infrastructure of acquisition and analysis of the necessary data. To provide a global overview of the smarBRAGA project, some outputs and results of this monitoring and information to the public platform will be presented.

3.1 The architecture of SmarBRAGA

The SmarBRAGA is an information system for the public and the urban environmental monitoring system of the city of Braga. With this platform, Braga owns an innovative service of information to the public that can be considered pioneer in Portugal.

The SmarBRAGA project aims to inform the population about the central aspects of the urban environment, such as: noise, air quality and meteorology, as well as to create an infrastructure of acquisition, storage, processing and communication of data related with the urban environment. In a more detailed level, the specific objectives are to monitor the urban noise, the air quality and the meteorological data in the city, plus to store the digital cartography of the urban noise and of atmospheric pollutants, as well as to generate scenarios of noise and air pollution predictions, and finally to maintain a system of alert and information to the public on noise and on air quality.

For the development work, three infrastructures (Figure 1) were created, namely a web platform that integrates the services of acquisition and analysis of data, and information to the public; a technological solution of acquisition of environmental data constituted by two mobile units of acquisition and collection of data; and an analysis centre for the development and deployment of scenarios for long term periods.

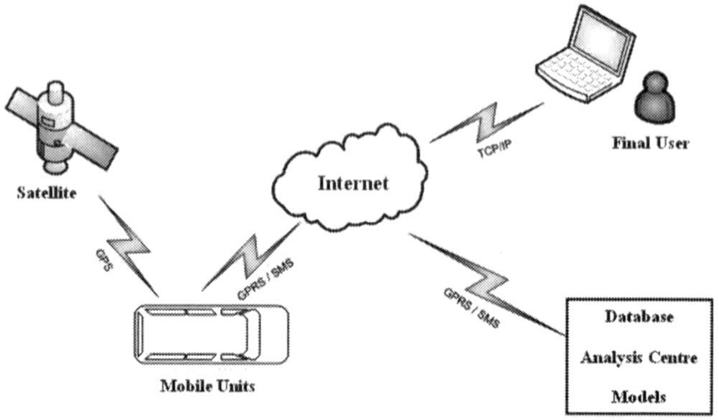

Figure 1: Overall functioning scheme of SmarBRAGA.

The SmarBRAGA project elapsed throughout 2007 and 2008, and it is foreseen that the availability of the service to the general public will start by the Summer of 2009. The global process of the project was established in four group stages, namely the equipment and hardware acquisition, the development work, the testing stage and the training and start-up.

3.2 The data acquisition and analysis infrastructure

Besides the human resources, in this project the following material resources were used: modified and adapted mobile units, GPS receptors, digital sound measurements, meteorological stations, particles and other air pollutant measuring equipments, simulation software for air pollution and noise prediction, among others.

The mobile units (Figure 2) take measurements in a specific location of the city (GPS technology), and transmit them through wireless communication for the web platform. On board of the mobile units are installed the modules for noise, air pollution, meteorological, multimedia, communication and positioning, even though this complete set is not always necessary for a particular measurement.

The aim of the analysis centre is to develop and provide the long term scenarios (in the form of maps) of the noise and the atmospheric pollution

a) Mobile unit in operation b) Measurement of particles *in loco* c) Recharging the batteries of several sensors

Figure 2: Examples of the functioning of data acquisition infrastructure.

through the web platform, as well as traffic statistics. For this purpose, the corresponding technological solution includes: i) the software for noise prediction with a traffic module based on the Directive 2002/49/CE and with the capacity to deal with mid/large cities; ii) the software for air pollution simulation, which integrates with the noise software; iii) and two high performance computers.

The software used in this case was CadnA, which integrates both components (noise and air pollution) and thus allows to rationalise the effort associated with data maintenance related with three-dimensional urban characteristics and traffic information, usually necessary to built urban databases for modelling noise and atmospheric pollution. Figure 3 shows some pictures of the analysis centre.

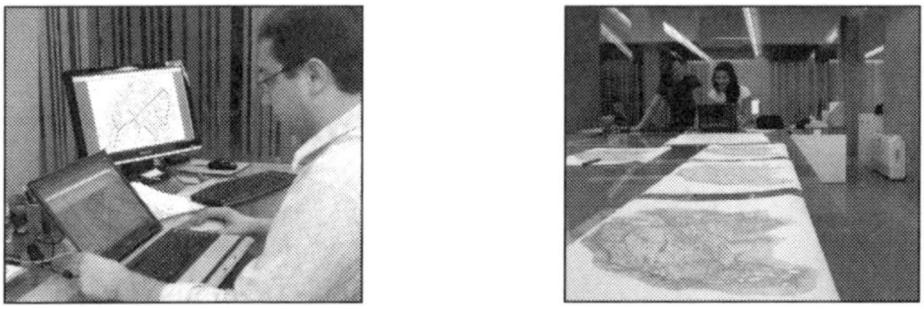

a) Simulation software in use b) Layout of information produced

Figure 3: Analysis centre in operation.

3.3 Examples of results

The results of SmarBRAGA project are mainly available on the web platform of information to the public. This platform aims to automatically process and provide information, integrating data acquired in real or differed time, as well as the long term simulations, through a user-friendly interface and innovative technologies of representation and mapping. This is structured in five channels, as follows:

The Homepage of SmarBRAGA. Includes a synthesis of up-to-date information, through a system of classification and information to the public based on colour quality labels, ranging from green to red (Figure 4).

Figure 4: SmarBRAGA: homepage.

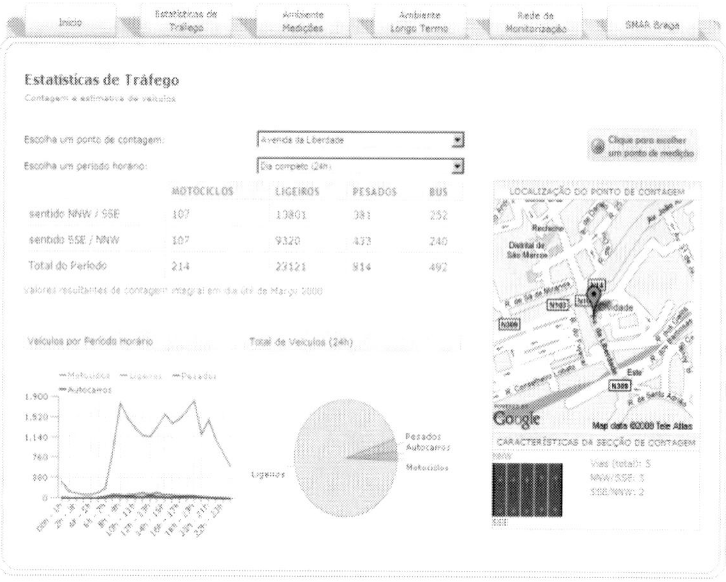

Figure 5: SmarBRAGA: traffic statistics.

Traffic statistics. The available data resulted from a traffic counting campaign (by category of vehicles) carried out in 2008 on the road network of Braga. These data are associated with specific periods of time; however the platform can be developed to integrate on-line information of traffic (Figure 5).

Long-term environment. Long-term environmental maps of noise and air pollution are available together with navigation tools developed over Google Maps technology (Figure 6).

Environmental measurements. The measurements of environmental and meteorological parameters are integrated automatically after its acquisition through the mobile units in the monitoring network points.

Monitoring network. Points of measurement that are part of the monitoring network are represented in a map with navigation functionalities and satellite image reference, as well as cross-links for information of measurements and environmental quality.

Figure 6: SmarBRAGA: Long-term PM_{10} concentrations map.

4 Conclusions

The aim of a monitoring and information system is to support the formulation of control strategies and, additionally, to inform the population about the urban environmental situation. This paper focused on the information objective.

A conceptual model for the urban environmental information system was presented. The critical aspects identified for such an information system were: i) the quality label-based information; ii) the long-term environmental maps versus the point-measurements; iii) the importance of motor traffic as the main pollution source; iv) the selection of air pollution and noise descriptors; and v) the algorithm which generates environmental quality labels.

Based on this conceptual outline, a web-based information system was developed for the city of Braga, in Portugal. The so-called SmarBRAGA project aims at informing the population about the central aspects of the urban environment, such as: noise, air quality and meteorology. For this purpose an infrastructure of acquisition, storage, processing and communication of data was created.

The web-based platform SmarBRAGA includes a home page where almost-real time environmental information is presented and classified through quality labels. Dedicated channels in this platform include information on traffic, point-measurements, long-term pollution maps, and the monitoring network.

References

[1] OECD, (eds). Roadside Noise Abatement, Organisation for Economic Co-operation and Development Publications: Paris, France, 1995.
[2] Silva, L.T., Avaliação da Qualidade Ambiental Urbana. PhD Thesis, Universidade do Minho, Braga, Portugal, 2008.
[3] Mendes, J.F.G.; Silva, L.T., Evaluating critical noise disturbance zones in a mid-sized city. International Conference on Sustainable Planning & Development, 2003.
[4] Mendes, J.F.G.; Silva, L.T., Evaluating conflict zones of air pollution in a mid-sized city. International Conference on Sustainable Planning & Development, 2007.
[5] Highways Agency. Design Manual for Road and Bridges Volume 11, Section 3, Part 1 – Air Quality, UK, 2003
[6] Silva, L.T., Mendes, J.F.G. Determinação do índice de qualidade do ar numa cidade de média dimensão. Engenharia Civil. 27, p. 63-74, 2006.
[7] Decreto-Lei nº 9/2007 de 17 de Janeiro. Diário da República, I Série-A, Lisboa, Portugal, 2007.

Guiding principles for creating environmental regulations that work

T. S. Mullikin
Government, Policy and Regulatory Affairs,
Moore & Van Allen PLLC, USA

Abstract

Creating sensible policies and regulations that reduce, stop and ultimately reverse the upward trend in greenhouse gas emissions are the topics of heated debate worldwide. By applying five guiding principles based upon years of experience formulating policies and regulations in a legal/legislative setting, workable environmental regulations can be developed and implemented.

Five guiding principles require that policies and regulations be: 1) flexible in application 2) global in scope 3) market-driven in implementation 4) firmly anchored in economic reality 5) successful in demonstrating tangible environmental benefits.

Keywords: government policies, regulations, climate change, environmental legislation, workable solutions, greenhouse.

1 Regulations must be flexible in application

Scientists do not deal in truths and untruths so much as they explore truths and "new" truths arising from additional research data. The constantly evolving knowledge from scientific methods and measurements requires that lawmakers and regulators be flexible when attempting to regulate man-made activities that result in a significant effect upon the Earth's environment.

As early as 1896, Nobel prize-winning chemist Svante Arrhenius of Sweden helped demonstrate how difficult it is to predict the effects of humankind's activities on our environment. Based upon his calculations, Arrhenius predicted that doubling the amount of carbon dioxide covering the Earth would result in a 5-6 degree Centigrade increase in the Earth's surface temperature [1]. But he also thought it would take 3,000 years to occur.

WIT Transactions on Ecology and the Environment, Vol 123, © 2009 WIT Press
www.witpress.com, ISSN 1743-3541 (on-line)
doi:10.2495/AIR090111

In reality, it has taken only 100 years for levels to increase by 30 percent. Fortunately, scientists can gauge with increased accuracy the temperature effects of greenhouse gases. Instead of the temperature rising 5-6 degrees as earlier predicted, scientists now estimate temperature increases at about half that, 2-3 degrees Centigrade [1].

While information is continuously changing and evolving, one constant that remains is that policies and regulations must be able to take into account new science and scientific data and be prepared to speed up or slow down regulatory induced controls as necessary.

For example, research in the Gubantonggut Desert in China indicates that deserts may take up carbon dioxide at the same rate as some temperate forests. Research in the American Mojave Desert resulted in similar findings.

Scientists need to do much research yet in this area, but if the initial findings hold true, then deserts and semi-arid regions of the world may be absorbing 5.2 billion tons of carbon a year. That amount is roughly half the amount of carbon released annually from the burning of all fossil fuels [2].

It is imperative, then, that enacted rules, regulations and policies be adaptable to changing conditions and new scientific findings.

2 Regulations must be global in scope

Early environmental legislation in the United States – such as Clean Air and Clean Water Acts – frequently addressed pollution from a local, state or even regional level. Nevertheless, in the 1970s and 1980s, another pollution problem called "acid rain" began receiving great attention. Forests in high mountain areas – particularly in the Eastern U.S., such as the Catskills, Adirondacks and Blue Ridge – were showing signs of damage and decay.

In Germany, trees damaged by acid rain appeared in nearby Eastern bloc countries. Scientists and politicians suddenly became aware that pollution knows no state or national boundaries.

Finally, the defining event that made the world aware of the global nature of the earth's physical environment occurred April 28, 1986, when technicians at Sweden's Forsmark Nuclear Power Plant outside Stockholm detected radiation levels inside the plant that were four to five times normal readings.

The mysterious radiation was not coming from Swedish reactors on site. Eventually, investigators concluded that prevailing winds were carrying dangerous radiation from the former Union of Soviet Socialist Republics (USSR).

Over the next several days, the world gradually learned that an explosion and fire at the Chernobyl nuclear power plant in northern Ukraine was spewing tons of dangerous radioactive isotopes into the atmosphere every day the fire raged out of control.

Even though Stockholm is 1,000 miles from Chernobyl, it received a significant amount of radioactivity from the accident. Chernobyl opened the world's eyes to the fact that environmental events in one country affect nearly every other country throughout the world. Thus, to be effective, we must implement pollution control measures on a worldwide basis.

Historically, nations have successfully banded together globally to solve environmental problems. One such example involves a family of chemicals known as chlorofluorocarbons – or CFCs for short. While these chemicals have been around since the 1930s, they did not achieve widespread use until the 1950s and 60s when they served as refrigerants in air conditioning compressors and as an inert propellant in aerosol sprays.

By the 1970s, research satellites already were beginning to pick up signs that the earth's ozone level was starting to diminish. Slowly the issue took on an international urgency. Various nations agreed to ban CFCs in aerosols but industry fought banning valuable CFCs in other applications.

In the early 1980s, models of the atmospheric chemistry involving CFCs became more and more complex, and various challenges arose to the validity of the science. Apparently, the industrial nations of the world required a great shock before they could commit to the uncomfortable task of phasing out CFCs.

That shock came in a 1985 field study by J. C. Farman, B. G. Gardiner and J. D. Shanklin of the British Antarctic Survey. Their study, published in Nature, May 1985, summarized data showing that ozone levels had dropped to 10% below normal January levels for Antarctica. A hole was developing in the ozone layer over the South Pole [3].

Just as importantly as the discovery of the hole in the ozone, these scientists demonstrated in a startling way that chemicals released by industrialized nations in major population centers of Europe and North America could gather over the poles and collectively change the atmosphere on a global scale.

September 2007 marked the 20th anniversary of the Montreal Protocol on Substances that Deplete the Ozone Layer. Initially, 55 countries and the European Economic Commission negotiated this historic international agreement. These countries banded together in the mid-1980s to set a schedule for freezing and then phasing out the production of CFCs. Eventually 191 countries joined the agreement including the United States under President Ronald Reagan. The United States ended CFC production 12 years ago.

Due to its widespread adoption and implementation, the Montreal Protocol has been regarded as an example of exceptional international co-operation by world leaders, including Kofi Annan, United Nations Nobel Peace Prize-winning Secretary General from 1997 to 2006. He said that the agreement is "Perhaps the single most successful international agreement to date...." And scientific evidence shows that CFCs in the lower troposphere are decreasing, a change that must occur before the ozone layer can regenerate in the upper levels of the stratosphere.

The significance of the ozone protection initiative is that action took place on a global scale even though there were challenges to the science involved. While eliminating greenhouse gases presents much tougher challenges than replacing ozone-depleting substances, the experience illustrates that nations can come together to achieve positive results even when there is disagreement about the science or politics involved. As illustrated in the ozone situation, remedies must be approached on a global basis or emissions will continue to emanate from other countries.

Just as electricity seeks the path of least resistance, industry locates where conditions are more favorable to economic success. If climate change issues are not addressed globally, then the emissions will migrate from advanced industrialized nations to other areas of the world where environmental controls are less restrictive.

For example, if U.S. climate change regulations place a burdensome tax or fee on carbon emissions, a business may be forced to relocate to a country where these regulations do not exist or are only minimally enforced. Which means that the U.S. may decrease its total carbon emissions when the polluting business moves overseas. The problem does not go away. It has just moved to another part of the world where there is less chance the problem will be mitigated.

3 Regulations must be market driven

Another lesson learned from the ozone experience is this: one of the key factors to change is motivating industry – especially large corporations – to develop the technologies that make change happen. Nothing drives change like economic success. One of the most notable achievements of the Montreal protocol was that the chemical companies quickly recognized the market opportunities created by the agreement.

With the discovery of a hole in the ozone, political forces began pushing for regulation of ozone-depleting chemicals. Chemical giant DuPont, which produced more than half the CFCs, then assumed a leadership role, immediately stepping up efforts to develop ozone-safe substitutes for CFCs.

The firm spent $500 million to develop substitutes and build production facilities. Within a few years they were supplying refrigeration and air-conditioning manufactures with profitable CFC substitutes [4].

DuPont created a new model for corporate behavior by teaming up with environmentalists to ban CFCs. DuPont correctly recognized the new markets that needed CFC replacements and leveraged their research and development strengths to fill diverse market niches.

The phase out of CFCs was assisted by the simple and unambiguous timeframe set out in the Montreal Protocol for the end to CFC production. By setting reasonable schedules, companies were able to rationally predict and develop markets for alternatives.

The CFC phase out teaches lessons today about the challenges created by greenhouse gas emissions. The speed with which CFC substitutes were rushed to market was based upon prospects of a lucrative new market for companies such as DuPont. Therefore, the effectiveness of any strategy on global climate change will depend upon how well it creates new markets.

It is estimated that the market for low-carbon energy products will reach $500 billion by 2050, according to the Stern Review on the Economics of Climate Change, released last year by the British Treasury. The report states that the transition to a low-carbon economy will bring challenges for competitiveness, but also opportunities for growth [5].

Market-driven incentives to reduce emissions work best. All regulations that affect individuals and industry must be crafted carefully to avoid creating burdensome requirements. The most effective policies will encourage the marketplace by offering tax rebates, development funds and other economic incentives to jump start technology and implementation into commercial production. For example, consumers could be offered tax credits to purchase alternative fuel vehicles or to replace inefficient home heating systems with solar-powered systems. Once the consumer demand has been established, then the private sector should enthusiastically manufacture the products required and assist in building the infrastructure. If hydrogen fuel vehicles become popular, then energy companies will have the incentive to create the fuel and distribution centers where motorists can conveniently refuel. Free enterprise is a powerful force when the driving factors are economic.

State governments in the United States are also taking steps to increase the amount of energy consumed from renewable resources. North Carolina, for example, the state in which our firm operates, has authorized tax credits for solar power generation. Legislation also requires that the state generate at least 12.5 percent of its power from renewable resources by 2021. This step establishes North Carolina as the first state in the Southeastern U.S. to adopt renewable energy standards.

The state has enacted additional legislation providing tax credits to both residential and corporate customers in North Carolina. The credit covers up to 35% of the cost of a solar electricity or other renewable energy system, capped at $10,500 for residential systems and a hefty $2.5 million for commercial and industrial systems.

Federal tax credits provide homeowners 30 percent of the cost of a project up to a maximum credit of $1,500 on a broad range of home energy improvements, such as energy saving doors, windows, insulation, roofing, energy efficient heating, air conditioning and ventilation systems and water heaters.

These state and federal tax rebates provide attractive incentives for homeowners to voluntarily make significant long-term infrastructure improvements to their dwelling

On the other hand, legislated regulatory systems to reduce emissions – such as cap and trade – are tricky and can result in an increase in government spending and a decrease in revenues, according to an April 2007 report by the U.S. Congressional Budget Office report. Such changes would come about due to rising energy costs for government, which would also be affected by the price of energy and other carbon-intensive goods and services. Businesses that are negatively affected by carbon costs create less revenue, which means less tax money will be paid to government.

If the cap-and-trade system is implemented, it may bring about economic changes too quickly and lead to an economic slowdown. Once again, government suffers because of falling tax revenues from tax-paying businesses and corporations [6].

4 Regulations must be anchored in economic reality

The world economy has an environment of its own, just as sensitive to change as our physical environment. For almost every action in the world, there is a measurable economic reaction. Therefore, any policy changes and regulations by a sovereign government also result in economic impacts. The key is in moving forward, making the world a better place to live without stopping economic growth. Economic realities cannot be ignored.

One of the most urgent questions that must be answered in reducing greenhouse emissions is, "Who will pay the bill?" Will costs be handled on an individual level, by passing costs on to consumers who create the demand for energy and other carbon-producing products? Carbon taxes on working households living from paycheck to paycheck may be saddled with the burden of extra costs. On the other hand, direct costs can create dramatic changes in behavior.

Individuals and families may have a significant economic incentive to reduce carbon consumption if charged for it. Likewise, through higher taxes coupled with manufacturer rebates and government tax incentives, a family may find it much more cost efficient to trade in an old gas-guzzler automobile for a new rechargeable electric, a gas hybrid or a fuel cell-powered car.

On the other hand, a large producer of carbon, such as a coal-fired electric power plant, may be in a better position to capture and sequester emissions on a commercial scale and share the costs with consumers in the form of higher electric rates. Here again, the consumer would have an incentive to reduce electric consumption to save money. Economics would drive conservation.

Cap-and-trade systems have the added drawback of significant administrative costs. To reduce emissions through a formal system of incremental reductions requires measures, databases and personnel and systems for operation. In other words, an expansion of government. While carbon credit auctions generate government revenue, money for investing in renewable energy projects or alternative fuels can be significantly reduced due to excessive administrative costs, a constant concern for local government.

And what other effects would carbon controls create? If a consumer has less money to spend, will it damper economic growth? If consumers stop buying, will that reduce jobs, further slowing down the economy? Will the economic stagnation affect large industrialized nations the most or will emerging industrial nations of Asia and South America be affected equally?

Past experience in the U.S. in dealing with environmental challenges involved reducing sulfur emissions, which combined with rain to produce an "acid rain." The threat from acid rain covered many areas of the economy, such as damage to public property and landscapes, harmful effects on wildlife in streams and ponds and deforestation. Sulfur dioxide emissions also endangered human lungs, particularly Americans who were already suffering chronic lung disease.

5 Regulations must result in real environmental benefits

One of the most significant lessons that can be learned from the efforts to eradicate CFCs is that policies and regulations to accomplish environmental changes must actually do what they are intended to do. For example, scientists as well as policy makers recognized that the only way to protect the ozone layer was to eliminate CFCs.

Simply reducing the amount would not end the environmental threat. Neither would a cap-and-trade system that has been used successfully in other environmental applications, such as sulfur dioxide emissions.

It is also important to separate similar objectives, such as increasing energy independence and reducing greenhouse gas emissions. These separate objectives are obviously not the same. So policy changes and economic incentives should be aimed at reducing carbon through clean fuels – such as hydrogen, nuclear, solar and wind-generated electricity – and other energy sources that do not produce carbon as a waste product.

On the other hand, making a country energy self-sufficient may involve intense use of ethanol, natural gas or other combustible fuels that equally produce greenhouse gas emissions.

Ultimately, the cleanest, most efficient electric generation plant is the one that did not have to be built. Increased energy efficiency and improved technology are the quickest ways to effect this type of greenhouse gas reductions. Some industries have already taken these steps. The American steel industry is a prime example.

According to the International Iron and Steel Institute, American steelmakers have greatly increased steel production while proportionately reducing carbon dioxide emissions. In 1990, American manufacturers produced 88.7 million metric tons of steel and 85 million metric tons of carbon dioxide as a by-product. In 2005, steel producers turned out 93.9 million metric tons of steel and reduced carbon dioxide emissions to 45.2 million metric tons. That reduction exceeds the 1997 Kyoto Protocol target by 900 per cent.

In addition to the use of more energy efficient electric arc furnaces, other manufacturing technology breakthroughs, such as the transition from ingot casting of steel to continuous casting and strip casting, have reduced energy requirements.

The changing economics of the iron and steel industry essentially drove progress in energy efficiency. A by-product of that change has been a dramatic reduction in greenhouse gas emissions. The recent increase in energy prices has the potential to spur the same kind of reductions in other sectors of the economy, arguably far faster and more cost effectively than through a government implemented environmental program.

One of government's most important roles is to identify proposed environmental policy objectives. Are they to reduce use of fossil fuels or to encourage energy efficiency and conservation? Is the objective to improve the quality of air or to spur consumers to use an alternative fuel?

Another way in which government can help is by ensuring that if a market-based approach is adopted, it is working. Government intervention has the power to provide incentives that remove barriers to environmentally positive behaviors.

For example, government tax policy can provide incentives that encourage gasoline service stations to augment their traditional petroleum fuel offerings with hydrogen fueling facilities. When hydrogen fuel is easier to obtain, motorists will be more inclined to buy and use hydrogen-powered vehicles.

6 Summary

Effective regulations must take into account the needs of all stakeholders on a global basis. Global environmental issues need global environmental solutions. Real leadership requires a more complex investigation into issues and solutions affecting global climate change issues.

Policies should be carefully aimed at achieving a particular objective without creating a burdensome regulatory environment. Private enterprise and market forces provide the greatest impetus for sustainable changes.

Most people have the potential to understand any issue once it is expressed in terms of dollars and cents. Policies that work leverage this principle by couching behavioral change—and acceptance of new energy sources and technology—in practical, economic terms.

References

[1] Sample, I. "The Father of Climate Change," The [Manchester] Guardian, June 30, 2005, http://www.guardian.co.uk/environment/2005/jun/30/climatechange.climatechangeenvironment, accessed September 30, 2008.

[2] Stone, R. "ECOSYSTEMS: Have Desert Researchers Discovered a Hidden Loop in the Carbon Cycle?" Science June 13, 2008: Vol. 320. no. 5882, pp. 1409 – 1410, DOI: 10.1126/science.320.5882.1409, accessed December 3, 2008.

[3] Farman, J. C., B. G. Gardiner, and J. D. Shanklin. 1985. Large losses of total ozone in Antarctica reveal seasonal ClOx/NOx interaction. Nature 315: 207-10.

[4] Rotman, D. "Remembering the Montreal Protocol," Technology Review, published by MIT, January 1, 2007, <http://www.technologyreview.com/energy/17994/page2/>, accessed October 3, 2008.

[5] STERN REVIEW: The Economics of Climate Change, Publication of HM Treasury, Office of Climate Change, <http://62.164.176.164/d/Executive_Summary.pdf>accessed October 10, 2008.

[6] "Trade-Offs in Allocating Allowances for CO2 Emissions." Report by U.S. Congressional Budget Office, April 25, 2007.

Atmosphere environment improvement in Tokyo by vehicle exhaust purification

H. Minoura[1], K. Takahashi[2], J. C. Chow[3] & J. G. Watson[3]
[1]*Toyota Central R&D Labs., Inc., Nagakute, Japan*
[2]*Japan Environmental Sanitation Center, Kawasaki, Japan*
[3]*Desert Research Institute, Reno, Nevada, USA*

Abstract

In Japan, PM regulations in diesel automobile emissions started in 1994. Long-term measurements of suspended particulate matter (SPM, <7 μm), PM_{fine} (<2.1 μm), and PM_{coarse} (2.1 to 7 μm) were obtained from an urban Kudan site in downtown Tokyo from 1994 to 2004 to evaluate the effects of emission reduction measures. A remarkable PM mass downward trend was found from 1996 onwards, especially in the PM_{fine} fraction, which decreased at a rate of 2.09 μg m^{-3} yr^{-1}. The decrease in PM_{fine} is attributable to the decreases in elemental carbon (EC) at the rate of 0.82 μg m^{-3} yr^{-1}. PM_{fine} EC concentrations at the roadside Noge site shows a threefold faster downward trend, at the rate of 2.56 μg m^{-3} yr^{-1}. This decrease is consistent with fleet penetration of engines and fuels that complied with a stringent Japanese emission reduction limit which began to take effect in 1994. It is apparent that vehicle emission reduction contributed to air quality improvement in Tokyo. The levelling off of the EC reductions since 2005 may be explained by the results from the carbon isotope (^{14}C) analysis, which suggested contributions from biomass combustion sources in addition to vehicle emissions in downtown Tokyo.

Keywords: Japan, Tokyo, trends, particulate matter, chemical composition, elemental carbon, biomass combustion.

1 Introduction

Atmospheric particulate matter (PM) is important for health, visibility, and climate [1–3]. Vehicle emissions are major contributors to PM, irrespective of recent advances in emission reduction technology [4]. Figure 1 shows a trend in

WIT Transactions on Ecology and the Environment, Vol 123, © 2009 WIT Press
www.witpress.com, ISSN 1743-3541 (on-line)
doi:10.2495/AIR090121

the number of vehicles and fuel consumption over the last 50 years. Passenger and light car numbers increased rapidly from the 1960s to the early 1990s, while truck numbers leveled after 1980. Passenger car and truck numbers decreased after 2000, but fuel-use decreased more rapidly after 2000, especially for diesel fuel. Diesel fuel consumption increased from 1987 to 2000, irrespective of the steady truck numbers, implying the increase of average kilometers traveled by each vehicle, which corresponds to the growth in trip length per year [5].

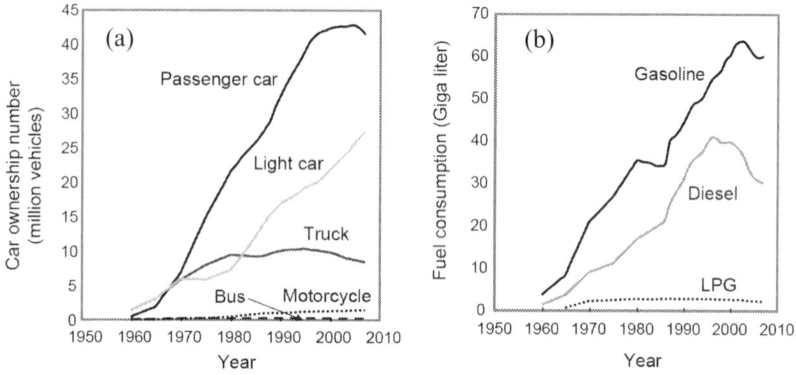

Figure 1: Yearly trends of Transport in Japan (a) the car ownership number for each sector, (b) the fuel consumption (data source: Ministry of Land, Infrastructure, Transport and Tourism).

The ratio of diesel passenger cars in Japan is very low. Figure 2 shows a trend for nitrogen oxides (NO_x) and PM emissions from on-road vehicles in the Tokyo area [6]. This decreasing trend suggests the effectiveness of the action plan: "NO diesel car operation," established in 1999 by the Governor of Tokyo, which prohibits the operation of old vehicles (trucks >7 years) in Tokyo [7].

Diesel engines emit high NO_x and low PM during high-temperature combustion with the reverse pattern at low temperatures [8]. NO_x limitations were more stringent than those for PM owing to high ambient nitrogen dioxide (NO_2) levels detected in the 1970s and early 1980s. PM emission limits were lowered in the 1990s, with new limits for long-term regulation (the most stringent in the world) anticipated since 2005 [9]. Figure 3 shows that the Japanese government proceeded to more stringent NO_x than PM regulations, while historically more stringent PM than NO_x were regulated in the U.S. and Europe. Rapid decrease in PM emissions (2003 – 2005), shown in Figure 2(b), reflect: 1) the new short term regulation, enforced in 2003; and 2) the decrease in on-road truck numbers (Figure 1(a)).

SPM (suspended particulate matter) refers to the Japanese PM standard for particles sampled through an inlet having 0% transmission for particles with 10 μm aerodynamic diameter; equivalent to a 50% cut-point of PM_7. Annual average SPM in Tokyo has exceeded 50 $\mu g\, m^{-3}$ from 1990–1997. Approximately 80% of the 49 monitoring stations exceeded the 24-hour standard

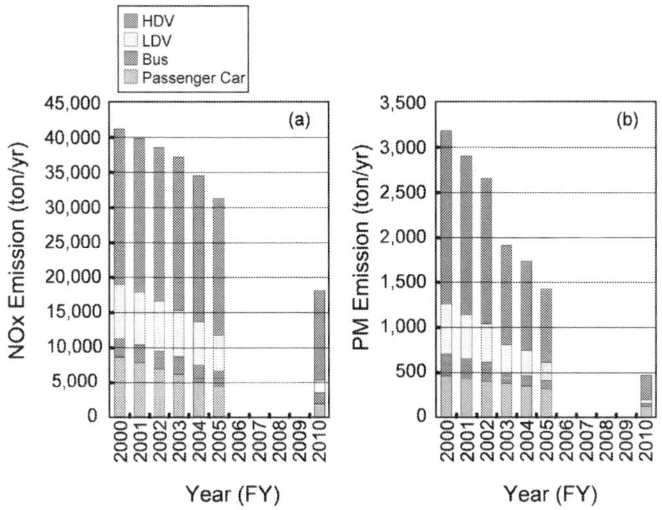

Figure 2: Vehicle emission trends for: (a) NO_x, expressed as NO_2; and (b) PM_{coarse}(?) for the Tokyo metropolitan area. HDV: Heavy-duty diesel vehicles; LDV: Light-duty diesel vehicles. Values before 2005 was based on the actual situation, and value of 2010 was estimated in consideration of automobile substitution.

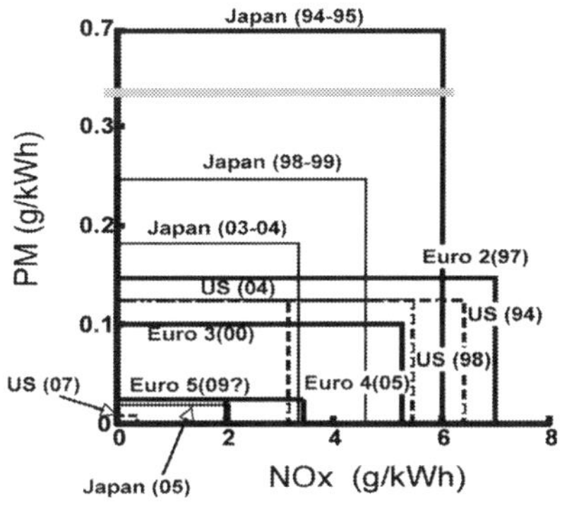

Figure 3: World emission regulation trend for heavy duty vehicle.

of 100 μg m^{-3} (98th percentile) from 1973 when monitoring was started until 1998. If engine exhaust is a large contributor to SPM, SPM levels in Japan would decrease as vehicle emission limits took place.

Ten years (1994–2006) of PM mass and chemical concentrations obtained from two sites in Tokyo traffic-dominated environment are examined. Special attention is given to the organic carbon (OC) and elemental carbon (EC) fractions (major PM constituents in vehicular emissions [10, 11]), along with six water-soluble ionic species. The objectives of this paper are: 1) to evaluate changes in vehicle contributions; and 2) to establish relationships between vehicle emissions and long-term trends in ambient concentrations.

2 Experimental method

As shown in Figure 4, the urban Kudan monitors were placed on the rooftop of a 10-story building about 30m above ground level in downtown Tokyo, with heavily travelled roadways nearby. The site is surrounded by office buildings and large open spaces, including the Yasukuni Shrine and the Imperial Palace. The nearby Chiyoda (1.9 Km) and Hibiya (2.4 Km) sites are compliance monitors operated by the Tokyo metropolitan government. The Noge monitors were located at the busy roadside of Ring 8 with a traffic volume of ~92,000 vehicles/day on weekdays, including 13% heavy-duty diesel-fuelled vehicles.

Table 1 shows the measurements for both in situ continuous hourly SPM by beta attenuation monitor (BAM) and two-week integrated SPM filter measurements by Andersen Cascade impactor acquired at the Kudan site. There may be some differences when comparing OC and EC between the Kudan and Noge sites, since two different carbon analysis protocols are used, but it will not affect the 10-year trend. All thermal carbon methods are operationally defined. Watson et al. [12] summarize nearly 20 different methods and 40 different intercomparison studies that show varying degrees of agreement worldwide.

Water-soluble chloride (Cl^-), nitrate (NO_3^-), sulphate ($SO_4^=$), sodium (Na^+), potassium (K^+), and ammonium (NH_4^+) were measured by ion chromatography after extraction of a portion of the quartz-fiber filter in distilled-deionized water [13]. Five field blanks were analyzed for carbon and ions with each batch of 100 ambient samples. Blank values were averaged and subtracted from the ambient samples prior to normalization to the sample volume.

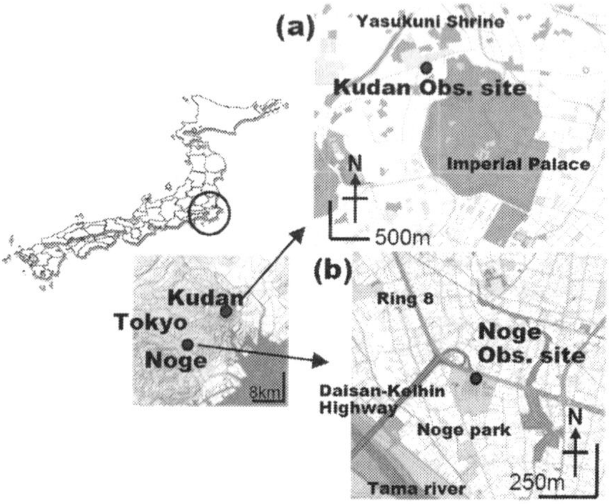

Figure 4: Sampling locations in Tokyo for the: (a) urban site at Kudan; and (b) roadsite at Noge.

Table1: Summary of ambient measurements at the urban Kudan and roadside Noge sites in Tokyo between 1994 and 2004.

Sampling Site	PM Size	Instruments	Observables	Measurement Method	Sampling Duration and Frequency	Measurement Periods
Kudan (urban background)	SPM (~PM_7)	Beta attenuation monitor (BAM; Model DUB-12, Towa DKK Corp. Ltd., Tokyo, Japan)	Mass	beta attenuation	1 hr, 24 samples/day, 365 days/yr	6/1/1994 through 3/31/2006
	PM_{fine} (< 2.1 μm) and PM_{coarse} (2.1-7 μm)	Andersen Cascade Impactor Sampler (Model AN-200, Tokyo Dylec, Tokyo, Japan) with quartz-fiber filters (2500 QAT-UP, Pall Sciences, Ann Arbor, MI, USA)	Mass	gravimetric analysis	2 weeks, 26 samples/ yr	6/1/1994 through 3/31/2006
			Organic carbon (OC), Elemental Carbon (EC), and Total Carbon (TC)	OC: 100% He at 650 °C for 8 minutes; TC: 90% He/10% O_2 at 950 °C for 5 minutes		
			Chloride (Cl^-), Nitrate (NO_3^-), Sulfate ($SO_4^=$), Sodium (Na^+), Potassium (K^+), and Ammonium (NH_4^+)	ion chromatography		3/31/1997 through 3/31/2006
Noge (roadside)	$PM_{2.5}$	Ambient carbon particulate monitor (Model 5400, Rupprecht & Patashnick, Albany, NY, USA)	OC and EC	OC: 340 °C for 13 minutes; EC: 750 °C for 8 minutes	2 hr, 12 samples/day, 365 days/yr	12/8/2002 through 9/5/2004

3 Results

3.1 PM mass concentrations

Figure 5 shows that before 2000, elevated SPM concentrations were frequently observed in summer (June – August) and winter (December – February) at the urban Kudan site. High concentrations in summer may be attributed to the photochemically generated $SO_4^=$. Elevated SPM were also found during winter due to the shallow surface layer that persists for several hours after sunrise. Annual SPM mass concentrations are consistently in the range of 51 – 53 μg m^{-3} between 1994 and 1997, with a 25% decrease from 47 μg m^{-3} in 1998 to 36 μg m^{-3} in 1999. This decrease was also found at the 27 roadside monitoring sites

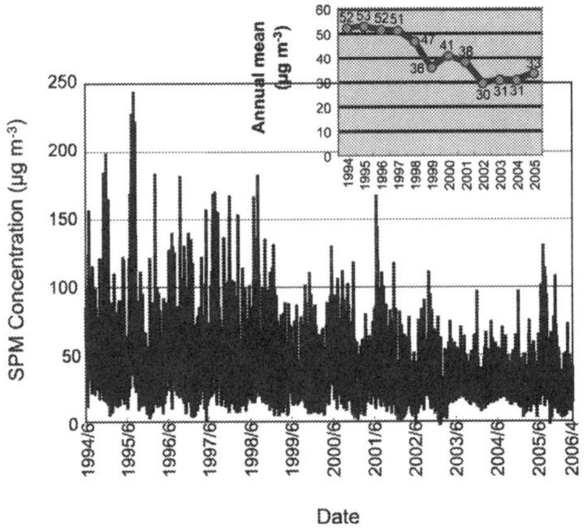

Figure 5: Hourly and annual (inset) arithmetic average SPM mass concentrations by beta attenuation monitor (BAM) at the urban Kudan site from 1994 through 2006.

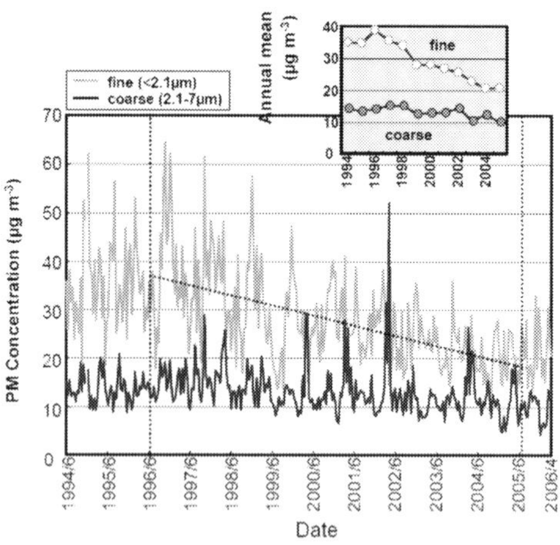

Figure 6: Two-week and annual average (inset) PM_{fine} (< 2.1 μm) and PM_{coarse} (2.1-7 μm) mass concentrations at the Kudan site from 1994 to 2006. The dashed line is a least-squares fit to the annual averages for nine years, which shows a decrease of 2.09 μg m^{-3} yr^{-1} for PM_{fine} starting September 1996.

operated by the Tokyo metropolitan government [13]. Mean wind speeds increased during summer and winter of 1999, and it is possible that there was greater dispersion.

As shown in Figure 6, the long-term improvement in SPM levels is due to the reductions in the PM_{fine} fraction at a rate of 2.09 μg m^{-3} yr^{-1}. PM_{coarse} concentrations did not change over the 10-year period, reflecting similar influence from road dust or marine aerosol. A few spikes in PM_{coarse} were found during springtime, probably reflecting contributions from Asian dust storms [14–16], which are common occurrences in Japan.

3.2 PM carbon trends

Figure 7 shows that EC was much higher than OC in PM_{fine} than PM_{coarse} fractions, suggesting the dominance of diesel engines (MOE manual, 1997). Average EC to total carbon (TC; sum of EC and OC) ratio was 0.76 in PM_{fine} and 0.62 in PM_{coarse}. The ratio of EC to TC also decreased, indicating that these primary emissions were decreasing more rapidly than OC, some of which comes from conversion of gases to particles. The high PM_{fine} EC/TC ratio is similar to other measurements reported in Japan [17, 18] and Hong Kong, China. Compared to the thermal/optical method (Chow et al., 1993), Takahashi et al. [19] found overestimation of EC by 41% in PM_{fine} and by 104% in PM_{coarse}, but no difference was found for TC.

The decrease in EC/TC ratio is consistent with diesel vehicles becoming more efficient combustors. PM_{fine} carbon started to decrease after 1996 at a rate of 0.20 μg m^{-3} yr^{-1} for OC and 0.82 μg m^{-3} yr^{-1} for EC based on linear regression.

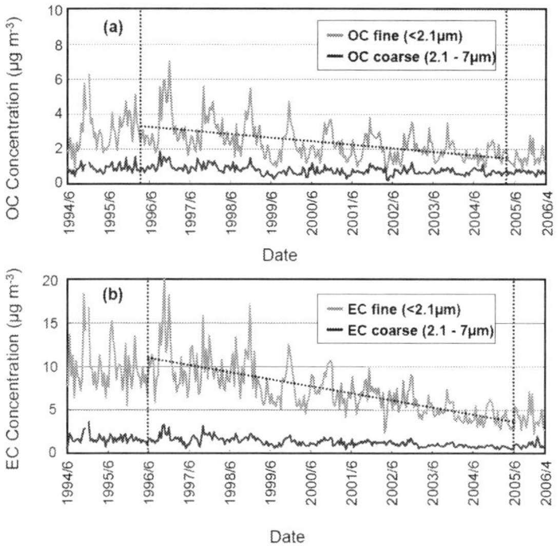

Figure 7: Temporal variations in OC and EC concentrations at the Kudan site. The reduction rate of each component since 1996, shown by the dashed line, is 0.20 µg m-3 yr^{-1} for OC and 0.82 µg m-3 yr^{-1} for EC.

These decreases reflect the fleet penetration of lower emitting diesel engines mandated by the 1994 for NO_x and PM.

The long-term EC trend at the urban Kudan site is consistent with the shorter-term trend found at the roadside Noge site. The Noge site reported a threefold higher EC reduction rate of 2.56 µg m^{-3} yr^{-1} (Figure 8), as compared to the Kudan site (0.82 µg m^{-3} yr^{-1}). Even though this rapid decrease may reflect the

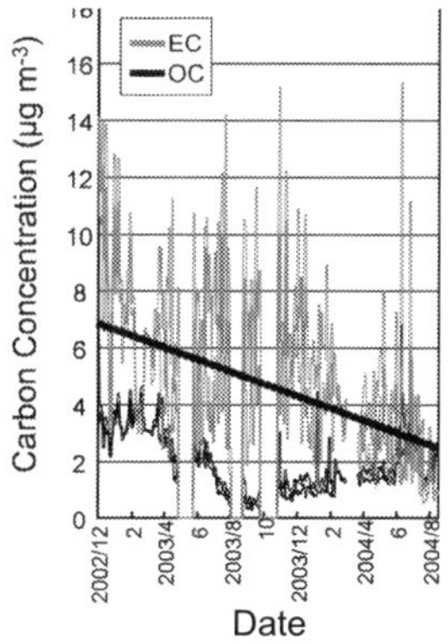

Figure 8: Two-hour averaged OC and EC concentrations from the R&P5400 carbon monitor at the Noge site from 12/1/2002 through 8/31/2004. The dotted line shows EC concentrations decreasing at a rate of 2.56 µg m^{-3} yr^{-1}.

WIT Transactions on Ecology and the Environment, Vol 123, © 2009 WIT Press
www.witpress.com, ISSN 1743-3541 (on-line)

phase in of low-emitting diesel vehicles, this value may be biased owing to the shorter measurement period, as well as different carbon measurement method used at the Noge site.

3.3 Chemical composition

Figure 9 shows that water-soluble ions contributed 38% (including 32% from NO_3^-, $SO_4^=$, and NH_4^+) to PM_{fine} for the nine years from 1994 – 2005, with 26% attributed to TC. Long-term trends in PM_{fine} components are more apparent for EC, to a much lesser extent for OC, NO_3^-, NH_4^+, and Cl^-, but are unclear for $SO_4^=$. A large change in the EC reduction rate was found beginning 2002 to 2004, which varied from -0.49 μg m^{-3} yr^{-1} to -1.09 μg m^{-3} yr^{-1}. As expected, material balance in PM_{coarse} shows 9% for sea salt (Na^+ and Cl^-), 21% for other ions (i.e., NO_3^-, $SO_4^=$, NH_4^+, and K^+), 14% for TC, and 56% in the "other" category. This "other" category can probably be attributed to the abundant geological material (i.e., metal oxides) found in PM_{coarse}.

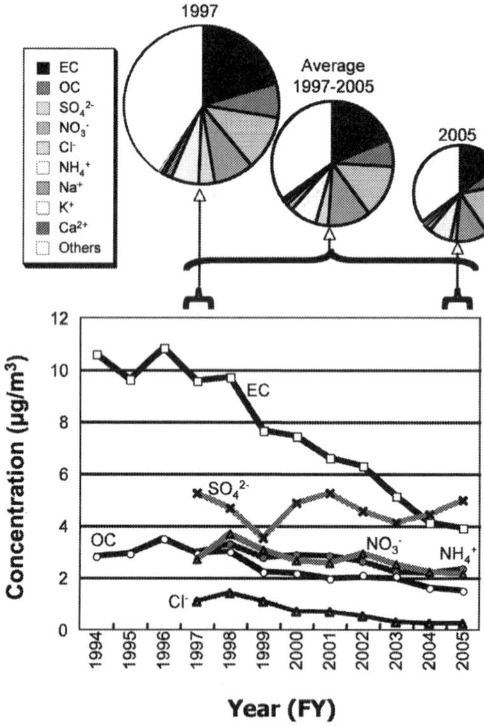

Figure 9: Annual variations of PM_{fine} composition for two-week impactor samples at the urban Kudan site from 1994 through 2005. The "others" category in the pie charts includes unmeasured species, such as mineral oxides in fugitive dust and oxygen and hydrogen associated with organics.

3.4 Other carbon sources

Seasonal variations in the PM_{fine} EC were examined to investigate the decrease in EC reduction ratios during 2004 – 2005. Figure 10 shows a monthly increase in EC concentrations from October to December, partially attributed to the atmospheric stability during winter.

Because ^{14}C is only present in contemporary carbon, not in fossil fuels, the influence of biomass burning is investigated [20]. Figure 11 shows that elevated

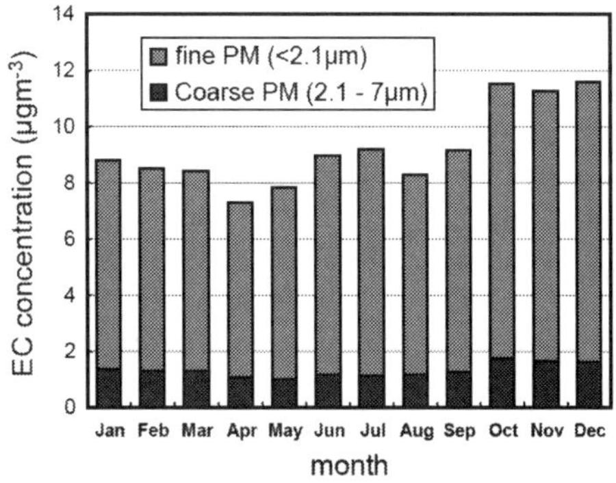

Figure 10: Seasonal variation in PM_{fine} and PM_{coarse} EC concentrations for two-week impactor samples at the urban Kudan site from 1994 through 2005.

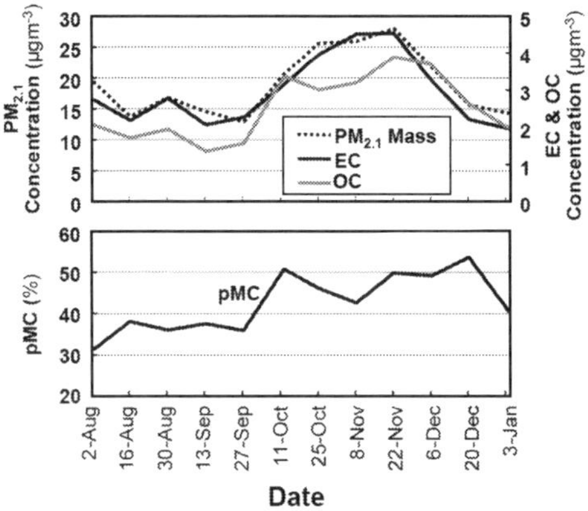

Figure 11: Temporal variation in the fine particle ($PM_{2.1}$) mass, EC, and OC concentrations (upper), and the percent of modern carbon, "pMC" (lower). The pMC was obtained on the basis of the isotope ratio of $^{14}C/^{12}C$. A pMC value of 100% means that all of the carbon is of biomass origin, and a pMC value of 0% means that all of the carbon is of fossil fuel combustion. The data for both charts was obtained from two-week impactor samples at the Kudan site in 2004.

PM_{fine} mass and carbon concentrations between October and November correspond to an increasing trend of pMC (the percentage of modern carbon in total carbon), suggesting the influence of vegetative burning. During 2004, as major improvements in vehicle engine technology and emission reduction was recognized [21], it became clear that vegetative combustion may account for 30–50% of carbonaceous aerosol in Tokyo. Tobacco smoke and burning of solid wastes in open dumps may be major sources, but the investigation of the sources of vegetative burning are underway.

4 Conclusions

Ambient SPM (~PM_7) concentrations have decreased from 1994 to 2006 at the urban Kudan site, near heavily travelled roadways in Tokyo. This trend is mainly due to reductions in the PM_{fine} (<2.1 μm) size fraction, that decreased at a rate of 2.09 $\mu g\ m^{-3}\ yr^{-1}$ after 1996. The decrease in PM mass concentrations corresponded to the introduction of new diesel engine technology in 1994 with stringent limits on PM emission rates. Most of the reductions can be attributed to the decrease in PM_{fine} EC (0.82 $\mu g\ m^{-3}\ yr^{-1}$) and OC (0.20 $\mu g\ m^{-3}\ yr^{-1}$). It became clear that vehicle emission reduction contributed to air quality improvement in Tokyo. By examining a carbon isotope ($^{14}C/^{12}C$) ratio on selected samples, it is found that biomass combustion sources could contribute to the slowed decreasing trend in EC from 2004 to 2005. More research is underway to identify and quantify PM source contributions from biomass and fossil fuel combustion.

References

[1] Vedal, S. 1997. Critical Review — Ambient particles and health: Lines that divide. Journal of the Air & Waste Management Association 47 (5), 551-581.

[2] Watson, J.G. 2002. Visibility: Science and regulation. Journal of the Air & Waste Management Association 52 (6), 628-713

[3] Jacobson, M.Z. 2002. Analysis of aerosol interactions with numerical techniques for solving coagulation, nucleatin, condensation, dissolution, and reversible chemistry among multiple size distributions. Journal of Geophysical Research 107 (D19), AAC 2-1-AAC 2-23, DOI:10.1029/2001JD002044.

[4] Chow, J.C., Watson, J.G. 2002. $PM_{2.5}$ carbonate concentrations at regionally representative Interagency Monitoring of Protected Visual Environment sites. Journal of Geophysical Research 107 (D21), ICC 6-1-ICC 6-9, doi: 10.1029/2001JD000574.

[5] MLITT, http://www.mlit.go.jp/statistics/index.html

[6] Tokyo Environmental White Paper 2004, 2005. Ed. The Bureau of the Environment, Tokyo Metropolitan Government, pp.140.

[7] Environment of Tokyo, http://www2.kankyo.metro.tokyo.jp/jidousya /diesel/

[8] Lloyd, A.C., Cackette, T.A. 2001. 2001 Critical Review — Diesel engines: Environmental impact and control. Journal of the Air & Waste Management Association 51 (6), 809-847.

[9] Karim and Ohno, 2000, www.env.go.jp/air/car/gas_kisei.html

[10] Watson, J.G., Chow, J.C. 2001. Ambient air sampling. In *Aerosol Measurement: Principles, Techniques, and Applications, Second Edition,* 2nd Edition, Baron, P., Willeke, K., editors. John Wiley & Sons, New York, NY, pp. 821-844.

[11] Watson, J.G., Chow, J.C., Fujita, E.M. 2001. Review of volatile organic compound source apportionment by chemical mass balance. Atmospheric Environment 35 (9), 1567-1584.

[12] Watson, J.G., Chow, J.C., Chen, L.-W.A. 2005. Summary of organic and elemental carbon/black carbon analysis methods and intercomparisons. Aerosol and Air Quality Research 5 (1), 69-102.

[13] Chow, J.C., Watson, J.G. 1999. Ion chromatography in elemental analysis of airborne particles. In *Elemental Analysis of Airborne Particles, Vol. 1*, Landsberger, S., Creatchman, M., editors. Gordon and Breach Science, Amsterdam, pp. 97-137.

[14] Fu, F.F., Watanabe, K., Yabuki, S., Akagi, T. 2004. Seasonal characteristics of chemical compositions of the atmospheric aerosols collected in urban seaside area of Tokaimura, eastern central Japan. Journal of Geophysical Research 109 (D20212), doi:10.1029/2004JD004712.

[15] Iino, N., Kinoshita, K., Tupper, A.C., Yano, T. 2004. Detection of Asian dust aerosols using meteorological satellite data and suspended particulate matter concentrations. Atmospheric Environment 38 (40), 6999-7008.

[16] Ma, C.J., Tohno, S., Kasahara, M. 2005. A case study of the size-resolved individual particles collected at a ground-based site on the west coast of Japan during an Asian dust storm event. Atmospheric Environment 39 (4), 739-747.

[17] Kadowaki, S. 1990. Characterization of carbonaceous aerosols in the Nagoya urban area 1. Elemental and organic carbon concentrations and the origin of organic aerosols. Environmental Science & Technology 24, 741-744.

[18] Ohta, S., Okita, T. 1994. Measurements of particulate carbon in urban and marine air in Japanese areas. Atmospheric Environment 18, 2439-2445.

[19] Takahashi, K., Minoura, H., Sakamoto, K., 2008. Chemical composition of atmospheric aerosols in the general environment and around a trunk road in the Tokyo metropolitan area. Atmospheric Environment 42, 113-125.

[20] Takahashi, K., Hirabayashi, M., Tanabe, K., Shibata, Y., Nishikawa, M., Sakamoto, K., 2007. Radiocarbon Content in Urban Atmospheric Aerosols. Water Air and Soil Pollution, 185, 305-310.

[21] Ministry of the Environment, The study meeting report of gross weight reduction measures environmental improvement effect, http://www.env.go.jp/press/press.php?serial=6432

Managing air pollution impacts to protect local air quality

C. Grant, R. Bloxam & S. Grant
Ontario Ministry of the Environment,
Environmental Sciences and Standards Division, Canada

Abstract

Local industrial/commercial sources of air pollution in Ontario, Canada have been regulated for almost four decades using air emission estimating, atmospheric dispersion models and point of impingement (POI) standards. Historically, the provincial Ontario Ministry of the Environment (MOE) set standards that considered technical, economic and scientific issues. Compliance assessment used air emission inventories and atmospheric dispersion models originally developed in the 1960s. The challenges of this type of approach included:

- A cumbersome standard-setting process that produced few standards – often dictated by technical/economic considerations.
- Inaccuracies in air emission inventories.
- Dispersion models that tended to under-predict impacts.

In August 2005, Ontario announced a significant overhaul of the local air pollution regulation that included:

- Air standards that are now set to protect against health and environmental impacts.
- Phase-out of current dispersion models and replacement with the more accurate dispersion models from the United States Environmental Protection Agency (US EPA).
- Rigorous air emission estimating rules including the use of a combination of dispersion modelling and ambient monitoring as a more accurate emission estimating technique for a wide variety of sources (including fugitives).
- Technical/economic considerations that are now addressed through a publicly transparent alternative air standards process that promotes continuous improvement. Site specific alternative standards represent the lowest technically and/or economically feasible levels that a specific facility could achieve. Decisions often hinge on the technology benchmarking report, which is similar to the US EPA "top-down" analysis.

This paper outlines key challenges and policy decisions during the development of the regulation; experiences in introducing more stringent scientific-based standards, including standards for lead and vinyl chloride, which are among some of the most stringently regulated standards in the world; and lessons-learned in the use of the combined monitoring and modelling emission estimating tool in the new alternative standards process.

Keywords: local air quality, air toxics, dispersion modelling, monitoring, technical and economic barriers, air standards, fugitive emissions.

WIT Transactions on Ecology and the Environment, Vol 123, © 2009 WIT Press
www.witpress.com, ISSN 1743-3541 (on-line)
doi:10.2495/AIR090131

1 Historical background

Ontario, Canada is home to a wide range of heavy industries including sectors such as iron and steel, petroleum refining, chemical production, pulp and paper, mining and primary metal smelting, as well as a large automotive sector. The Ontario Ministry of the Environment (MOE) has regulated local air pollution since the late 1960s. It was one of the first provinces in Canada to regulate air pollution. The province sets contaminant-specific Point of Impingement (POI) air standards to manage air pollution from industrial and commercial sources. Mathematical air dispersion models have been the primary tools used to assess compliance with air quality standards along with occasional ambient monitoring in specific communities. The initial regulation set out simple Gaussian air dispersion models, which were considered state-of-the-art at that time. Outputs from the models are used to compare POI concentrations to the air standards: ambient monitoring could also be used. Ontario's reliance on POI air standards is somewhat unique where many other jurisdictions emphasize air emissions or best available control technology standards. For example, since the Clean Air Act Amendments of 1990, the United States has used a combination of Maximum Achievable Control Technology (MACT) standards and state-specific requirements to regulate air toxics (US EPA [1]). As other jurisdictions moved to regulating technology or emission standards – as opposed to concentration based POI standards – Ontario began a review of its regulatory framework and considered similar approaches. However, proposed changes created uncertainty for industry and raised questions about technical barriers to compliance and costs. In 2001, Ontario began a new consultation process. The key drivers for change were based on criticisms that the current air standards were over 20 years old and were not protective of health and environmental impacts and that current air dispersion models were over 30 years old and could be underestimating concentrations. For many years, Ontario had challenges in setting and updating air quality standards and updating its air dispersion models since:

- a change in the air standard for one substance could affect a variety of sectors in different ways;
- more stringent air standards can create technical and economic implementation issues for industry;
- changes to the air dispersion models, used to assess POI concentrations, could lead to potential compliance issues;
- uncertainty in assessing compliance could result in permitting delays which could lead to uncertainty in making business decisions;
- the past practice of setting an air standard that considered economics, technology and science was complex; lacked public transparency; resulted in less protective standards; and took significant time and resources for each standard.

1.1 A new legal framework to protect local air quality

In August 2005, Ontario announced a significant overhaul of the local air pollution regulation – Ontario Regulation 419/05: Air Pollution – Local Air

Quality (hereafter referred to as the "Regulation"): further amendments were announced in 2007 (MOE [6]). The new regulatory framework combined science-based air quality standards with the use of technology standards (similar to the US EPA [1]) as an alternative if compliance with air standards could not be achieved in the short term. This Regulation is the primary tool used to protect local air quality by enforcing air quality standards. One of the key policy shifts that occurred as a result of the Regulation was a move towards provincial air quality POI standards that were set to protect against health and environmental effects – as opposed to setting air standards that considered technical and economic concerns (MOE [3]). Under the new regime, the past practice of relaxing standards based on the concerns of one industry or a specific sector would not continue. This Regulation now includes:

- 59 new or updated air standards set to protect against health and environmental impacts - new or more stringent standards are phased-in.
- Phase-out of current atmospheric air dispersion models and replacement with the more accurate dispersion models from the United States Environmental Protection Agency (US EPA).
- Rigorous emission estimating rules that include a combination of dispersion modelling and ambient monitoring as a more accurate method to assess emission rates for a wide variety of sources.
- A new risk-based decision making process that considers technical and economic barriers through a publicly transparent alternative air standards process, which promotes continuous improvement. Site specific alternative standards represent the lowest technically (and/or economically) feasible levels that a specific facility could achieve. Decisions often hinge on a technology benchmarking report, which is similar to the US EPA "top-down" analysis (US EPA [4]).

Implementation issues that result from the updating of air quality standards and air dispersion models can be addressed by allowing a facility to request an alteration of the standard to address site specific technical and/or economic concerns through a publicly transparent process. Ontario's regulatory framework has evolved into an effective hybrid of the POI approach using effects-based standards and a process that allows for site specific technology standards if needed to address compliance issues and promote continuous improvement.

2 Emission estimating methods

One of the items that the Regulation addressed was uncertainties in emission estimation methods that are inherent in any air dispersion modelling approach. The regulatory framework requires a facility to determine, for any given source, an emission rate that represents the highest emission that a facility is capable of based on a given operating condition. In order to expedite compliance assessments, a facility may initially use a "conservative" emission estimate to determine the maximum off-property POI concentration. If this maximum POI concentration complies with the air standard, no further assessment is required. If a facility exceeds the standards at those levels, then they are required to "refine" their emissions to be as accurate as possible. Refinement of emissions must be done in one of two ways:

- Estimating emissions based on a combined modelling and monitoring analysis; or
- Source testing across a range of operating conditions may also be considered but only if it is specifically approved as a more accurate or effective approach than a combined modelling and monitoring analysis.

Alternatively, the regulation allows a facility to move directly towards abatement improvements instead of complex studies to determine emission rates.

2.1 The combined modelling and monitoring method

Typically, source testing is conducted at a maximum operating condition that gives rises to a conservative emission rate with the premise that other operating conditions would result in lower emission rates (and hence a lower POI concentration). However, source testing is often conducted under optimal conditions and these emission rates may not be representative. In addition, source testing for fugitive emissions, which are often underestimated, is not possible. A more accurate assessment of emissions uses a combination of air dispersion modelling along with ambient air monitoring. Monitors are strategically located to determine emissions rates from key dominant sources of that contaminant for each monitoring period. It is important to locate monitors close enough to the source to capture all emissions, in particular fugitive emissions. However, if there is any uncertainty in source release parameters care should be taken not to locate monitors too close to the source. Verification of source release parameters for fugitive sources requires careful location of monitors. Siting considerations for the monitors is important and will include nearby proximity to the most significant sources of air emissions; site-specific geometry that may affect the dispersion of the contaminants from the source; obstructions to the monitor that may affect collection of the ambient air sample; and the predominant wind direction. This type of ambient monitoring program is often focused on collecting samples of contaminants immediately downwind of the key sources of contaminant and generally ends after collection of approximately thirty samples that are significantly above "background" levels of the contaminant. The approach is particularly well suited for estimating fugitive air emissions from sources of both particulate matter and volatile organic compounds but can also be used for other sources. The results of the ambient monitoring are used in combination with an iterative approach to revise the air emission rates for the dominant sources until the dispersion modelling results match the ambient monitoring results. In summary, a combined monitoring and modelling analysis recommends the following steps (MOE [5]):

i) Approval by the Ontario MOE of a combined monitoring and modelling plan that includes pre-agreement of the appropriate source of meteorological data for the dispersion modelling; appropriate siting of monitors and recording of relevant facility operating parameters during the monitoring periods (e.g., generally, a 24-hour monitoring period).

ii) Installation of monitors and gathering of samples on a regular basis or when meteorological forecasting indicates that the monitors will be downwind of the main sources of a contaminant.
iii) A preliminary analysis of the monitoring data, the predominant wind directions and wind speeds for the monitoring period and the facility operating conditions to select the most significant measurements for further analysis and refinement of air emission estimates.
iv) Application of the atmospheric dispersion models for each of the monitoring locations (plus four locations in the immediate vicinity of each monitor to average any spatial anomalies between the monitor and modelling locations) using the meteorological data inputs for the specific monitoring period.
v) Results that are paired in time can be plotted on logarithmic paper of modelling versus monitoring data. Air emission estimates, for the most dominant and uncertain sources, are modified by iteration that can be focused on the most significant emission sources where the estimates are uncertain.
vi) Monitoring and modelling data is then re-organized, unpaired in time, from highest to lowest for each of the monitoring and modelling data.
vii) The revised emission estimates provides an indication of the variation in air emissions, from the most significant contributors to the measured concentrations, during the monitoring study. For log-normal-like distributions of data, the use of the mean plus one standard deviation for the most significant sources identified from the study in combination with average emissions from all other sources is recommended.

Field costs for a combined monitoring and modelling analysis can be comparable or slightly less than the costs for a comprehensive source testing program. Computer run times for the dispersion modelling are relatively short but the required analysis between iterations can be resource intensive. To test this methodology, MOE conducted a study of lead emissions from a large brass and bronze foundry in southern Ontario. Figure 1 provides a plot of the initial unpaired monitoring versus modelling data for measurement of lead concentrations in the immediate vicinity the foundry. The modelled data was derived using an existing air emission inventory for the facility based primarily on source testing. This initial plot shows a trend towards the right-hand side of the plot, which has been interpreted by MOE as a possible under-estimating of POI concentrations and air emissions at the higher measurements (MOE [6]). Other combined analysis projects are underway for chemical plants, integrated iron and steel mills, smelting operations, and petroleum refineries.

3 Implementing air quality standards

Many jurisdictions recognize the need to protect human health and the environment while acknowledging the importance of goods and services provided by industry. The risk-based alternative standards process provides a mechanism to balance these interests in a publicly transparent manner. Ontario's

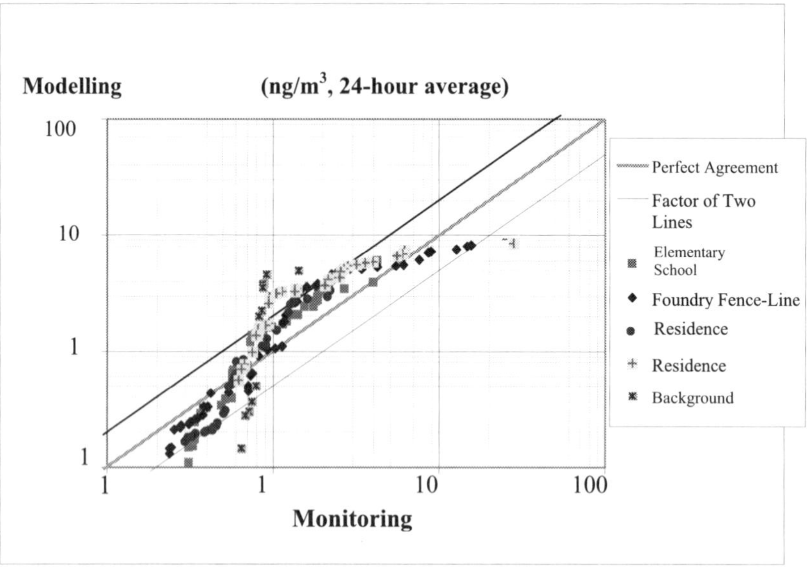

Figure 1: AERMOD model vs. monitoring for lead emissions.

risk-based framework for air quality standards is presented in Figure 2. The inverted triangle represents a measure of "risks". Risks generally increase at higher POI concentrations or exposures - the frequencies of those exposures may also be a consideration (MOE [3]). Provincial air quality standards that are set to protect against health and environmental effects are represented by the lower level line, below which risks are considered generally acceptable (Zone 1: "Broadly Acceptable Region"). The MOE's objective for air standards for carcinogenic effects is to set the standard that corresponds to an incremental lifetime risk of 1 in a million (10^{-6}) (i.e. the risk of one person in a population of a million who may develop some form of cancer). Standards for non-carcinogenic effects are based on a Reference Concentration (RfC) that is derived from a threshold toxicological endpoint (the most sensitive) and the use of uncertainty factors to account for gaps in the data. Generally, the objective for non-carcinogenic risk is to set the standards based on a Hazard Quotient (HQ) of one, which is the ratio of the concentration of a contaminant to the corresponding air standard or RfC. The framework also defines an "Upper Risk Threshold" (shown as the upper level line – Zone 3). Concentrations in this region require timely action to assess and if necessary, to reduce contaminant concentrations as soon as possible. URTs for carcinogens are generally set at a 1 in 10,000 risk level (10^{-4} or 100 times the standard); URTs for non-carcinogens are generally set at a HQ of 10 (or 10 times the standard). POI concentrations between the upper and lower levels are in the "Region of Concern" (Zone 2). Facilities operating in this Region are required to take all reasonable steps to get into compliance with the effects-based standard by the phase-in date. If compliance with a standard is not possible by the phase-in date, these facilities may be eligible to request a site specific alternative standard – which is a risk-based approach (CSA [7], McColl *et al* [8]).

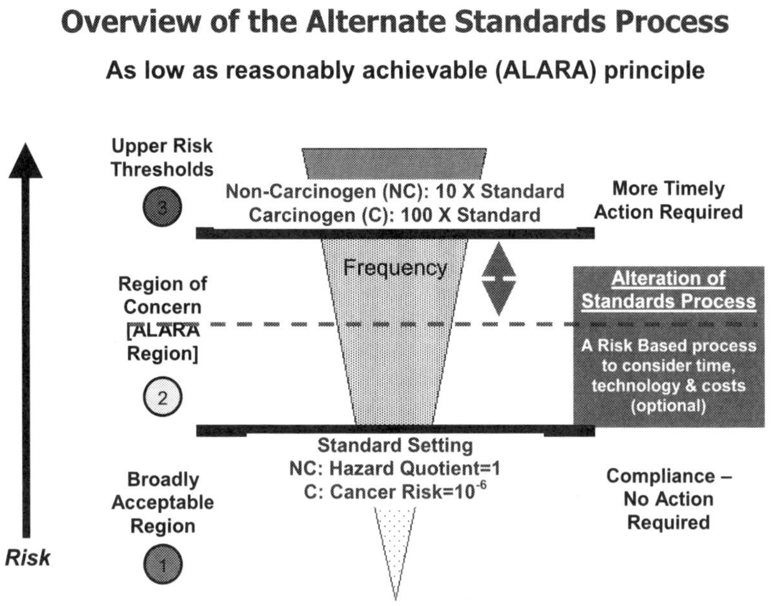

Figure 2: Risk framework for air quality standards.

3.1 The alternative standards process

Introduction of effects-based air standards, and more advanced air dispersion models, meant that not all facilities would be able to achieve compliance immediately due to technical limitations, economic realities or the need for more time to assess, plan, and if necessary, finance and install new equipment or processes to come into compliance. In these cases, the goal is to strive for reducing risk to "As Low As Reasonably Achievable" (ALARA). These issues could be reviewed as part of the site specific alternative standards process set out in the Regulation. An alternative standard will not be considered if the URT is exceeded at receptors such as daycares, schools, dwellings, hospitals or homes for the elderly. The frequency of exceedances is also a consideration (MOE [3]).

For most facilities, a phase-in period provides time to assess, plan, budget and implement technical solutions to ensure compliance with the air standards. Phase-in periods for new or more stringent standards are normally 3 to 5 years, but can vary based on the contaminant. If a facility can identify feasible technical solutions that can be implemented within the phase-in period to achieve compliance, then it should proceed to do so. For other facilities, compliance with air standard(s) might not be achievable within the phase-in period. If the technical solutions are not readily available to allow a facility to achieve compliance before the end of the phase-in period, these facilities may consider requesting Director's Approval for a site specific alternative standard. An alternative standard would establish an interim site specific standard with the goal of continuous improvement toward achieving the effects-based standard over time. When making a request for an alternative air standard, industry must include in their request:

- An Emission Summary and Dispersion Modelling (ESDM) report using emission rates determined from a modelling and monitoring analysis;
- A technology benchmarking report, comparing the facility to others within the sector to ensure they are doing the best they can;
- Summary comments from a public meeting held in the community; and,
- An action plan to minimize POI concentrations.

If approval is granted, the decision would be periodically reviewed to ensure that the technical (or economic (optional)) issues considered at the time are still relevant for that particular facility. The Director may approve a site specific alternative standard for a period of up to 5 years (up to 10 years in extenuating circumstances). A facility is eligible to re-apply but the Director must consider the number of times a request has been made for an alternative standard and the subject of the request. This will be considered on a case-by-case basis.

3.2 Technical and economic considerations

The alternative standards process allows for consideration of technical and economic feasibility in separate analyses. Technical feasibility is a mandatory component of the process and is documented in a "Technology Benchmarking" report. Economic feasibility is optional. Any information submitted as part of the request must also be shared with the local community if requested.

3.2.1 Technology benchmarking reports

A Technology Benchmarking Report is a key document submitted to support a request for an alternative standard. A list of process and site-specific air pollution control strategies are assessed in a prescriptive approach that is based upon the US EPA New Source Review "top-down" analysis (US EPA [4]). Various pollution control options are ranked from most to least effective at controlling emissions and POI concentrations. In practice, the technology benchmarking process may be simplified by focusing on the priority contributors to POI concentrations identified through the combined monitoring and modelling analysis. MOE is the midst of reviewing alternative standard requests for two polyvinyl chloride (PVC) chemical plants in southern Ontario. MOE's review of the technology benchmarking reports included the following components:

- A review of requirements for similar facilities in other jurisdictions (e.g., the U.S, Europe and Australia);
- A review of the performance of the Ontario facilities (grams of total/fugitive pollutant emitted per tonne of product produced) relative to information published on similar facilities in other jurisdictions; and

- Third party technical experts to assess feasible technical methods for the key sources identified through the combined monitoring and modelling analysis.

Figure 3 provides a summary of the outcome of the Ontario MOE review of the technology benchmarking report and request for alternative standard for one facility. The graphic includes the relative performance (in grams per tonne of product produced) of the facility from 1994-1999 and in 2007.

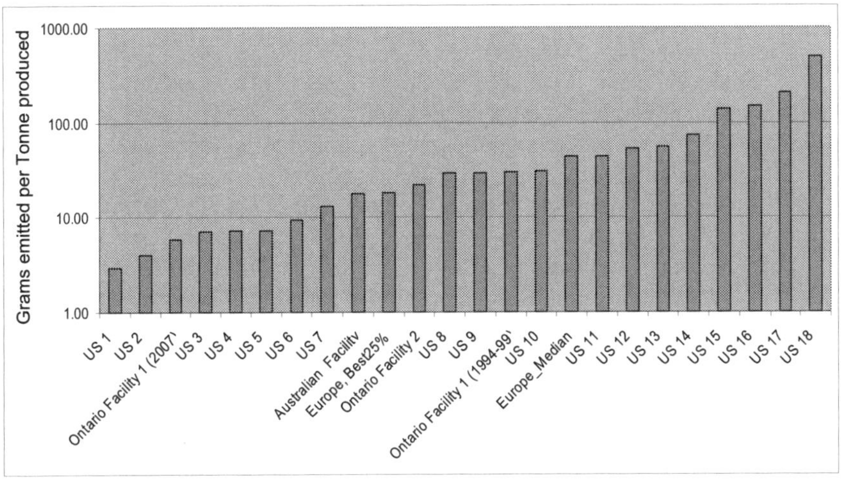

Figure 3: Benchmarking of Ontario facility.

3.2.2 Financial hardship and cost effectiveness

The alternative standards process also allows for the consideration of economic feasibility. MOE guidance includes economic ratios to assist in determining financial hardship. MOE is also considering development of cost effectiveness indicators which contemplates using traditional parameters, such as a cost per unit emission rate, and traditional pollution control costing techniques, such as those recommended by the US EPA (US EPA [9]), that are modified to account for relative toxicity of the contaminant; type of emission source (e.g., point, volume or area fugitive sources) and the magnitude and frequency of exceedances.

3.2.3 Ongoing challenges to implementation

The Regulation is primarily designed to address air toxics from larger industrial facilities. Smaller to medium sized facilities may find the regulatory approach burdensome. Alternatively, sector specific approaches are being considered for sectors that have common air pollution issues. Other on-going challenges include the need to address concerns about the cumulative effect of multiple facilities and contaminants within an air shed.

References

[1] US EPA Clean Air Act Amendments, 1990 http://www.epa.gov
[2] Ontario Regulation 419/05:Air Pollution–Local Air Quality www.ene.gov.on.ca
[3] MOE. 2005. Guideline for the Implementation of Air Standards in Ontario, Toronto, Ontario. Queen's Printer for Ontario, PIBs #5166e http://www.ene.gov.on.ca/envision/gp/5166e.pdf
[4] New Source Review Workshop Manual, Prevention of Significant Deterioration and Nonattainment Area Permitting, Draft, October 1990" Chapter B, Best Available Control Technology (top down analysis) www.epa.gov/ttn/nsr/gen/wkshpman.pdf
[5] MOE. 2005. Procedure for Preparing an Emission Summary and Dispersion Modelling (ESDM) Report, Toronto, Ontario. Queen's Printer for Ontario, PIBs #3614e02 http://www.ene.gov.on.ca/envision/gp/3614e02.pdf
[6] MOE. 2007. Draft Report for an Air Emissions Study and Development of "An Accurate As Possible" ESDM Report for a Brass and Bronze Foundry with contributions from Senes Consultants and Ortech Environmental
[7] CSA. 1997. Risk Management: Guideline for Decision Makers. Canadian Standards Association, CAN/CSA-Q850-97 (R2002)
[8] McColl S, Hicks J, Craig L, and Shortreed J. 2000. Environmental Health Risk Management - A Primer for Canadians. Network for Environmental Risk Assessment and Management (NERAM). Waterloo, ON N2L 3G1
[9] US EPA Control Cost Manual, Office of Air Quality Planning and Standards, Research Triangle Park, NC 27711, EPA 453/b-96-001

Section 3
Emission studies

Application of mineral magnetic concentration measurements as a particle size proxy for urban road deposited sediments

C. J. Crosby[1], C. A. Booth[1], A. T. Worsley[2], M. A. Fullen[3], D. E. Searle[1], J. M. Khatib[1] & C. M. Winspear[1]
[1]SEBE, University of Wolverhampton, UK
[2]NGAS, Edge Hill University, UK
[3]SAS, University of Wolverhampton, UK

Abstract

The application of mineral magnetic concentration parameters (χ_{LF}, χ_{ARM} and SIRM) as a potential particle size proxy for urban road deposited sediment collected from Scunthorpe, North Lincolnshire, U.K. has been investigated. Correlation analyses between each magnetic parameter and traditional particle size classes (i.e. sand, silt and clay) and respiratory health related size classes (i.e. PM_{10}, $PM_{2.5}$ and $PM_{1.0}$) are reported. Significant relationships (p <0.01; n = 35) exist between clay content and two of the magnetic concentration parameters (χ_{ARM} and SIRM). This is also the same for each PM_{10}, $PM_{2.5}$ and $PM_{1.0}$ sizes. Of the three magnetic parameters, χ_{ARM} displays the strongest correlation (r = 0.45; p <0.01; n = 35) values and is the most significant parameter, which is consistent with class sizes of each approach. In doing so, these associations indicate mineral magnetic measurements have considerable potential as a particle size proxy for determining urban roadside particulate matter concentrations. Given the speed, low-cost and sensitivity of the measurements, this suggests magnetic techniques could potentially be used as an alternative and/or complementary exploratory technology for pilot particulate pollution investigations. Furthermore, in certain instances, it could be useful for examining linkages between respiratory health and particulate pollution and vehicle emissions.

Keywords: environmental magnetism, particle size, urban street dust, built environment, epidemiology, public health.

WIT Transactions on Ecology and the Environment, Vol 123,
www.witpress.com, ISSN 1743-3541 (on-line)
doi:10.2495/AIR090141

1 Introduction

Sediments within urban environments originate from a wide range of sources, both natural and anthropogenic [1]. Numerous studies [2,3] have shown that street sediment is composed of a wide range of particle sizes, commonly biased towards coarse material. That said, the descriptive term of street dust is not always appropriate and, as a consequence, road-deposited sediment (RDS) is a more preferable terminology for studies of urban particulate matter (PM) [4,5].

Urban PM pollution continues to be intensively studied because the finest sediment particles can exert a potentially harmful influence on public health, especially those susceptible to respiratory illness [6]. McConnell [7] found strong linkages with child residence near to major sources and an associated risk of asthma (95%) due to road traffic sources. Maher [8] analyzed PM on tree leaves at different heights, which showed that metal-rich particulate pollution concentrations are highest at ~0.3m (i.e. small child height) and at 1.5–2m (adult head height) above ground level. Infant exposure to air pollution is a special concern because their immune system and lungs are not fully developed when exposure begins, raising the possibility of different responses to those seen in adults [6]. It is customary for air quality studies to analyse particulates of two size fractions; the coarse fraction PM_{10} ($d_a \leq 10$ μm) and fine fraction $PM_{2.5}$ ($d_a \leq 2.5$ μm). More recently, the adopted changeover point between coarse and fine particles occurs at $PM_{1.0}$ ($d_a \leq 1.0$ μm). The success or failure of the respiratory defence systems partly depends upon the size of the particulates inhaled and the depth of their penetration into the respiratory tract [9]. A recent assessment by the 'EPAQS' [10] concluded that both coarse and fine fractions represent health risks, although the disease outcomes may differ for the two size fractions [11].

Since 1997, UK local authorities have been reviewing and assessing air quality in their area, to work towards meeting the national air quality objectives. There are over 1500 monitoring sites across the UK, which monitor air quality (monitoring PM_{10} concentrations, amongst those of several other pollutant measurements). In 2003 the UK Government set lower limits for fine particle concentrations, which need to be reached by 2010. Currently, the main source of airborne fine particulates in the UK is road traffic emissions [12]. However, it is estimated, by 2010, that road transport emissions of fine particles will fall by two-thirds of those a decade ago [13]. Moreover, the European Commission has proposed that the legislation on particulate matter should be supplemented by setting a limit value of 35 $\mu g/m^3$ for $PM_{2.5}$ particles and an interim reduction target of 20% to be attained between 2010 and 2020 [12]. Assessment of the extent and severity of urban dust concentrations requires thorough investigation before Air Quality Management Plans and remediation can be instigated, which means that there is scope for new PM monitoring technologies.

It is, therefore, timely for innovative PM technologies to be considered as an alternative, or in tandem, to those already employed to determine $PM_{2.5}$ and PM_{10} concentrations. Ideally, they need to be rapid, reliable and inexpensive. However, to assess the suitability of any analytical technique as an efficient particle size proxy it is necessary that the nature of the relationship between the proposed parameters and particle size follow a predictable pattern. Many studies have previously explored relationships between mineral magnetic measurements and the physico-chemical properties of soils, sediments and dusts. Based on these investigations, from host environmental settings (e.g. soils, deserts, glacial, lakes, coastal and marine), magnetic measurements have previously been identified as a suitable proxy for geochemical, radioactivity, organic matter content and particle size data [14–17].

Two hypotheses are tested here: Firstly, the extent to which particular magnetic concentration parameters can be used as a particle size proxy for urban RDS; and, secondly, whether theses data associations follow the predictable trends of other environmental studies.

2 Case study

Air quality PM_{10} measurements for the UK reveal a host of towns and cities that exceed the permitted European concentration levels [18]. This includes the town of Scunthorpe, which is the administrative centre of North Lincolnshire and home to an estimated resident population of ~72,500. It was founded in Roman times but the town mainly grew, from 1859 onwards, when local iron ore was discovered and mined. This resulted in the development of a nearby iron and steel industry and a concomitant rapid population growth. Today, the steel industry (Corus, an Indian-owned firm) still remains (on the eastern side of the town (national grid reference: SE 911 104)) and is a major employer in the town (4,500 personnel in 2008). Nationally, this industry is acknowledged to be a potential major contributor to poor air quality [19]. As a consequence, the local authority (North Lincolnshire council) operates a series of air sampling stations, which are used to monitor levels of particulates in the area (Figure 1). These sources are mainly industry and traffic related [19].

Figure 1: Scunthorpe scenes: (a) the main A18 dual-carriageway road facing west (NGR: SE 902 092) and (b) the air sampling station near to the town centre crossroads (NGR: SE 891 108).

The heart of the town is close to the crossroads of the main A18 road (Figure 1) (Brigg is ~12km to the east and Hatfield is ~24km to the west) and the A159 road (Figure 1(b) (Gainsborough is ~23km to the south). The limits of the town are constrained b)y the M180 and the M181 motorways (highways) to the south and west, respectively.

3 Materials and methods

RDS samples were collected from both main and residential roads within the town of Scunthorpe (July 2008).

3.1 Sample collection and preparation

RDS were collected from the surface of roads, gutters and pavements (sidewalks) of the town's roads. Typically, 30–50 g sediment samples were collected (from ~1 m^2) by sweeping with a small hand-held fine-bristle brush. Sediment was then transferred to clean, pre-labelled, self-seal, airtight plastic bags. In the laboratory, samples were visibly screened to remove macroscopic traces of hair, animal and plant matter [20].

3.2 Mineral magnetic measurements

All samples were subjected to the same preparation and analysis procedure. Samples were dried at room temperature (<40 °C), weighed, packed into 10 ml plastic pots and immobilized with clean sponge foam and tape prior to analysis. Initial, low–field, mass–specific, magnetic susceptibility (χ) was measured using a Bartington (Oxford, England) MS2 susceptibility meter. By using a MS2B sensor, low frequency susceptibility was measured (χ_{LF}). Anhysteretic Remanence Magnetisation (ARM) was induced with a peak alternating field of 100 mT and small steady biasing field of 0.04 mT using a Molspin (Newcastle–upon–Tyne, England) A.F. demagnetiser. The resultant remanence created within the samples was measured using a Molspin 1A magnetometer and the values converted to give the mass specific susceptibility of ARM (χ_{ARM}). The samples were then demagnetized to remove the induced ARM and exposed to a series of successively larger field sizes up to a maximum 'saturation' field of 1000 mT, followed by a series of successively larger fields in the opposite direction (backfields), generated by two Molspin pulse magnetisers (0-100 and 0-1000 mT). After each 'forward' and 'reverse' field, sample isothermal remanent magnetisation (IRM) was measured using the magnetometer [21].

3.3 Laser diffraction measurements

All samples were subjected to the same textural preparation and analysis procedure, using sieving (2000 μm aperture) followed by laser diffraction analysis. Low Angle Laser Light Scattering (LALLS), using a Malvern (Malvern, England) Mastersizer Long-bed X with a MSX17 sample presentation unit, enabled rapid measurement of particle sizes within the 0.1-2000 μm range. Macroscopic traces of organic matter were removed from representative sub-samples before being dampened by the dropwise addition of a standard chemical solution (40 g/l solution of sodium hexametaphosphate ($(NaPO_3)_6$) in distilled water) to help disperse aggregates. To ensure complete disaggregation, each slurry was then subjected to ultrasonic dispersion in a Malvern MSX17 sample presentation unit. For greater precision, the mean of five replicate analyses was measured with a mixed refractive indices presentation setting. A standard range of textural parameters was calculated, including the percentage of sand, silt and clay class sizes and their sub-intervals. The Malvern instrumentation was regularly calibrated using latex beads of known size [22].

4 Results

Particle size data (n = 35) indicates samples are dominated by sand (~77%), silt (~19%) and clay (~3%), in respective orders (Table 1). From a respiratory-health perspective, PM_{10} grains represent ~8%, $PM_{2.5}$ ~3% and $PM_{1.0}$ ~2% of the sediments. Once suspended, particles <10 μm are able to remain airborne for hours or days and, in some cases, even weeks [11]. Therefore, the presence of PM of these sizes on pavement surfaces indicates either the particles have not been disturbed recently or they have only just settled-out. That said, since the town centre pavements normally receive frequent and heavy foot-traffic, it is assumed the time of sampling (04:00 – 08:00) and the weather conditions (warm, dry and still) have permitted sizeable PM accumulations.

Table 1: Summary particle size properties of urban RDS: (a) traditional sediment size fractions and (b) respiratory health-related size fractions (n = 35 samples).

(a)	**Mean (%)**	**Maximum (%)**	**Minimum (%)**	**Standard Deviation**
Sand (63-2000 μm)	77.23	88.18	56.38	7.80
Silt (2-63 μm)	19.86	39.95	9.36	7.42
Clay (<2 μm)	2.89	7.01	1.04	1.58
(b)	**Mean (%)**	**Maximum (%)**	**Minimum (%)**	**Standard Deviation**
$<PM_{10}$	8.06	16.18	3.21	3.65
$<PM_{2.5}$	3.35	7.76	1.40	1.72
$<PM_{1.0}$	1.92	5.23	0.73	1.20

Table 2 summarizes the mineral magnetic characteristics. χ_{LF} is roughly proportional to the concentration of ferrimagnetic minerals within the sample, although in materials with little or no ferrimagnetic component and a relatively large antiferromagnetic component, the latter may dominate the signal. χ_{ARM} is particularly sensitive to the concentration of magnetic grains of stable single

www.witpress.com, ISSN 1743-3541 (on-line)

domain size, e.g. ~0.03-0.06 μm. SIRM is related to concentrations of all remanence-carrying minerals in the sample, but is also dependent upon the assemblage of mineral types and their magnetic grain size. These data indicate the samples contain moderate to very high magnetic concentrations. Yet, compared with previous urban magneto-dust studies, the mean values are more than double those of Liverpool (23.7 $\times 10^{-7}$ m^3 kg^{-1}) [23] and Shanghai (29.9 $\times 10^{-7}$ m^3 kg^{-1}) [24].

Table 2: Summary magnetic properties of urban RDS (n = 35 samples).

	Units	Mean	Maximum	Minimum	Standard Deviation
χ_{LF}	10^{-7} m^3 kg^{-1}	60.86	123.34	20.87	28.32
χ_{ARM}	10^{-7} m^3 kg^{-1}	0.180	1.09	0.03	0.21
SIRM	10^{-5} Am^2 kg^{-1}	1270.39	9686.42	194.58	1682.48

Table 3 shows the Pearson's correlation coefficient values (*r*) between the mineral magnetic concentration parameters and particle size parameters, grouped according to traditional sediment size fractions and respiratory health-related size fractions. Significant relationships (p <0.01; n = 35) exist between clay content and both the χ_{ARM} and SIRM magnetic concentration parameters. This is also the same for with all the respiratory class sizes of the finer fraction $<PM_{10}$ (p <0.01). Therefore, this indicates that χ_{ARM} and SIRM magnetic concentration parameters could be used as a particle size proxy, particularly if the kinship is required with particle sizes $PM_{1.0}$, $PM_{2.5}$ or PM_{10}. However, it is also noteworthy that χ_{LF} displays no significant relationship throughout all size fractions.

Table 3: Pearson's correlation coefficients (r) between mineral magnetic concentration and particle size parameters for urban RDS: (a) traditional sediment size fractions and (b) respiratory health-related size fractions (n = 35 samples).

(a)	**Clay <2 μm**	**Silt 2-63 μm**	**Sand 63-2000 μm**
χ_{LF}	0.30	-0.12	0.05
χ_{ARM}	0.45**	0.0	-0.09
SIRM	0.43**	0.05	-0.13
(b)	**$<PM_{1.0}$**	**$<PM_{2.5}$**	**$<PM_{10}$**
χ_{LF}	0.31	0.29	0.26
χ_{ARM}	0.42**	0.45**	0.44**
SIRM	0.40*	0.43**	0.43**

Note: Significance levels: p <0.05 = *; p <0.01 = **.

5 Discussion

Previous magnetic studies have noted significant correlations between χ_{LF}, χ_{ARM}, SIRM and particle size. To date, Oldfield *et al.* [25] has identified that anhysteretic remanent magnetisation (ARM) measurements can be used to reflect the concentration of fine-grained magnetite (<0.1 μm) in the clay fraction and low-frequency magnetic susceptibility (χ_{LF}) measurements can be used to infer the presence of coarser multi-domain magnetite (>1.0 μm) in sands and coarse silts. Clifton *et al.* [15] found χ_{LF} was strongly associated with sands and medium silts, susceptibility of ARM (χ_{ARM}) was strongly associated with clay and fine silts, and saturated isothermal remanent magnetisation (SIRM) was strongly associated with very fine to medium silts. Zhang *et al.* [26] suggested that both percentage frequency-dependent magnetic susceptibility ($\chi_{FD\%}$) and χ_{ARM} can be used as a proxy for clay content. The potential application of mineral magnetic measurements as a proxy for urban PM was highlighted by Booth *et al.* [17], who noted that χ_{LF}, χ_{ARM} and SIRM parameters correlate with all class sizes. Moreover, Booth *et al.* [16] suggested χ_{LF}, χ_{ARM} and SIRM have potential as a particle size proxy for several sedimentary environments, but highlights the importance of fully determining the nature of the relationship between sediment particle size and magnetic properties before applying mineral magnetic data as a size proxy. More recently, this was further explored by Booth *et al.* [27] who discussed potential problems of employing this technology.

These studies illustrate sand correlated negatively with χ_{LF} ($r = -0.94$), χ_{ARM} ($r = -0.96$) and SIRM ($r = -0.91$); silt correlated positively with χ_{LF} ($r = 0.96$), χ_{ARM} ($r = 0.96$) and SIRM ($r = 0.96$); and clay correlated positively with χ_{LF} ($r = 0.82$), χ_{ARM} ($r = 0.94$) and SIRM ($r = 0.81$). They also illustrate PM_{10} correlated positively with χ_{LF} ($r = 0.69$); $PM_{2.5}$ correlated positively with χ_{LF} ($r = 0.70$), χ_{ARM} ($r = 0.30$) and SIRM ($r = 0.33$); and $PM_{1.0}$ correlated positively with χ_{LF} ($r = 0.66$), χ_{ARM} ($r = 0.41$) and SIRM ($r = 0.32$). When data presented here are compared to these earlier investigations, it is apparent that the trends observed are mostly similar to previous studies (e.g. clay and χ_{ARM} ($r = 0.45$) and SIRM ($r = 0.43$); PM_{10}, $PM_{2.5}$ and $PM_{1.0}$ with χ_{ARM} ($r = 0.45$) and SIRM ($r = 0.43$)) but, unlike other work, χ_{LF} displays no significant relationship. The magneto-associations with each of the traditional sediment class sizes highlights the potential use of mineral magnetic data as a means of normalizing compositional analytical data (i.e. geochemical) for particle size effects.

The significant correlations between the magnetic parameters and the respiratory-health related size classes is perhaps of greater importance, because this highlights the technique as a possible alternative PM monitoring tool, which could be linked to both health and pollution studies. Given the combination of low-cost and sensitivity of the method, it can be argued that mineral magnetic measurements have considerable potential to act as a reliable particle size proxy. The method is also rapid; bulk samples require little preparation and individual measurements of magnetic susceptibility (χ_{LF}) can be made in ~1 minute, in either a laboratory or field setting. Therefore, it is feasible that mineral magnetic measurements could be a dependable and rapid exploratory technology for pilot urban roadside PM investigations.

6 Ongoing related work

This work forms part of a wider investigation extended to other UK towns and cities (Runcorn, Salford, Dumfries, Oswestry, Norwich, Wolverhampton and London), which is attempting to answer the same hypotheses posed in the case study presented in this work. In doing so, it is anticipated it will offer better insights of the reliability of mineral magnetic technologies as an alterative PM monitoring approach.

7 Further work

Although not presented here, when the magnetic concentration data for this case study is entered into a GIS and the information presented spatially, it reveals a distinct and noteworthy pattern that illustrates an east-to-west magneto-gradient from low to high concentration values. At this time, it is perceived that this trend can be attributed to emissions from industrial activities on the eastern side of the town and, therefore, proffers that magnetic technologies may be a suitable technology for establishing sources linkages or distinguishing between vehicular and industrial pollution signals of urban sediments.

8 Conclusions

Analyses indicate magnetic concentration parameters could be reliably employed as a suitable particle size proxy for urban RDS. Of the three magnetic parameters, χ_{ARM} and SIRM has the strongest and most significant correlation values ($P < 0.01$) with all respiratory class sizes <10µm. In most cases, these data associations follow the predictable trends of other environmental studies. Therefore, given the speed, low-cost and sensitivity of the measurements, this suggests magnetic techniques could be used as a rapid alternative exploratory technology for pilot particulate pollution investigations.

Acknowledgements

All authors thank the School of Applied Sciences for unlimited access to analytical facilities. The first author also gratefully acknowledges the receipt of a doctoral studentship hosted by the School of Engineering and the Built Environment at the University of Wolverhampton.

References

[1] Robertson, D.J., Taylor, K.G., & Hoon, S.R. 2003. Geochemical and mineral magnetic characterisation of urban sediment particulates, Manchester, UK *Applied Geochemistry*, 18, 269-282.

[2] Lau, W.M., & Wong, M.H. 1983. The effect of particle size and different extractants on the contents of heavy metals in roadside dusts. *Environmental Research*, 31, 229-242.
[3] Fergusson, J.E., & Ryan, D.E. 1984. The elemental composition of street dust from large and small urban areas related to city type, source and particle size. *Science of the Total Environment*, 34, 101-116
[4] Herngren, L., Goonetilleke, A., & Ayoko, G.A. 2006. Analysis of heavy metals in road-deposited sediments. *Analytica Chimica Acta*, 571, 270-278.
[5] Sutherland, R.A. 2003. Lead in grain size fractions of road-deposited sediment. *Environmental Pollution*, 121, 229-237.
[6] Schwartz, J. 2004. Air pollution and children's health. *Pediatrics*, 113, 1137-1043.
[7] McConnell, R., Berhane, K., Yao, L., Jerrett, M., Lurmann, F., Gilliland, F., Kunzli, N., Gauderman, J., Avol, E., Thomas, D., & Peters, J. 2006. Traffic, susceptibility and childhood asthma. *Environmental Health Perspectives*, 114, 766-772.
[8] Maher, B.A., Moore, C., & Matzka, J. 2008. Spatial variation in vehicle derived metal pollution identified by magnetic and elemental analysis of roadside tree leaves. *Atmospheric Environment*, 42. 364-373.
[9] Yeh, H.C., G. M. Schum & Duggan, M.T. 1979. Anatomical models of the tracheobronchial and pulmonary regions of the rat. *Anat. Rev.* 195, 483 492.
[10] EPAQS. 2001 Airborne particles, Expert Panel on Air Quality Standards, London: HMSO.
[11] Harrison, R.M. 2004. Key pollutants – airborne particles. *Science of the Total Environment*, 334, 3-8.
[12] www.defra.gov.uk/environment/airquality/aqeg
[13] www.environment-agency.gov.uk
[14] Bonnett, P.J.P., Appleby, P.G., & Oldfield, F. 1998. Radionuclides in coastal and estuarine sediments from Wirral and Lancashire. *Science of the Total Environment*, 70, 215-236.
[15] Clifton, J., McDonald, P., Plater, A., & Oldfield, F. 1999. Derivation of a grain-size proxy to aid the modelling and prediction of radionuclide activity in saltmarshes and mud flats of the Eastern Irish Sea. *Estuarine, Coastal and Shelf Science*, 48, 511-518.
[16] Booth, C.A., Walden, J., Neal, A., & Smith, J.P. 2005. Use of mineral magnetic concentration data as a particle size proxy: a case study using marine, estuarine and fluvial sediments in the Carmarthen Bay area, South Wales, U.K. *Science of the Total Environment*, 347, 241-253.
[17] Booth, C.A., Winspear, C.M., Fullen, M.A., Worsley, A.T., Power, A.L., & Holden, V.J.C. 2007. A pilot investigation into the potential of mineral magnetic measurements as a proxy for urban roadside particulate pollution. In: *Air Pollution XV*, (Editors) C.A. Borrego & C.A Brebbia, WIT Press, 391-400.
[18] www.bv-aurnsiteinfo.co.uk

[19] www.northlincs.gov.uk
[20] Shilton, V.F., Booth, C.A., Giess, P., Mitchell, D.J., & Williams, C.D. 2005. Magnetic properties of urban street dust and its relationship to organic matter content in the West Midlands, U.K. *Atmospheric Environment*, 39, 3651-3659.
[21] Booth, C.A., Walden, J., Neal, A., Smith, J.P., & Morgan, E. 2004. A comparison of inter-unit, intra-site and intra-sample variability in environmental magnetic data: an example based on the Gwendraeth Estuary, South Wales, U.K. *Journal of Coastal Research*, 20, 808-813.
[22] Booth, C.A., Fullen, M.A., Smith, J.P., Hallett, M.D., Walden, J., Harris, J., & Holland, K. 2005. Magnetic properties of agricultural topsoils of the Isle of Man: their characterisation and discrimination by factor analysis. *Communications in Soil Science & Plant Analysis*, 36, 1241-1262.
[23] Xie, S., Dearing, J.A., & Bloemandal, J. 2000. The organic matter content of street dust in Liverpool, UK and its association with dust magnetic properties. *Atmospheric Environment*, 34, 269-275.
[24] Shu, J., Dearing, J.A., Morse, A.P., Yu, L., & Yuan, N. 2001. Determining the source of atmospheric particles in Shanghai, China, from magnetic geochemical properties. *Atmospheric Environment*, 35, 2615-2625.
[25] Oldfield, F., Richardson, N., Appleby, P.G., & Yu, L. 1993. ^{241}Am and ^{137}Cs activity in fine-grained saltmarsh sediments from parts of the N.E. Irish Sea shoreline. *Journal of Environmental Radioactivity*, 19, 1-24.
[26] Zhang, W., Yu, L., & Hutchinson, S.M. 2001. Diagenesis of magnetic minerals in the intertidal sediments of the Yangtze Estuary, China, and its environmental significance. *Science of the Total Environment*, 266, 160-175.
[27] Booth, C.A., Fullen, M.A., Walden, J., Worsely, A.T., Marcinkinos, S., & Coker, A.O. 2008. Problems and potential of mineral magnetic measurements as a particle size proxy: a case study using Manx topsoils. *Journal of Environmental Engineering and Landscape Management,* 3, 151-157.

Microbial and endotoxin emission from composting facilities: characterisation of release and dispersal patterns

L. J. Pankhurst[1], L. J. Deacon[1, 2], J. Liu[3], G. H. Drew[1], E. T. Hayes[4], S. Jackson[3], P. J. Longhurst[1], J. W. S. Longhurst[3], S. J. T. Pollard[1] & S. F. Tyrrel[1]
[1]*School of Applied Sciences, Cranfield University, UK*
[2]*Environmental Knowledge Transfer Network, UK*
[3]*Centre for Research in Biomedicine, UWE, Bristol, UK*
[4]*Air Quality Management Resource Centre, UWE, Bristol, UK*

Abstract

The potential risk to human health posed by exposure to bioaerosols released from composting is an important issue. Further growth in the number of composting facilities in the UK is anticipated as biodegradable waste is diverted from landfill. To date, studies of bioaerosol emission from composting have focussed on culturable bioaerosols. This paper describes both culturable bioaerosol and endotoxin release and dispersal from two large green waste composting facilities in the UK. *Aspergillus fumigatus*, actinomycetes, Gram-negative bacteria, and endotoxins were simultaneously and repeatedly sampled to describe the release and dispersal from these sites. Meteorological and site operational observations were recorded, allowing analysis of factors influencing bioaerosol release and dispersal. The highest measured concentrations of bioaerosols were associated with composting activities such as shredding and turning. Between release and 50–80m downwind bioaerosol concentrations reduced by 80-90%. An unexpected second peak was detected 100–150m downwind from source at both sites. Endotoxin dispersal patterns were site specific and showed some differences to dispersal patterns of culturable microorganisms.

Keywords: bioaerosol, aspergillus fumigatus, actinomycetes, gram-negative bacteria, endotoxins, composting, dispersal.

doi:10.2495/AIR090151

1 Introduction

Within the UK, increasing amounts of biodegradable waste are being diverted to composting facilities in order to meet targets set through the Landfill Directive (EC/31/99). In 2006/07 3.6 million tonnes of waste was composted in the UK, with 79% of this waste processed in open-air turned windrow facilities (The Composting Association [1]). The successful composting of this waste is dependent on a host of thermophilic and thermotolerant microorganisms, whose proliferation is encouraged. However, open-air windrow composting requires several high-energy processes, such as shredding, turning, and screening. While static compost windrows allow the aerosolisation of some composting microorganisms, these agitation processes increase their release dramatically. Past research has shown how activities result in a 3-log increase in bioaerosol emissions (Clark *et al* [2]; Taha *et al* [3, 4]).

Bioaerosols can be defined as any aerosol of biological origin (Swan *et al* [5]). In the case of composting the term refers specifically to microorganisms, their constituent parts, and by-products. Common composting bioaerosols of concern include *Aspergillus fumigatus*, actinomycetes, and endotoxins (Millner *et al* [6]). Exposure to these bioaerosols has been shown to result in a range of respiratory conditions, including asthma, mucosal membrane inflammation, and invasive aspergillosis (Dutkiewicz [7]; Swan *et al* [5]). Tentative, precautionary threshold levels of 1000 CFU m^{-3} for total bacteria and total fungi, and 300 CFU m^{-3} for Gram-negative bacteria, have been recommended; furthermore, facilities should preferably not be situated within 250m of sensitive receptors (Environment Agency [8]).

Despite the health risks and regulatory limits, the dispersal range of bioaerosols remains uncertain (Albrecht *et al* [9]). Many past studies present dispersal ranges based on few sampling occasions with limited sampling times (Swan *et al* [5]) and using mean rather than peak bioaerosol concentrations. In order to produce valid data, the episodic nature of bioaerosol emissions must be accounted for; therefore peak emission data should be enumerated and presented (Albrecht *et al* [9]). The physical properties of bioaerosols remain poorly characterised (Drew *et al* [10]; Swan *et al* [5]), hence behaviour in the atmosphere remains uncertain and atmospheric modelling programs are unable to accurately represent bioaerosol dispersal. In addition, sampling and enumeration methods used for bioaerosols remain underdeveloped compared to those used within soil and aquatic microbiology (Malik *et al* [11]; Sykes *et al* [12]). Bioaerosol concentrations may therefore have been consistently underestimated, as the non-culturable fractions remain poorly quantified (Swan *et al* [5]).

The above factors have contributed to gaps in our understanding of bioaerosol emission and dispersal from composting facilities. It has been suggested that current guidelines surrounding bioaerosols, and their sampling and enumeration, require updating (Environment Agency [13]). A thorough understanding of bioaerosol dispersal is required, given their potential human health impacts. This study enumerates culturable microorganism and endotoxin emission and dispersal from composting facilities through an extensive experimental program;

creating a validated dispersal profile for these bioaerosols, of use within academia, regulatory bodies, and industry.

2 Methods

Two composting facilities in the UK were sampled. Both sites are large-scale green waste composting facilities that receive approximately 25,000 tonnes of waste per annum from civic amenity sites and kerbside collection, and operate open-air turned windrow systems.

Site A is located on a closed and capped landfill facility. Bunding encircles the northern, western and southern sides, with hedgerow on all sides. The nearest residence is located 500m to the north of the site. Surrounding land is agricultural, with an area of woodland to the west. Site B is located on the south-west corner of an industrial estate; the northern edge of the site is surrounded by concrete bunding. The industrial estate extends approximately 630m to the north and 500m to the east of the facility, workplaces can be found 50m from the site to the edges of the industrial estate. A large river bounds the west side of the site; arable land surrounds the southern edge of the site and the industrial estate.

SKC personal aerosol filter samplers were used to collect bioaerosols, with a flow rate of 2.2 ± 0.1 L min^{-1}; sampling heads were elevated to 1.7m, and loaded with sterile polycarbonate filters, pore size 0.8 μm (SKC Ltd, UK) (Taha *et al* [3]). A Kestrel 4000 weather data logger (Nielsen Kellerman, PA, USA) was used to record meteorological conditions throughout sampling.

Filters for culture were placed into buffer solution (NaCl 1 g l^1 and 3 drops of Tween 80 L^{-1}, in 1 l sterilised ddH_2O) and stored at 4°C. Within 24 hours samples were processed under aseptic conditions. Samples were shaken to create a suspension, which was diluted to a common logarithm order (10^{-1} and 10^{-2}) and 100 μl of each dilution was transferred onto a range of media. Malt Extract Agar (MEA) was used to culture *Aspergillus fumigatus*, Compost Agar (CA) (Taha *et al* [4]) was used for actinomycetes, while Gram-negative bacteria were grown using MacConkey Agar (MAC). Plates were incubated at 37 ± 2 °C for 3-7 days (MEA and MAC) or 44 ± 2 °C for 7 days (CA). Colony forming units (CFUs) were visually enumerated and converted to CFU m^{-3} of air (Taha *et al* [3, 4]).

Samples for endotoxin assay were sealed with the cap and clip supplied (IOM multi-dust cassettes) and stored at -20 °C until extraction at UWE Bristol. Extraction solution was prepared in Limulus Amebocyte Lysate (LAL) grade water (Lonza Wokingham Limited, UK) containing 0.05% Tween 20. The extraction solution was centrifuged at 1000×g for 15 minutes, the supernatant was collected, vortexed and aliquots placed into two pyrogen-free glass tubes. Analysis was carried out using the PyroGene rFC Endotoxin detection kit (Lonza Wokingham Limited, UK) (Spaan *et al* [14]).

At Site A, a total of 17 sampling trips were made, from September 2007 until August 2008. Sampling at Site B commenced in August 2008, and is ongoing with 10 sampling trips reported. Samples were taken from on-site, to site boundary (edge of the composting area), and the farthest possible point downwind (Site A: 355m, Site B: 600m). Upwind samples were taken to evaluate ‘background’ concentrations. Samples were taken regardless of weather conditions or site activities.

3 Results

3.1 Aspergillus fumigatus

A. fumigatus concentrations and dispersal patterns show many similarities at Sites A and B (Figures 1-2). Downwind concentrations are characterised by a significant increase of 1-2 log from upwind to those found on-site during composting activities. Between this peak in concentration and 50–80 m downwind, concentrations decline by approximately 90%. However, a second peak in concentration can be observed 100m downwind. From this point a steady decline in *A. fumigatus* concentrations can be observed. In the case of Site B, a third peak in concentration is visible at 500–600m downwind.

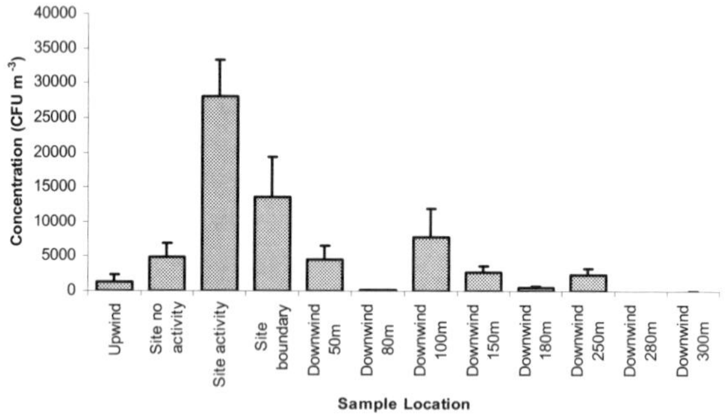

Figure 1: Mean Site A *A. fumigatus* concentrations presented in CFU m^{-3} represented by bar, standard error represented by whisker.

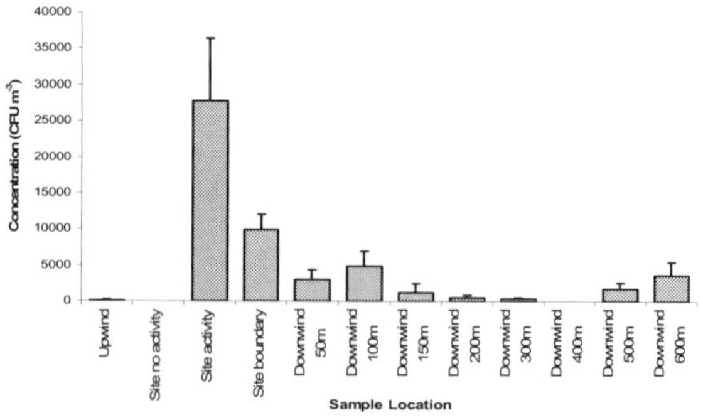

Figure 2: Mean Site B *A. fumigatus* concentrations presented in CFU m^{-3} represented by bar, standard error represented by whisker.

3.2 Actinomycetes

There are some site specific features for actinomycete release and dispersal patterns (Figures 3-4). For Site B, the highest concentrations are found at the site boundary, where they are 2-log above those found upwind and 1-log higher than the Site A peak, which was found on-site during activities. At both sites, concentrations are significantly above those found upwind at the on-site activity, site boundary, and downwind 100-150m locations. At Site A there is no significant difference between the concentrations found at any other sampling location and upwind. At Site B, a third peak in concentrations is shown 500–600m downwind, with concentrations significantly above those found upwind.

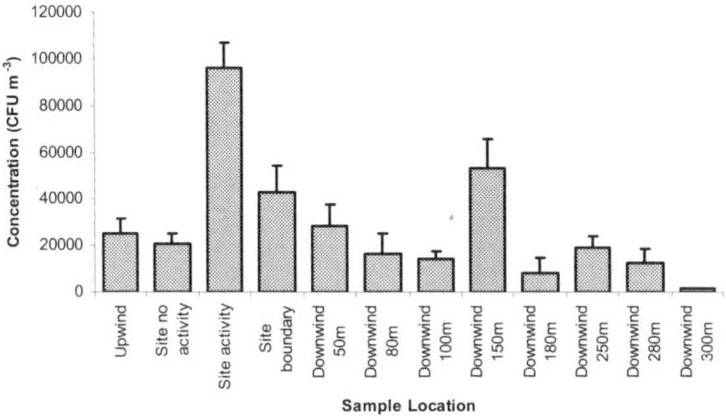

Figure 3: Mean Site A actinomycete concentrations presented in CFU m^{-3} represented by bar, standard error represented by whisker.

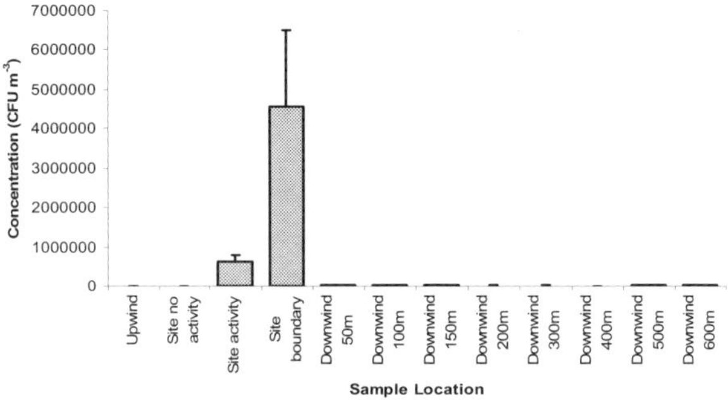

Figure 4: Mean Site B actinomycetes concentrations presented in CFU m^{-3} represented by bar, standard error represented by whisker.

3.3 Gram-negative bacteria

Gram-negative bacteria were found in the highest concentrations on-site during activity at Site A, and at the site boundary of Site B (Figures 5-6). These concentrations are up to 3-log higher than those found upwind. Downwind concentrations are characterised by a decline of over 90% between the peak and 50–80m downwind. At both sites a second peak, significantly different from concentrations found upwind, is seen at 150m downwind. This peak is more significant at Site A; nevertheless, concentrations reached those found upwind by 280m downwind. At Site B, concentrations do not decline to upwind concentrations by 600m from site boundaries.

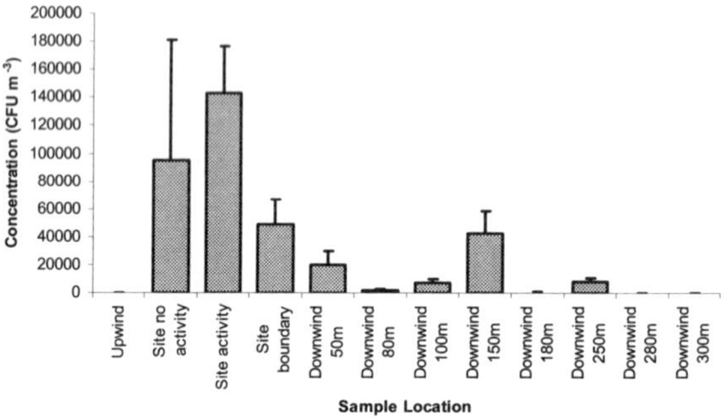

Figure 5: Mean Site A Gram-negative bacteria concentrations presented in CFU m^{-3} represented by bar, standard error represented by whisker.

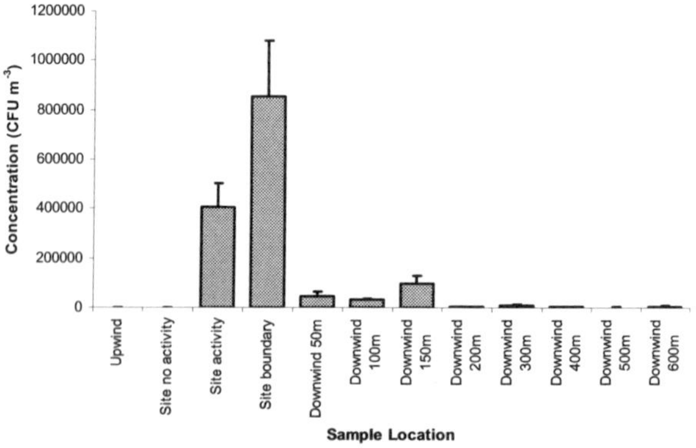

Figure 6: Mean Site B Gram-negative bacteria concentrations presented in CFU m^{-3} represented by bar, standard error represented by whisker.

3.4 Endotoxin

At both sites, a peak in endotoxin concentrations was found at the site boundary, with concentrations from on-site to downwind 280m at Site A, and 200m at Site B above those found upwind (Figures 7-8). Downwind concentrations at both sites are characterised by a decline of 90% from site boundary to 50m downwind. A second peak can be seen at 150m downwind for Site A, and 200 m downwind for Site B. However, at Site A endotoxins showed the highest concentrations 280m from site boundary, While at Site B by 300m downwind concentrations had returned to those found upwind.

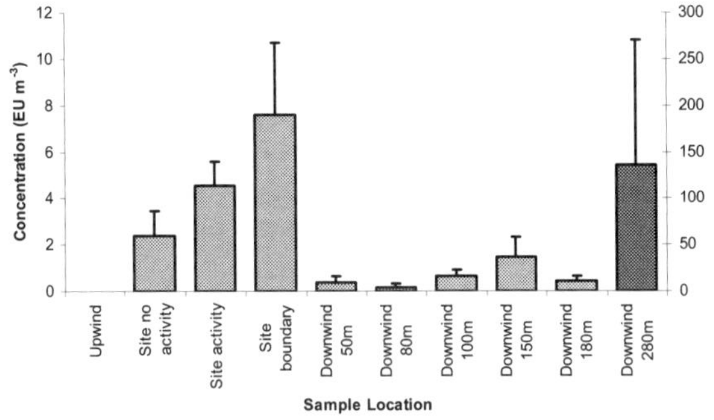

Figure 7: Mean Site A endotoxin concentrations presented in endotoxin units per cubed metre (EU m^{-3}) represented by bar, standard error represented by whisker. Diagonally shaded bar represented on secondary y-axis.

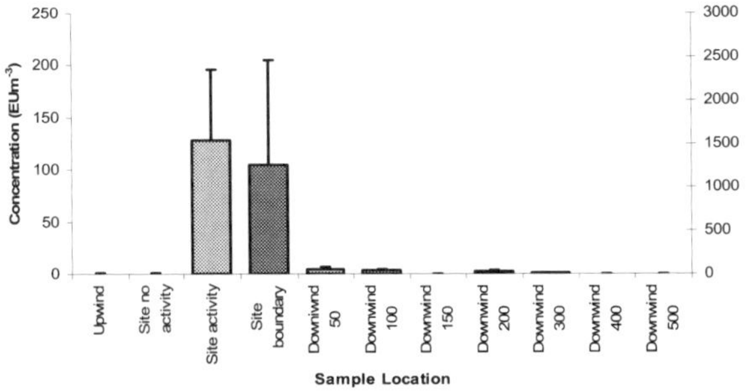

Figure 8: Mean Site B endotoxin concentrations presented in endotoxin units per cubed metre (EU m^{-3}) represented by bar, standard error represented by whisker. Diagonally shaded bar represented on secondary y-axis.

4 Discussion

All of the bioaerosols enumerated through culture showed similar patterns of release; characterised by peak emissions on-site (including site boundary). This is in agreement with Albrecht *et al* [9] and Taha *et al* [3]. At Site B, the highest concentrations of actinomycetes and Gram-negative bacteria were found at site boundaries. This may be due to an accumulation of bacteria here as several activities are carried out upwind. Dispersal patterns were also similar, culturable microorganisms showed an 80-90% reduction in concentrations between peak concentrations and 50–80m downwind. This feature of dispersal has previously been reported anecdotally (Environment Agency [15]). Between 100 and 150m downwind a second peak in concentrations was detectable; this may be due to buoyancy effects, causing some bioaerosols to rise above sampling height until cooled enough to sink back to sampling height. This pattern of dispersal was shown by both sites, suggesting that it is not due to site-specific features.

A. fumigatus reached concentrations similar to those found upwind, and below 1000 CFU m^{-3}, by 280m from Site A, and 200m from Site B. At Site B, however, *A. fumigatus* concentrations show a further peak 500–600m from site boundaries. It is currently unknown whether this peak is due to site emissions or a secondary source. Actinomycetes reached concentrations similar to those found upwind by 180m from Site A, and 300m from Site B. At both sites, actinomycetes were found in concentrations above the 1000 CFU m^{-3} limit for total bacteria, even at upwind locations. This may have implications for the future revision of threshold levels, as the relatively high background concentrations of actinomycetes needs to be accounted for. Gram-negative bacteria declined to upwind concentrations, and were below 300 CFU m^{-3}, by 280m from Site A. At Site B, concentrations did not decline to those found upwind at any sampling location.

While *A. fumigatus* was released in similar quantities at both sites; this was not the case with the bacteria measured. Actinomycetes peaked at quantities 1-log higher at Site B than Site A. However, up- and downwind concentrations were largely in the same order of magnitude at both sites. Site B Gram-negative bacteria also peaked at concentrations 1-log above those found at Site A. As with actinomycetes, other sampling locations showed concentrations largely within the same order of magnitude at both sites. The higher quantities of bacteria found at Site B may be due to differences in feedstock, as it receives more domestic biodegradable waste than Site A. Furthermore, concentrations at site boundary at Site B are typically higher than at Site A. This site-specific feature may be due to site design, and warrants further investigation. Despite these differences in concentration, dispersal patterns off-site remain similar at both sites.

Endotoxin dispersal patterns show several differences. At both sites, peak on-site concentrations are seen at the site boundary and are similar to those previously found (Liebers *et al* [16]). As with culturable microorganisms, this peak is followed by a 90% decline in concentrations. At Site B, this decline is followed by a further small peak at 200m downwind, after which concentrations are similar to those found upwind, or undetectable. At Site A, a second peak in concentration is seen at 150m downwind, as with culturable microorganisms. However, the highest peak was seen at 280m downwind, where concentrations were 2-log higher than those found on-site. This feature was not seen at Site B, despite the fact that emissions were 3-log higher on-site as compared to Site A.

5 Conclusions

This data-set demonstrates how bioaerosols do not follow a typical decay curve from emission to background concentrations. Peaks in concentration at 100–150m downwind suggest that buoyancy and air temperature impact upon their dispersal. The initial 80–90% decay in concentrations, followed by a small peak, appears to be a common feature of bioaerosol dispersal, despite the volume of bioaerosols released. This feature of dispersal was previously unquantified. However, the bioaerosols measured also showed some differences in dispersal. Actinomycete concentrations should be interpreted with care, as they were found above reference values at most locations. Furthermore, although Gram-negative bacteria reached background concentrations at Site A, they did not at Site B and remained above the threshold limit.

Bioaerosols may be found above threshold values at distances above the risk assessment limit of 250m, where sensitive receptors can be exposed, and at upwind locations. This suggests that caution is required in the interpretation of results, and that samples must be repeated in order to ensure valid results.

The creation of this data-set has provided validated dispersion profiles for these bioaerosols, and demonstrated how physical properties affect dispersal range. Further data collection and detailed analysis is continuing in order to add to the data-set. The information provided can be used to improve best-practice risk assessment of bioaerosols, as well as provide a basis for further investigations into bioaerosol dispersal.

Acknowledgements

The authors would like to acknowledge UWE Bristol for their analysis of samples for endotoxin content. The financial support of SITA part of the SUEZ Environment group and NERC (NERC grant NE/E008534/1) is also acknowledged.

References

[1] The Composting Association, The State of Composting and Biological Waste Treatment in the UK 2007/07, pp 4, UK, 2008.

[2] Clark, C. S., Rylander, R. and Larsson, L., Levels of Gram-negative bacteria, Aspergillus fumigatus, dust, and endotoxin at compost plants, Applied and Environmental Microbiology, 45 (5), pp. 1501-1505, 1983.

[3] Taha, M. P. M., Drew, G. H., Longhurst, P. J., Smith, R. and Pollard, S. J. T., Bioaerosol releases from compost facilities: Evaluating passive and

active source terms at a green waste facility for improved risk assessments, Atmospheric Environment, 40, pp. 1159-1169, 2006.

[4] Taha, M. P. M., Drew, G. H., Tamer-Vestlund, A., Aldred, D., Longhurst, P. J. and Pollard, S. J. T., Enumerating actinomycetes in compost bioaerosols at source – Use of soil compost agar to address plate 'masking', Atmospheric Environment, 41, pp. 4759-4765, 2007.

[5] Swan, J. R. M., Kelsey, A., Crook, B., and Gilbert, E.J., Occupational and Environmental Exposure to Bioaerosols from Composts and Potential Health Effects – A critical review of published data, Health and Safety Executive, UK, 2003.

[6] Millner, P. D., Olenchock, S. A., Epstein, E., Rylander, R., M.D., Haines, J., Walker, J., Ooi, B. L., Horne, E. and Maritato, M., Bioaerosols associated with composting facilities, Compost Science and Utilization, 2 (4), pp. 6-57, 1994.

[7] Dutkiewicz, J., Bacteria and fungi in organic dust as potential health hazard, Annals of Agriculture and Environmental Medicine, 4 (1), pp. 11-16, 1997.

[8] Environment Agency, Health Effects of Composting. A study of three compost sites and review of past data, Bristol, UK, 2001.

[9] Albrecht, A., Fischer, G., Brunnemann-Stubbe, G., Jäckel, U. and Kämpfer, P. Recommendations for study design and sampling strategies for airborne microorganisms, MVOC and odours in the surrounding of composting facilities. International Journal of Hygiene and Environmental Health, 211, pp. 121-131, 2008.

[10] Drew, G. H., Tamer, A., Taha, M. P. M., Smith, R., Longhurst, P. J., Kinnersley, R. and Pollard, S. J. T. Dispersion of bioaerosols from composting facilities, Waste 2006 Conference, 19-21st September, Stratford, UK, 2006.

[11] Malik, S., Beer, M., Megharaj, M., and Naidu, R. The use of molecular techniques to characterize the microbial communities in contaminated soil and water, Environment International, 34, pp. 265 – 276, 2008.

[12] Sykes, P., Jones, K. and Wildsmith, John. D. Managing the potential public health risks from bioaerosol liberation at commercial composting sites in the UK: An analysis of the evidence base. Resources, Conservations and Recycling, 52 (2), pp. 410-424, 2007.

[13] Environment Agency. Updated Review of Methods for Monitoring Compost Bioaerosols. Environment Agency, Bristol, UK, in press.

[14] Spaan, S., Wouters, I. M., Oosting, I., Doekes, G. and Heederik, D., Exposure to inhalable dust and endotoxins in agricultural industries, Journal of Environmental Monitoring, 8, pp. 63-72, 2005.

[15] Environment Agency. Monitoring the Environmental Impact of Waste Composting Plants. R&D Technical Summary P1-216, Environment Agency, Bristol, UK, 2001.

[16] Liebers, V., Brüning, T. and Raulf-Heimsoth, M., Occupational endotoxin exposure and possible health effects on humans, American Journal of Industrial Medicine, 49, pp. 474-491, 2006.

Annual study of airborne pollen in Mexicali, Baja California, Mexico

S. Ahumada-Valdez, M. Quintero-Nuñez, O. R. García-Cueto & R. Venegas
Engineering Institute, Air Quality Department, Autonomous University of Baja California, Mexico

Abstract

The purpose of the study is to present a preliminary survey of the presence of airborne pollen in the air of Mexicali, and to record the most common pollen particles present in its atmosphere. Samples were collected using Rotorod model 40, every 24 hours. The quantitative parameters measured were: diversity, richness, volume, and frequency, as well as its relation with meteorological factors, like average temperature, relative humidity average, and precipitation. The airborne pollen concentration was estimated as 1,973 grains/m^3 and 72 taxon pollen were detected outstanding March with 30 pollen types that correspond to 41.6% of the total detected. The pollen types were grouped according to their phenology. In spring a total of 44 types were classified (61%), for summer 23 types were listed (32%); in autumn 40 types appeared (55%), during the winter were identified 39 types (54%). Respect the shrubs 63 grains/m^3 was observed in August representing the 24.9%, all equivalent to 3.19% of the annual total. In relation to the weeds, were better observed in September with 295 grains/m^3, representing 30.87% of this group, equivalent to 14.95%, of the annual total. The most abundant group with greater biological diversity was the trees with a bulk that fluctuated between 103 to 146 grains/m^3 and richness identified with 32 pollen types. The Pearson correlation analysis demonstrated pollen grains do not have a significant linear association with the mean monthly temperature, but they have it, with the maximum relative humidity.

Keywords: aerobiological survey, airborne pollen, allergenic particles, Mexicali.

1 Introduction

Nowadays aeropalinologycal studies are of great importance at worldwide level, they give us the opportunity to develop in many areas and topics, such a climate change, archaeology and health, they contribute whit valuable data related with taxonomic spectrum of pollen in the atmosphere of different regions in the world (Kobzar [1]; Detandt and Nolard [2]; Gehrig and Peeters [3]; Rodriguez-Rajo *et al.* [4]). These studies have allowed one to establish the relation with meteorological parameters and have helped to develop dispersion models, that have been helpful in the diagnosis of health problems in respiratory tract (D'Amato *et al.* [5]; Trivi *et al.* [6]; Schueler and Schlunzen [7]; Egger *et al.* [8]; Assing *et al.* [9]). There is information about pollen with allergenic characteristics affecting the quality of life of sensitive people; the pollen particles are suspended in the atmosphere, therefore they are considered as part of the atmospheric pollutants and the diagnostic of respiratory diseases (Gonzalez-Lozano *et al.* [10]). A few studies have been realized in some regions around Mexico (Bronillet [11]), as well as studies related with the behavior of particles suspended in the atmosphere for different zones in Mexico city (Montes and Cisneros [12]; Ramirez and Rodriguez [13]; Cid del Prado [14]; Rosas *et al.* [15]; Gonzalez-Lozano *et al.* [10]); unfortunately for the city of Mexicali, BC, Mexico, there are not records for this kind of aerobiological analyses, being a city where the levels of atmospheric contamination have violated the permitted air quality many times (Reyna and Arriola [16]).

The palinologycs studies allow us to know the types, dominance, and behavior of pollen and its relation with some meteorological parameters. These studies could be also considered as a useful tool for the health sector to allow identify causal agents in patients with allergic problems, as well as the implementation of preventive measures for this kind of diseases in the respiratory tract (Riggioni *et al.* [17]; Ferreiro *et al.* [18]; Subiza [19]; Mandrioli and De Nuntiis [20]).

2 Materials and methodology

This study was performed in the city of Mexicali, B. C., Mex., located in northwest of Mexico with geographical coordinates North Latitude 32°40', West longitude 115° 28', (Figure 1) in the Subprovince of the Low Delta of the Colorado River belonging to the Great Physiographic Province of the Sonoran Desert, (Shreve and Wiggins [21]). Its basic characteristics, is almost flat area, expose to flood, conformed by very fine sediments like clays and silts (Venegas [22]). Most of the time these particles under certain atmospheric conditions remain suspended in the air, contributing to the atmospheric pollution. The climate conditions, warm and dry summer, winter rain regime, and mean annual precipitation slightly greater to 70 mm, wind direction in winter is NW-SE and summer the dominant wind is S-SE; the mean annual temperature is 22.4°C with extreme annual variations, during the summer the maximum temperature reach 50°C and in the winter the minimum can be lower than 0°C (García [23]).

The city has been built over the Delta Colorado river, it allows an intensive agricultural activity; in fact, two immense agricultural valleys surround the city; to the north the Imperial Valley, in the state of California in the United States of North America and to the south, the Mexicali Valley in the state of Baja California, Mexico, fig. 1.

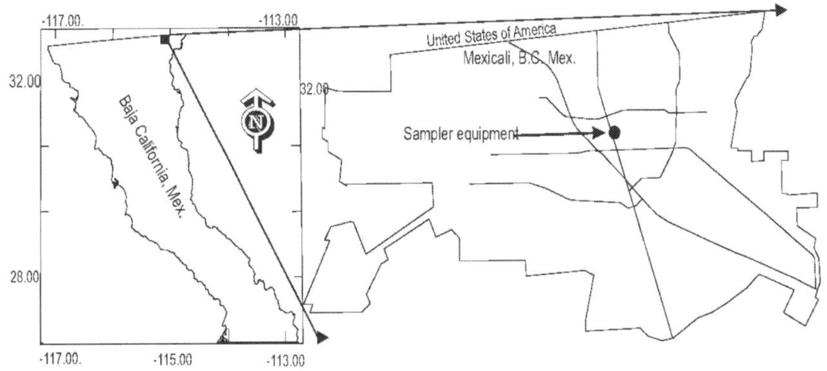

Figure 1: Localization of study area.

The farming activity in the Mexicali valley has two agricultural cycles, during spring-summer, the main ones are: cotton, sorghum, and corn; and during the autumn-winter: wheat, barley, safflower, and green onion. The total surface under irrigation in the Mexicali valley is 165.121.00 hectares.

In relation to the urban landscape, Venegas [24] identified 47 species of trees and shrubs the most common species used for landscaping. Only five of these, are regional or natives: *Populus fremontii* (fremont cottonwood), *Prosopis juliflora* (Mesquite), *Washingtonia robusta* (Mexican fan palm) and *Washingtonia filifera*, (California fan palm) *Atriplex lentiformis* (big saltbush). The majority of the species are considered non natives, they come mainly from tropical and subtropical ecosystems, like: *Bougainvillea spectabilis* (great bougainvillea), *Caesalpinia gilliesii* (bird of paradise), *Callistemon citrinus* and *C. viminalis* (lemon bottlebrush), *Casuarina equisitifolia* (she-oak, horsetail), *Eucalyptus camaldulensis* (river red gum), *Ficus benjamina* (Benjamin tree), *Ficus microcarpa* (Indian Laurel), *Jacaranda mimosifolia* (jacaranda), Melia azederach (chinaberry), Morus nigra (black mulberry), *Nerium oleander* (rose bay), *Phoenix dactilifera* (Date palm trees), *Pluchea sericea* (arrowweed).

The airborne pollen samples were collected using Rotorod model 40 equipment, with rotation and impact of particles in pre-lubricated rods, obtaining two simultaneous and equal samples (Frenz and Guthrie [25]). The equipment was installed on the roof of the Engineering Institute of the Baja California University (UABC), fifteen meters above ground, geography coordinates 32° 37' 52" N, 115° 26 ' 41.4" W.

The standard parameters for identification and quantification of pollen grains were adjustments to obtain sampling at 5% instead of 10%, as a result the volume of air sampling was 1,56 m^3 per period of sampling (Elander [26]). The collected samples were stained with solution of Calberla's, consisting of basic preparation of fuchsine solution, widely used for pollen identification under microscope.

Samples were collected every 24 hours, five days a week, according to the standard method (Brown [27]). The grains of pollen were identified following the criteria of Grant *et al.* [28], most of them at taxonomic level of Family, Genus and Species.

The identified airborne pollen types were grouped according to their biological form in trees, shrubs, and weeds. In the last group were included the grasses, herbaceous and all the plants that compete with the cultured species. The quantitative parameters used to measure the types of pollen were their diversity, richness, volume, and frequency, as well as its relation with meteorological factors, like mean temperature, relative humidity average and precipitation. To analyze the relationship between pollen data and meteorological variables, the Pearson correlation was used to determine the level of significance of each correlation. The analysis was made for all the period of capture with program Statistica V6.0.

3 Results

A total of 1,973 grains/m^3 of airborne pollen concentration (TP) were identified with a richness of 72 types of pollen during the year, fig. 2.

In relation to the richness, the months from January to June had a similar behavior with an average of 24 airborne pollen monthly; outstanding March with

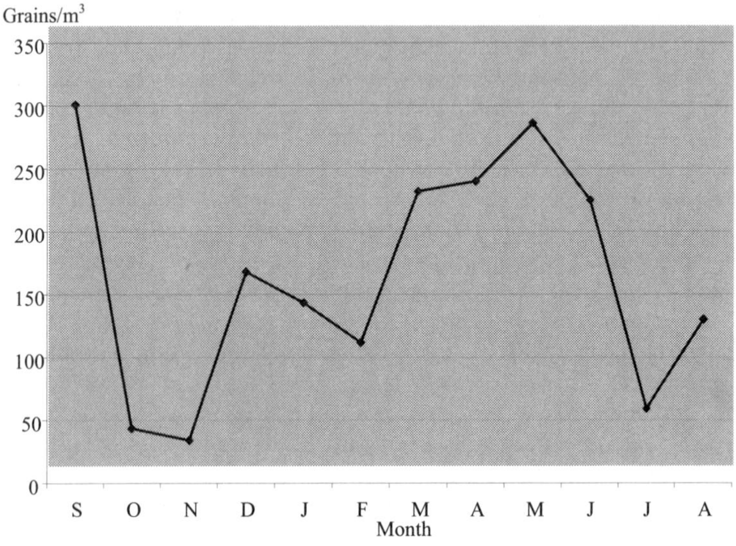

Figure 2: Annual pollen distribution.

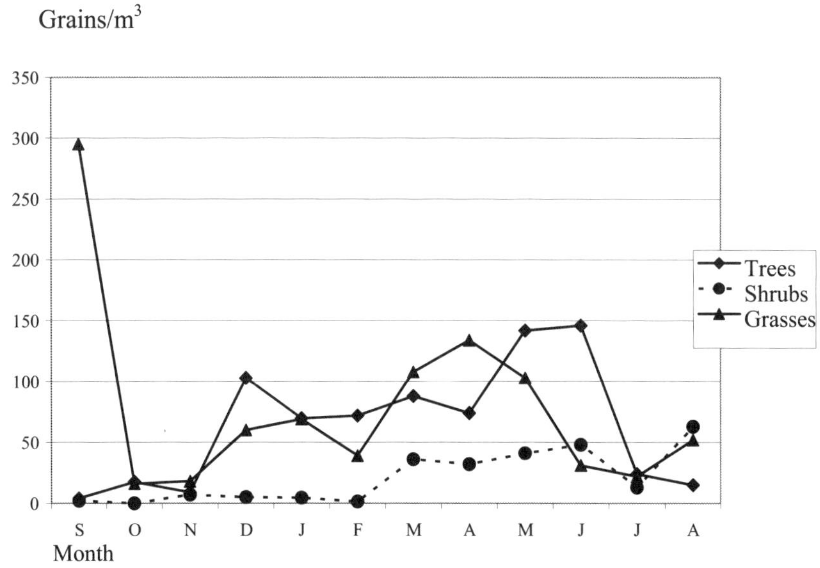

Figure 3: Trees, shrubs, grasses.

30 pollen types that correspond to 41.6% of the total identified for all the sampling year.

According to distribution by life forms, figure 3 shows its behavior during the sampling period.

Trees have the highest values from December to June; months of May and June outstanding 142 and 146 grains/m^3 respectively, these two months represent each one, the 18.58% and the 19.10% for this group of plants, and 7.2% and 7.40% for the same period in relation to the annual total.

Respect the shrubs the relevant data with 63 grains/m^3 is observed in August representing the 24.9% for this form of life, all equivalent to 3.19% of the annual total. In relation to all the plants that were included within the group of the weeds, relevant data is observed in the months of March 108 grains/m^3, April 134 grains/m^3, May 103 grains/m^3 and September with 295 grains/m^3, representing each month the 10.77%, 14%, 11.30%, and 30.87% respectively of this group of plants, equivalent to 5.22%, 6.8%, 5.47%, and 14.95%, of the annual total.

Within the weeds group all herbaceous are included, weeds that compete with the cultivated plants and plants used for landscaping in the city and private gardens. In this group the species belonging to the *Poaceae* Family represent the greater volume of pollen collected, but due to the difficulty of identifying them at taxonomic level of genus and specie, most of these are reported at taxa family; the distribution during the four seasons in the year is shown in table 1.

Trees prevail in all the four seasons; in spring season a total of 44 types were identified, prevailing *Castanea sativa* with 172 grains/m^3; for summer season

23 types were identified; in autumn 40 types appeared, prevailing genus *Casia* with 83 grains/m^3, during the winter were identified 39 types, with the family *Poaceae* showing dominance in volume with 102.23 grains/m^3.

Table 1: Types and volume.

	Trees	Types	Shrubs	Types	Grass	Types
Spring	362	44	121	9	268	13
Summer	43	23	78	6	369	7
Autumn	129.5	40	12	10	94	13
Winter	228.62	39	41.8	5	216	16

4 Discussion and conclusions

The density of pollen particles of families *Chenopodaceae-Amarataceae* during sampling months fluctuated between 1-362 grains/m^3; the greater volume was counted in September, which means that this month could be the one with greater production of pollen, but also is observed that it could be transported by winds NE-NW; other significant peaks with greater concentrations are for March, April, May and June with a mean of 245.75 grains/m^3, representing 12.46% of the annual total. Also from March to June, the greater velocity of wind coming from the NW was registered with an average of 2.05 m/s (personal communication García-Cueto [29]).

The meteorological conditions in Mexicali do not adjust to the four season characteristics; consequently the sampling period was divided in two stages (García-Cueto *et al.* [30]), warm season from April to September, and the cold one from October to March, tables 2 and 3. The monthly presence of pollen grains/m^3 does not have a correlation with the temperature, rainfall and the wind velocity. For July and August, the wind blow trend did not follow defined pattern for regular behavior S-SE, in the summer season.

March appears as the month with a great diversity, with 30 airborne pollen types; this matches the beginning of the flowering in spring, outstanding pollen pertaining to the Family *Poaceae* and the Genus *Ambrosia*. The presence of the Family *Poaceae* throughout the year and periods with greater abundance,

Table 2: Pollen in warm season.

Month	A	M	J	J	A	S
Grains/m^3	240	286	225	59	130	301
T°C	23.6	28.8	32.5	37.2	35.5	29.3
%HR	28.9	30.9	29.3	34.2	42.4	32.9
PP mm	10.8	0	0	0	43	0
Wind m/s	2.2	1.8	2.2	1.8	1.8	1.9
Wind direction	NW	N	N, NE	N	NW	SE

Table 3: Pollen in cold season.

Month	O	N	D	J	F	M
Grains /m^3	44	34	168	143	112	232
T°C	24.5	17.1	15	15.9	16.6	20.8
%HR	42.6	49	52.5	62.1	63.6	37.8
PP mm	26.5	13	8.75	27.7	22.5	10
Wind m/s	1.4	1.2	1.1	1.3	1.4	2
Wind direction	NW	NW	NW	NW	N	NW

concords with the agricultural cycles of the region, for example cycle autumn-winter sowing wheat, safflower, barley, and Rye grass. Cycle spring-summer sowing forage sorghum and corn; in addition to perennial cultured like the Bermuda grass.

The most abundant group with greater biological diversity was the trees with a bulk that fluctuated between 103 to 146 grains/m^3 and richness identified with 32 airborne pollen. In this plants group were identified pollen with allergenic characteristics according to the Department of Health Sciences of the University of Arizona, among them appear the Genus: *Washingtonia, Tamarix, Platanus, Quercus, Fraxinus, Ulmus, Juniperus, Populus, Ligustrum and Juglans* (Shumacher [31]).

Respect to the weeds group, they were also pollen with allergenic characteristics, like Genus *Ambrosia, Amaranthus, Chenopodium, Artemisia* (Shumacher [31]). In the Chenopodiacea-Amarantacea group the greater volume happens in September with 362.1 grains/m^3, the mean temperature for the month was 29.3°C, and it seems not to be influence by the environmental suspension and dispersion of particles, the mean relative humidity for this month was of 32.9% that could have favored the suspended particles. The Family *Poaceae* included densities that fluctuate between three and eighty grains/m^3, the richness consisted of nine types detected, and among them is *Sorghum halepense* as allergenic pollen (Shumacher [31]).

In relation to the months with highest relative humidity were December, January and February, with 52.4%, 60.6.1% and 63.4%, respectively, being observed an agreed relation with the volume of pollen; the same happens in relation to the average temperature.

Table 4 shows the results of the statistical analysis between meteorological variables and the presence or absence of pollen. The data correspond to monthly averages, the correlation of Pearson where R^2 is the coefficient of determination and R is the correlation coefficient, P is the significance associated to the value of R, whereas HRMX is maximum relative humidity; HRMN, minimum relative humidity; HRMED, mean relative humidity; TMAX, maximum temperature; TMED, minimum temperature; and PREC, pluvial precipitation.

This analysis show that the correlation coefficient between pollen and mean temperature, was low and non-significant $R = 0.13$, $P = 0.22$. The relationship between pollen and maximum relative humidity showed a better correlation in

Table 4: Pearson correlation.

Climatic variables	(R^2)	(R)
HRMX	0.3375	0.58
HRMN	0.1944	0.44
HRMD	0.2702	0.52
TMAX	0.024	0.15
TMIN	0.0089	0.09
TMED	0.0164	0.13
PREC	0.2182	0.47
TMED- HRMAX	0.36	0.60

relation to the previous, R = 0.58, P= 0.001; it means the coefficient indicates that the volume of pollen is directly proportional with the maximum values of relative humidity. Oliveira *et al.* [33] found positive correlation with relative humidity, Boral and Bhattacharya [34] found the same positive relation between meteorology variables like relative humidity, temperature and precipitation.

When the inclusion of an additional variable is made (mean temperature-maximum relative humidity), and it correlates with the temperature average and maximum relative humidity, R was considered = 0.60, whose result also is significant, which was the best one of the coefficients of found correlation; this means that the value of pollen can be explained up to a 36% according to the coefficient of determination (R^2) estimated.

Finally this first aerobiological study in the region provides interesting results with the meteorological variables, and its possible relation of species identified as allergenic. These results demonstrate the importance of continuing with this kind of aerobiology studies, to develop a pollen map of the region that could contribute to recognize the airborne allergenic particles and their influence in the individuals.

References

[1] Kobzar, V.N., Aeropalynological monitoring in Bishkek, Kyrgyzstan. Aerobiologia 15, pp. 149-153, 1999.

[2] Detandt, M., & Nolard, N., The fluctuations of the allergenic pollen content of the air in Brussels (1982 to 1997). Aerobiologia 16, pp. 55-61, 2000.

[3] Gehrig, R., & Peeters, A.G., Pollen distribution at elevations above 1000 m in Switzerland. Aerobiologia 16, pp. 69-74, 2000.

[4] Rodríguez-Rajo, Seijo, M.C., & Jato, V., Estudio aerobiológico de la atmósfera de A Guardia, NO de España. REA 7, pp. 7-15, 2002.

[5] D'Amato, G., Liccardi, G., & Frenguelli, G., Thunderstorm-asthma and pollen allergy. Allergy 62, pp.11-16, 2007.

[6] Trivi de Mondri, M. E., Burry, L.S., & D'Antoni H.L., Dispersión-depositación del polen actual en la Tierra del Fuego, Argentina. Revista Mexicana de Biodiversidad 77, pp. 89-95, 2006.

[7] Schueler, S., H.L. & Schlunzen K., Modeling of oak pollen dispersal on the landscape level with a mesoscale atmospheric model. Environmental Modeling and Assessment, 11(3), pp. 179-194, 2006.

[8] Egger, M., Mutschlechner, S., Wopfner, N., Gadermaier, G., Briza, P., & Ferreira F., Allergy 61, pp. 461-476, 2006.

[9] Assing, K., Bodtger U., & Poulsen, L. K., Seasonal dynamics of chemokine receptors and CD62L in subjects with asymptomatic skin sensitization to birch and grass pollen. Allergy 61 (6), pp.759-768,2006.

[10] González-Lozano, C., Cerezo-Moreno, A., González-Macías, C. & Salazar, C.L., Comportamiento de las partículas suspendidas y polen en la atmósfera de la región norte de la Zona Metropolitana de la Ciudad de México. Revista de la Sociedad Química de México, 43(5), pp. 155-164, 1999.

[11] Bronillet, T. I., An annual study of airborne in northern Mexico City. Aerobiologia 12, pp. 191-195, 1996.

[12] Montes, J., & Cisneros, P., Los polenes atmosfericos de la ciudad de Mexico. Alergia. 29(2) pp. 51-60, 1982.

[13] Ramírez, O., & Rodríguez, B., Estudio ilustrado de los pólenes del aire de México. Alergia 3, pp. 187-217. México, 1961.

[14] Cid del Prado, L., Lluvia de Polen de la Ciudad de Toluca. Revista de la Facultad de Medicina, Universidad Autónoma del Estado de México. 2(1), pp. 28-31, 1992.

[15] Rosas, I., Calderon, C., Ulloa, M., & Lacey, J., Abundance of airborne *Penicillium* CFU in relation to urbanization in Mexico City. Applied and environmental Microbiology. Aug., pp. 2648-2652, 1993.

[16] Reyna, C.M., & Arriola Z. H., Contaminación y medio ambiente en Baja California (Capítulo 6). El estudio de las principales enfermedades respiratorias y los contaminantes del aire en Mexicali, Baja California, ed. Porrua: México, pp.137-155, 2006.

[17] Riggioni, O., Montiel, M., Fonseca, J., Jaramillo, O., Carvajal, E., Rosencwaig, P., & Colmenares A., Los pólenes de gramíneas y su relación con manifestaciones alérgicas en Costa Rica. Rev. Biol. Trop., 42(I Supl.), pp. 41-45, 1994.

[18] Ferreiro, A., Núñez, O., Rico, D., Soto, M., & López, R., Pólenes alergénicos y polinosis en el área de La Coruña. Revista Española de Alergología e Inmunología Clínica 13(2) pp. 98- 101, 1998.

[19] Subiza J., Cómo interpretar los recuentos de pólenes. Alergología e Inmunología Clínica. 16, pp. 59-61, 2001.

[20] Mandrioli, P., & De Nuntiis, P., Forecasting tools for pollinosis: Aerobiology and Allergens. Allergy 57(s73), pp. 277, 2002.

[21] Shreve, F.& Wiggins, I.L., Vegetation and Flora of Sonoran Desert. Volume 1, Stanford University Press. Stanford California: USA, 1964.

[22] Venegas, F. R., Physical and Biological Features of the Colorado River. The U.S.-Mexican border environment: Lining the All-American Canal: Competition or Cooperation for Water in the U.S.-Mexican Border

COLEF-SCERP Monograph Series, Vicente Sanchez Munguía: Mexico, pp. 1-19 2006.
[23] García, E., Modificaciones al sistema de clasificación climática de Köppen, (Para adaptarlo a la República Mexicana). Instituto de Geografía, UNAM. pp. 250, 1981.
[24] Venegas F. R., Manual para el reconocimiento de los árboles y arbustos más comunes en la ciudad de Mexicali. UABC: México, 1991.
[25] Frenz, D., & Guthrie, B.L., A rapid, reproducible method for coating Rotorod Sampler collector rods with silicone grease, Annals of Allergy, Asthma & Immunology, 87, pp. 390- 393, 2001.
[26] Elander, J, Volumetric Calculations. Sampling Technologies, Multidata LLC, 2000.
[27] Brown, T., Frenz, D.A., &.Wimpsett, T. L., Operating Instructions for the Rotorod Sampler. Sampling Technologies, Inc., Minnetonka, 1993.
[28] Grant, S.E., Choice and use of equipment (Chapter 2). Sampling and Identifying Allergenic Pollen and Molds, ed. Blewstone Press: San Antonio Texas, pp. 7-13, 2000.
[29] García-Cueto O. R. Personal Communications, 20 July 2007, Head of Meteorology Department, Engineering Institute, UABC, Mexico
[30] García-Cueto, O.R., Martínez, C.A., Cervantes, P.J. & Tejeda N. D., Evaluación del bioclima humano en Mexicali, B. C. y su comparación con otras ciudades cálidas de la Republica Mexicana. Reporte técnico, Instituto de Ingeniería, UABC, 1996.
[31] Allergy and Asthma in the southwestern Unites States. Web site, Arizona, USA http;//allergy. peds.arizona.edu/southwest/trees_shrubs, content owner Schumacher, M. J., 2007.
[32] Allergy and Asthma in the southwestern Unites States. Web site, Arizona, USA http://allergy. peds.arizona.edu/southwest/weeds, content owner Schumacher, M. J., 2007.
[33] Oliveira, M., Ribeiro H., & Abreu, I., Annual variation of fungal spores in atmosphere of Oporto: 2003. Ann Agric Environ Med. 12, pp. 309–315, 2005.
[34] Boral, D., & Bhattacharya, K., Aerobiology, allergenicity and biochemistry of three pollen types in Berhampore town of West Bengal, India. Aerobiologia 16, pp. 417-422, 2000.

Impact of road traffic on air quality at two locations in Kuwait

E. Al-Bassam[1], V. Popov[2] & A. Khan[1]
[1]*Environment and Urban Development Division, Kuwait Institute for Scientific Research, Kuwait*
[2]*Wessex Institute of Technology, Southampton, UK*

Abstract

Kuwait having one of the highest GDP and the least fuel price provides ideal opportunity for ownership of motorized vehicles. Weather has also a major role in this issue where in long summer (lasting about nine months), temperature sores to nearly 50 °C very often and in short winter it drops to single digit value in early mornings and nights. The road transport is vital for the local inhabitants as the sole means of transport (commuting and transporting goods).

In the last decade, motorized road vehicle fleet has grown significantly bringing unprecedented mobility to the burgeoning population. With the growth of vehicles, the fuel consumption has also increased. Motor vehicles are a critical source of urban air pollution (PM_{10}, CO, CO_2, NOx, O_3, SO_2 and VOCs). Air pollution is a serious health problem and accounts for hundred of millions of dollars for health care and welfare cost.

This paper focuses on environmental impact of road transport on the air quality around two selected schools for a period of two weeks each using an air pollution monitoring station which continuously recorded various pollutants' concentrations and meteorological variables in five minute intervals. The results show that for both sites during the weekdays, the measured pollutants emitted from the road traffic next to the selected schools, such as carbon monoxide (CO) and nitrogen dioxide (NO_2), were always under the allowable limits for Kuwaiti air quality standards, except for a single occurrence for NO_2 concentration at morning hours for the governmental school. On the other hand, the values of non-methane hydrocarbon pollutants were found to be several times above the Kuwaiti air quality standards throughout the investigated period. The suspended particulates (PM_{10}) concentrations have twice exceeded the limits of Kuwaiti air quality standards.

Keywords: congestion, pollutants, air quality standards, schools.

WIT Transactions on Ecology and the Environment, Vol 123, © 2009 WIT Press
www.witpress.com, ISSN 1743-3541 (on-line)
doi:10.2495/AIR090171

1 Introduction

The transport sector is considered a major cause of the high level of air pollution in urban regions which affects the environment and ultimately, the human health. Air quality expert group has shown that pollutants emitted from the traffic have a big effect on human health [1]. A German paper mentioned that based on many studies it was shown that particulate matter causes damage to health and is responsible for increased rates of mortality and to lung cancer [2]. Amato et al. [3] mentioned that the frequency of allergic respiratory symptoms is higher in urban areas due to vehicle traffic than in rural areas. A Dutch study found that respiratory disorders of 632 children aged 7 to 11 worsened as air pollution increased [4]. Guo et al. [5] have reported that over 331,000 middle school children in Taiwan found that traffic related air pollution especially, carbon monoxide and nitrogen dioxides, were positively associated with the prevalence of asthma.

According to the Ministry of Education (MOE) in Kuwait statistical data and Ministry of Planning statistics, the school buses are serving approximately 17 to 18% of students in the government schools. Based on 2003/2004 statistics, there are 23,302 students using buses out of 131,597 total students. The rest of students mostly depend on private transportation.

2 Study area

This study deals with two schools located in flat and homogeneous terrain with out any major local air pollution emission sources except road traffic. One of these schools is located next to Muthana road at Hawalli Area and the other school is located next to Road 57 at Mishref Area. Muthana road is surrounded by many residential buildings, grocery stores, bookshops and schools. The school is private and has an area of 67,000 m^2. It is surrounded by four roads, from the east Muthana major road and the main gate for the school opens at this road for secondary and middle school students, teachers' cars and buses. At the north, there is a narrow road with one lane having traffic flow in both directions but from 6:00 to 9:00hr and 13:00 to 15:00hr, it accommodates only unidirection traffic flow. This road intersects with Muthana two-lane road without traffic signals. In this narrow road the gate of the school opens for elementary class children. There is a narrow lane in the west that has another gate available for elementary and kindergarten students. From south, there is another narrow lane parallel to a main freeway called Fourth Ring Road. The Fourth Ring Road is a vital and essential freeway consisting of three lanes for each side and the traffic volume of the road is extremely high at weekdays. The other school at Mishref is a governmental school with no school buses at morning or afternoon time for students. It is surrounded by Road 57 from north and from the east there is another school under construction and other governmental schools. From west and south, the school is surrounded by a street with two lanes having traffic in each direction. The school area is 18,000 m^2 and has a parking in the front of the entrance of the school gate.

3 Monitoring

Air quality and weather data were recorded at sampling intervals of 5 minutes from 18 February to 18 March 2006 for both sites by Kuwait Institute for Scientific Research (KISR), using its mobile air monitoring station. The measured data included the concentration of different pollutants such as methane, CH_4 and non methane hydrocarbons, nitrogen oxide (NO), nitrogen dioxide (NO_2), carbon monoxide (CO) and PM_{10}. In addition, the measured data included wind speed, wind direction, solar radiation, humidity and ambient temperature. The monitoring station was parked on Muthana road almost 100 m from the main gate of the selected private school. Later it has been moved to the other location at Mishref area next to the governmental school entrance. A road side counter was used to register the number of cars on the roads around the chosen schools for every 15 minutes throughout this study period. The measurements were taken during the months of February and March including the weekdays, weekends and national holidays.

4 Methodology

Air quality and weather data were recorded at sampling intervals of 5 minutes by Kuwait Institute for Scientific Research (KISR) air monitoring station as shown in figure 3 for two weeks. The measured data included the concentration of different pollutants such as carbon monoxide (CO), carbon dioxide (CO_2), methane (CH_4) and non-methane hydrocarbons, nitrogen oxides (NO_x), nitrogen dioxide NO_2, and suspended particulates (PM_{10}). In addition, the measured data included wind speed, wind direction, solar radiation and ambient temperature. The monitoring station was parked in Hawalli area next to the private school for two weeks approximately then it moved to Mishref area next to the governmental school entrance. A traffic counter was used to record the number of cars in the roads next to the selected schools for every 15 minutes throughout the study period. The measurements were taken during February and March 2006 including weekdays and weekend holidays.

5 Discussion and results

5.1 Traffic

The hourly average weekday and weekend traffic flow profiles are shown in Figures 1 and 2 for the roads next to the selected schools. The profile of the traffic indicates two distinct peaks for both sites during the working days which are related to the schools opening time and employees going to work and noon time returning from schools and work for the students and employees respectively. At weekends there are no conspicuous peaks, and traffic flow gradually increases followed by slight decrease at afternoon time, then with minor increase in the evening.

During peak traffic times, there was larger traffic flow near the governmental school than private school due to the occurrence of many schools and less traffic network cluster in Mishref area.

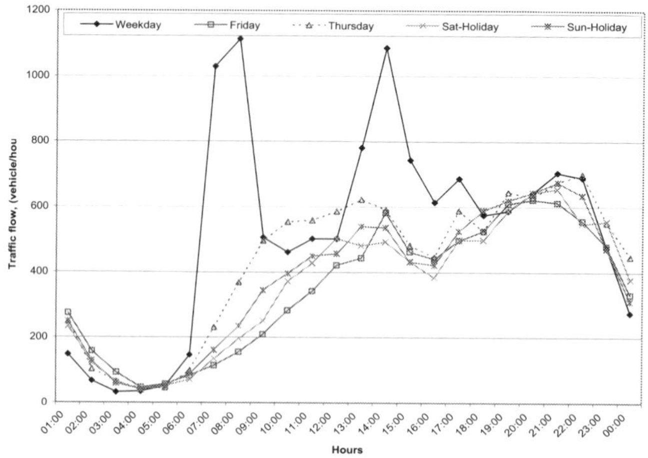

Figure 1: Hourly traffic flow for weekdays and weekend days next to the private school.

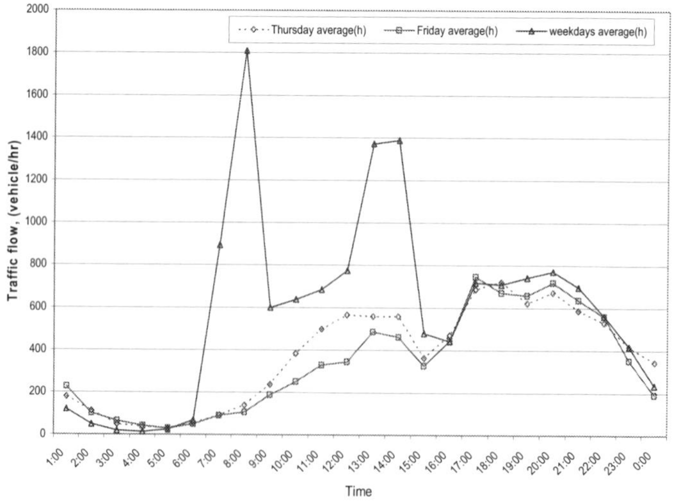

Figure 2: Hourly traffic flow for weekdays and weekend days next to the governmental school.

5.2 Measured pollutants

The concentrations of primary pollutants such as CO, NO_2 and PM_{10} were recorded from 18 February to 18 March 2006 for both sites. The profile for CO concentration has distinct variation, the highest concentration recorded in the morning hours and the second highest in the afternoon hour as shown in Figs. 3 and 4 for private and governmental school respectively.

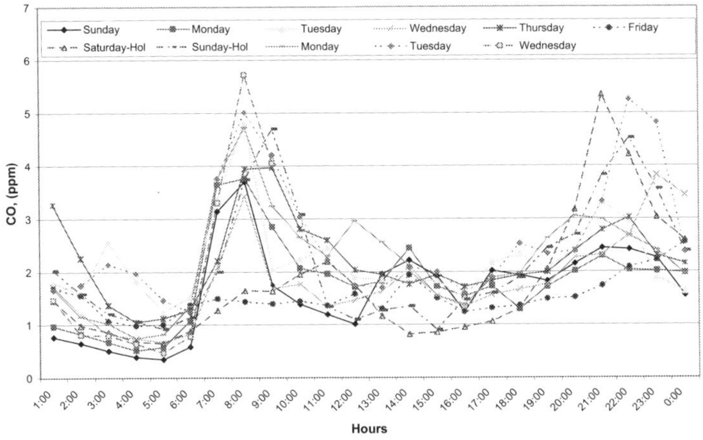

Figure 3: Average daily variation of CO emissions next to the private school from 19 February to 1 March 2006.

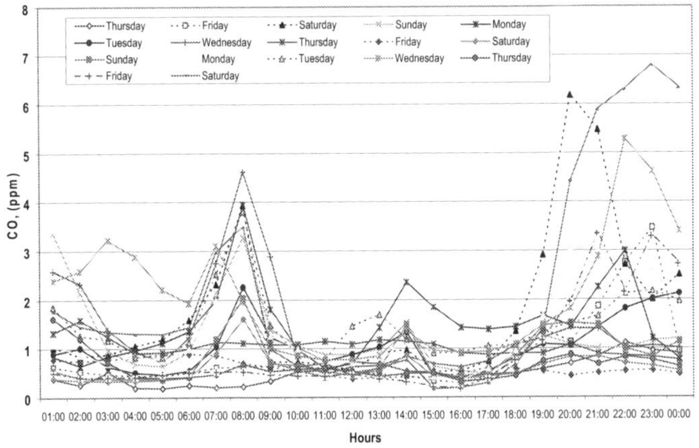

Figure 4: Average daily variation of CO emissions next to the governmental school from 2-18 March 2006.

On the contrary to traffic flow, CO concentrations were slightly higher at private school than the governmental school in morning opening hours and noon closing hours with the exception of late night hours, where high concentrations were recorded at the governmental school. The high concentrations were due to traffic congestion, road layout, local terrain and prevailing meteorological conditions. There is a conspicuous weekend effect reflecting low concentrations on Thursdays and Fridays at both sites.

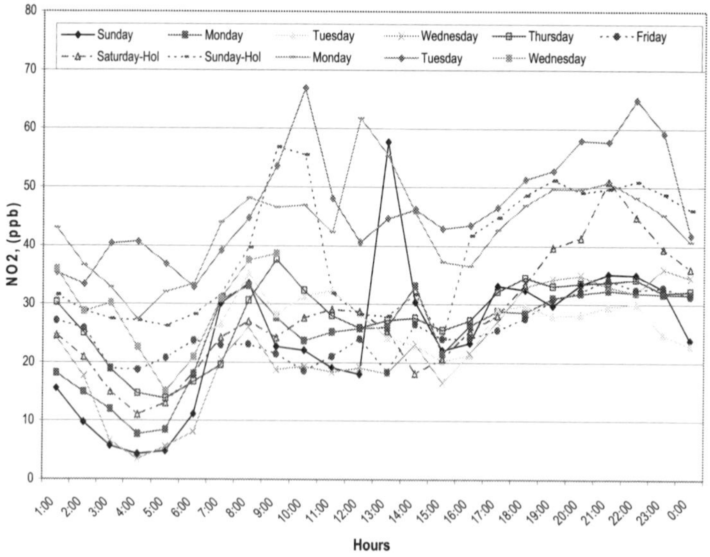

Figure 5: Average daily variation of NO_2 emissions next to the private school from 19 February to 1 March 2006.

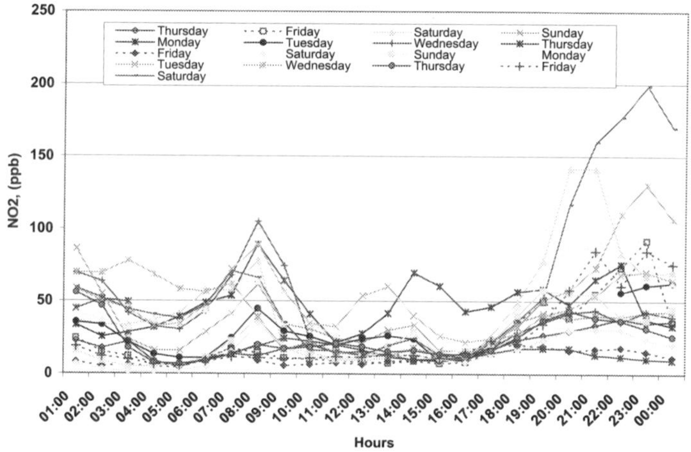

Figure 6: Hourly variation of NO_2 for different days near the governmental school.

Figs. 5 and 6 show the NO_2 profile for the two sites. The lowest concentrations of NO_2 were recorded on Thursday and Friday clearly indicating low traffic flow on weekend at both sites. NO_2 is a major constituent of NOx and NOx profiles are identical to NO_2 with a multiple factor. NO_2 and NOx concentrations were higher at government school than private school, reflecting the traffic density, where the influence of traffic flow was not that much pronounced (Lee et al., 2003). The diurnal trend of NO_2 was obvious from Figs. 5 and 6 showing an increase in morning hours followed by another smaller peak in the afternoon and later another increase in evening, depicting shopping and other recreational activities at both sites.

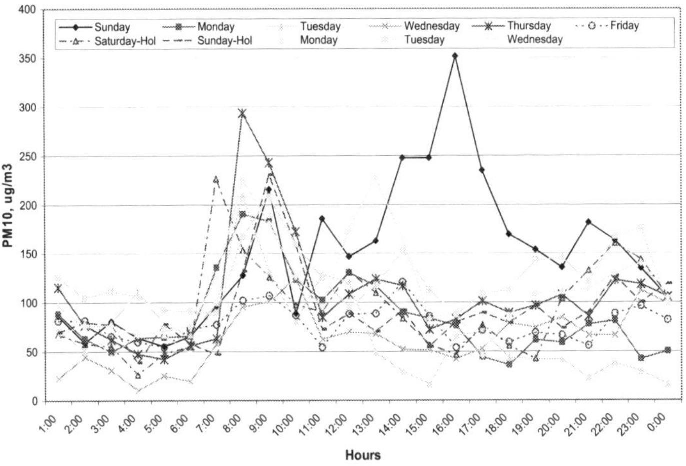

Figure 7: Hourly variation of PM_{10} for different days near the private school.

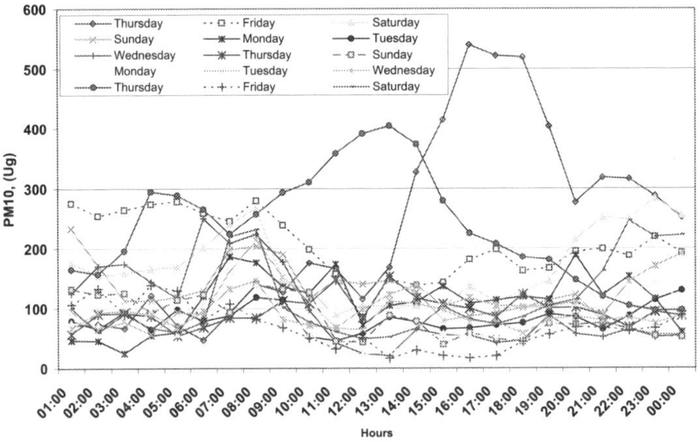

Figure 8: Hourly variation of PM_{10} for different days near the governmental school.

The concentration of particulate matter, PM_{10} is presented in Figs. 7 and 8 for private and government schools. PM_{10} concentration are dependent on prevailing meteorological conditions due to arid climate in the State of Kuwait and existence of high frequency of dust storms that overwhelmingly suppressed the contribution of traffic emission as obvious from Fig. 8. Fig. 7 shows the diurnal variation of PM_{10} concentrations representing high activities in morning hours of week days.

5.3 Air quality

All the measured pollutants' concentrations in the vicinity of the selected schools for a period of four weeks were compared with the allowable levels according to Kuwait's air quality standards. The Air Quality Standards (A.A.Q) in the residential areas for Kuwait, Federal US and California states are presented in Table 1. The concentration level of CO, NO_2, and PM_{10} pollutants are shown in Figures 3 to 8. The CO concentrations are always under the allowable limits. The average NMHC concentrations are always above the specified limits as shown in figure 9. NO_2 concentration had exceeded the allowable limits 15 times during the study period near the governmental school. The NO_2 exceedances are mainly due to road traffic since these values were associated with the increase of CO levels. Regarding PM_{10} levels it has exceeded the limits of A.A.Q (on two occassions) during the time of recording.

Table 1: The hourly air quality standards for Kuwait, Federal US and California State.

Pollutant	Kuwaiti Standards	Federal Standards	EU standards
Ozone (ppb),[$\mu g/m^3$]	80, [157]	120, [235]	90, [180]*
CO [ppm]	30	35	10** mg/m^3
NO_2 [ppb]	100	-	200*** $\mu g/m^3$
PM_{10} [$\mu g/m^3$]	350 (24 hours)	150 (24 hours)	50 (24 hours)
NMHC [ppm]	0.24 for a period of 3 hours (6-9 AM)	-	-

Source: [6].

6 Conclusion and recommendation

The air quality in the vicinity of two schools: one governmental and the other private school have been assessed for a period of four weeks using KISR's Mobile Air Quality Monitoring Station. It is found that the air quality in the vicinity of both schools is not adequate for the children due to high levels of non methane hydrocarbon almost all the times above the Kuwaiti EPA limit and also high NO_2 levels showing exceedence sometimes. These pollutants are mainly due to oil related activities, oil production, refineries and other petrochemical industries, power station, fuel pumping stations, road traffic etc. It is important to maintain high standards of air quality around the schools in order to reduce the effect of traffic pollutants on health of children and their performance. High levels of pollution and traffic conjunctions are recognized as health risk. The Kuwait government should consider public transportation for the governmental schools students to abate traffic conjunction and associated air pollution problems in the country.

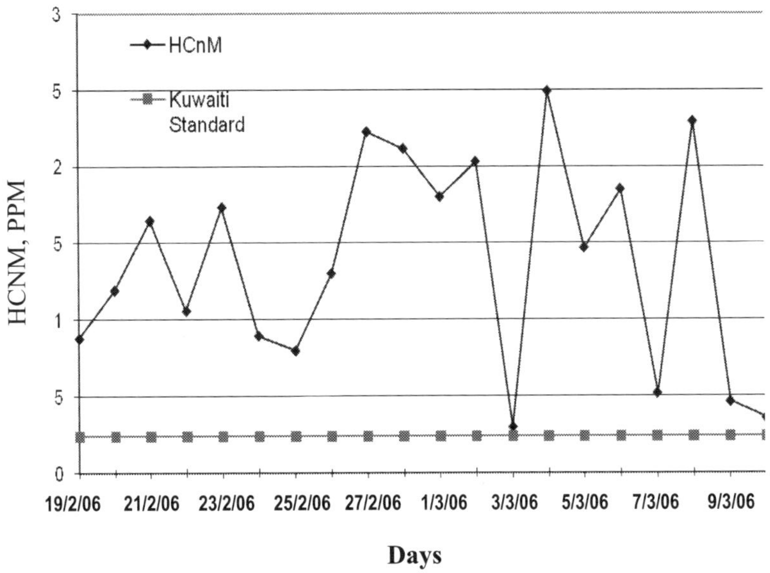

Figure 9: Daily variation of the measured non-CH4 hydrocarbon around the private and government schools compared to Kuwaiti standards.

References

[1] Air quality expert group, 2007. "Air quality and climate change-UK Perspective Summery". Published for Environment, Food, and Rural Affairs; Scottish Executive: Welsh assembly Government and Department of the Environment in North Ireland. Crown copyright 2007.

[2] Uwe Lahl, Wilhelm Steven, 2005, "Traffic guidance and restrictions-Legally permissible options for air quality control policy". Federal Ministry for the Environment, Nature Conservation and Nuclear Safety (BMU), Bonn, Germany.

[3] G. D'Amato, G. Liccardi, M. D'Amato and M. Cazzola, 2002. European Respiratory Journal, 2002; 20:763-776.

[4] BBC NEWS. Health: Medical notes. "Exhaust emissions". 6 May, 1999.

[5] Guo,Y.L. et al., 1999. Climate, traffic-related air pollutants, an asthma prevalence in middle-school children in Taiwan. Environ Health Prospect, 107(12):1001-6.

[6] Al-Kuwait Al-Youm. 2001. Annexure No. 533, 2 October 2001, year 47. Ministry of Information (Arabic).

Remote sensing study of motor vehicles' emissions in Mexican Cities

A. Aguilar, V. Garibay & I. Cruz-Jimate
DGICUR, Instituto Nacional de Ecología, Mexico

Abstract

According to the North American Free Trade Agreement (NAFTA), signed by Canada, the USA and Mexico in 1992, beginning on January 1st 2009, Mexico may not maintain restrictions on the importation of used cars older than ten years. The increased influx of used vehicles might cause significant changes in the composition of the country's vehicle fleet and this might increase its contribution to air emissions. Due to this situation and the lack of reliable information for decision-making, in 2007 the National Institute of Ecology of Mexico carried out studies of emissions, activity and composition of the vehicular fleet in the Mexican cities of Mexicali and Tijuana, in Baja California State, which share a border with the USA. Measurements were carried out with an AccuScan RSD3000 remote sensing system for on-road vehicle emissions, to obtain concentrations of carbon monoxide, hydrocarbons and nitric oxide (CO, HC and NO) from exhaust fames. To determine the activity, composition and technological characteristics of vehicles, surveys were conducted and analyses of databases were made. The results show that the average of CO and HC from Mexicali and Tijuana are lower than in the Mexico City Metropolitan Area (MCMA), while the average for NO is higher. This study is the first effort conducted by Mexican environmental authorities to document the impact of imported uses cars in emissions.
Keywords: remote sensing, exhaust emission, vehicle fleet characteristics.

1 Introduction

Emissions inventories are a basic tool to identify the sources of emissions that impact air quality and therefore, these tools are used to design suitable strategies to tackle air quality deterioration [3]. According to the National Emissions

WIT Transactions on Ecology and the Environment, Vol 123, © 2009 WIT Press
www.witpress.com, ISSN 1743-3541 (on-line)
doi:10.2495/AIR090181

Inventory of Mexico-1999, transport is the main source of anthropogenic emissions of nitrogen oxides (NOx) and volatile organic compounds (VOCs), which are ozone precursors [2]. These sources also release particulate matter (PM) and carbon monoxide. Although the emissions come from fossil fuel (e.g. petrol, gas and diesel) use, there are other factors that determine the amount of emissions from a vehicle such as the technology, the use and driving modes as well as maintenance [5].

Additionally, according to the North America Free Trade Agreement (NAFTA), appendix 300-A.2, starting in January 2009, Mexico may not maintain a ban or restriction on the importation of vehicles manufactured in Canada or the U.S. which are at least 10 years old [14]. When restrictions are removed, it is expected that these vehicles produce significant changes in the vehicular fleet in Mexican cities and, consequently, increase their emissions. Vehicle sales reported by the Mexican association of Vehicles Distributors (AMDA) and reports from the Mexican Tax Management Service (SAT), belonging to the Ministry of Treasury and Public Credit (SHCP), show that from October 2005 to January 2007, in average, two million imported vehicles from Canada and U.S. were legalized. This yields a ratio of two second-hand imported vehicles for each new car sold in Mexico. These vehicles are additional to the non-registered vehicles that were introduced into Mexican territory before 2005 and that are called "*chocolate*", from which there is no information available [11].

Based on this, the Ministry of the Environment and Natural Resources of Mexico (SEMARNAT) is implementing a strategy to reduce the impact of the emissions coming from these vehicles on the air quality of border cities. This strategy includes the vehicular emission verification program, aimed at controlling and reducing on-road vehicle emissions. However, information on vehicle fleet characteristics and emissions levels is necessary to implement these programs.

Due of the lack of reliable information for decision-making in this regard, in 2007 the National Institute of Ecology of Mexico carried out a study to measure on-road emissions from vehicles in Mexicali and Tijuana. A remote sensing device (AccuScan RSD3000) was used to monitor and measure gas concentrations of CO, HC and NO, as well as speed and acceleration. This equipment also takes a picture of the license plate, which allows further vehicle identification and characterization [4]. Finally, an integral analysis of the emissions and driving modes were carried out [6].

2 Experimental section

2.1 Vehicular emissions

Measurements were carried out in 2007, from the 4th to the 10th of October in Mexicali and from the 11th to the 17th of October in Tijuana, monitoring from 9:00 to 14:00 hrs. using an AccuScan RSD3000 system from Environmental System Products, CT, USA. The instrument is built up of a non-dispersive infrared component for CO and HC detection, and a dispersive ultraviolet spectrometer for NO measurements. The RSD process of remotely measuring emissions begins with the light source projecting infrared (IR) and ultraviolet (UV) beams across the road (approximately 15-28 ft) and the Corner Cube Mirror (CCM) returning the transmitted light to a series of detectors. Fuel specific concentrations of HC, CO and NO, as well as smoke in the vehicle exhaust are calculated based on the absorption bandwidth of 3.3, 4.6 and 4.3μm. of IR/UV light [4,10]. During this process, the camcorder system captures and stores the image of the license plate, which can be used to obtain the technical and technological data from the monitored vehicles (e.g. model year, brand, sub-brand, motor size, etc.) [4]; simultaneously, the speed/acceleration sensors (S/A) record the speed and acceleration of the vehicle. Vehicles passed through the selected sites by restricting circulation to one lane, to obtain vehicle speeds of 30 to 40 km/h, [6].

The gas analyzer was calibrated daily with a mixture of certified gases, (CO, propane and NO). The HC measurements were expressed in terms of the "n-hexane ppm equivalent". The CO tolerance was 10% or 0.25% (whichever was greater) for all expected concentrations below 3.0%, and 15% for all CO expected concentrations above 3.0%. In the case of HC, the tolerance was 150ppm or 15% of the expressed HC concentration (whichever was greater) throughout the range of HC concentrations. The NO tolerance was 250ppm or 15% of the expected NO concentration (whichever was greater) throughout the range of the NO concentration [5,6].

To determine the activity, composition and characteristics of the vehicles, surveys were applied. Ten sites were selected in both cities to ensure that they are representative of different socioeconomic strata and diversity of land use.

2.2 Vehicular characteristics and vehicular activity

The aim of the surveys was to determine the size of the vehicular fleet in use and its composition in both cities, distinguishing them by their origin (*national* or *imported)*, as well as their characteristics (model year and vehicle type). The surveys were applied to the drivers at fuel stations, and direct counting was carried out on roads. Complementary data was collected from sales statistics. Surveys were carried out from the 5th to 9th of December 2007 in Mexicali and from the 10th to the 13th of December 2007 in Tijuana, area over the premise that those sites would be in the surrounding were the RSD3000 was installed.

3 Results and discussion

Table 1 shows the total number of readings, as well as the valid readings and readings found per site and per city. 25 369 readings of vehicular emission in total were collected in Mexicali, 5 692 were valid readings (22%). While in Tijuana, 19 951 were collected, from these readings, 4 871 (24%) were valid.

The valid readings were considered those readings that contained data from emissions, speed and acceleration.

Table 1: Number of readings per sampling site per city.

City	Date	Site	Readings	Valid readings All	Valid readings With plate	Valid readings Without plate
Mexicali	04-Oct-07	Av. Colón	4 035	685	470	215
	05-Oct-07	Av. Universidad	3 504	583	385	198
	06-Oct-07	Av. Fco. Montejano	3 177	761	521	240
	07-Oct-07	Blvd. Anáhuac	3 408	849	611	238
	08-Oct-07	Av. Río Nuevo	3 628	1 007	654	353
	09-Oct-07	Calle Novena	3 605	666	454	212
	10-Oct-07	Av. Colón	4 012	1 141	766	375
		Total	25 369	5 692	3 861	1 831
		Readings (%)		22	15	7
		Valid readings (%)			68	32
Tijuana	11-Oct-07	Vía Rápida Poniente	5 010	1 092	700	392
	12-Oct-07	Blvd. Casa Blanca	3 928	1 195	735	460
	13-Oct-07	Blvd. Aeropuerto	2 436	529	271	258
	14-Oct-07	Vía Rápida Poniente	1 480	160	83	77
	15-Oct-07	Libramiento Sur	5 091	1 573	877	696
	16-Oct-07	Paseo Ensenada	2 006	322	186	136
		Total	19 951	4 871	2 852	2 019
		Readings (%)		24	14	10
		Valid readings (%)			59	41

Ideally, the measurement of vehicular emissions should be carried out when the vehicle experiences a light increment in the speed, preferable on a road whit a slight slope [8]. In the case of CO and especially for HC, emissions can increase with the load on the engine from moderate to high, that is, when the vehicle increases its speed from moderate to high. HC emissions can also increase during decelerations [9]. NO is primarily formed in the post-flame of the gases during the combustion process into the cylinder engine. The kinetic of this reaction is highly dependant of the temperature of the gas; high temperatures favour the rate of formation of NO [7]. Nitrogen oxide emissions can also be increased when there is a load on the engine and with a high air/fuel mixture [1].

Regarding pollutants emissions, the average CO volume percentage was 0.70% for Mexicali and 0.83% for Tijuana while HC emissions for Mexicali were 132 ppm and 130 ppm for Tijuana. On the other hand, the behaviour of NO emissions for both cities was, for Mexicali 1,200 ppm and Tijuana 966 ppm. Table 2 shows the summary of the average, standard deviation and confidence interval of the emissions data as well as speed and acceleration for both cities.

On the emissions report the vehicular fleet was classified based on the technological stratum that represents the technologic characteristics and emission control system. For the 1981-1990 stratum, these vehicles did not have emission control systems; while for stratum 1991-1992, when legislation for control emission started, these vehicles had two-way catalytic converters installed.

Table 2: Summary of speed, acceleration and emissions per city.

City (No. valid data)	Parameter	CO (vol %)	HC (ppm)	NO (ppm)	Speed (km/h)	Accel. ($km\ h^{-1}\ s^{-1}$)
Mexicali (5 692)	Average	0.70	131.85	1,200.86	29.65	2.52
	SD	1.39	170.44	1,170.33	6.08	1.19
	±95% CI	0.12	14.17	97.27	0.51	0.10
Tijuana (4 871)	Average	0.83	129.84	965.79	28.61	2.07
	SD	1.60	209.33	892.80	6.55	1.14
	±95% CI	0.14	18.81	80.21	0.59	0.10

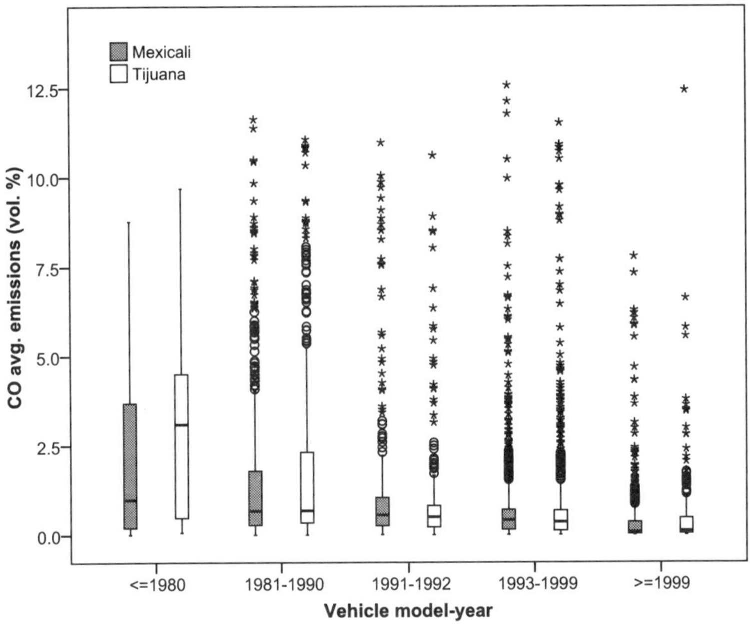

Figure 1: CO emission per vehicular stratum and per city.

Finally, from the 1993 model year vehicles were equipped with three-way catalytic converters and electronic fuel injection.

Based on this classification, emissions of CO (% vol.), HC (ppm) and NO (ppm) per vehicular stratum and per city are presented in fig. 1-3. It is observed that the mean vehicular emissions per stratum show a decrease trend that can be explained by the improvement in the emission control systems. The average values of CO and HC were very similar in Mexicali and Tijuana, while NO was higher in Mexicali.

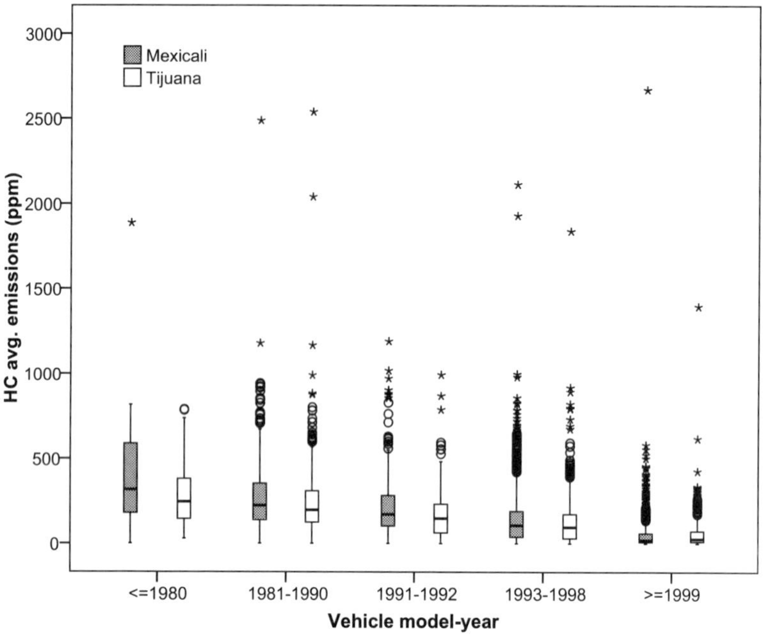

Figure 2: HC emission per vehicular stratum and per city.

Regarding data dispersion, observed even in the technological strata (1983-1999 y >=1999) that has the best control emission technologies, the major data dispersions are found in the model year 1998 and earlier, that is more than 10 years of age. That largest data dispersion can be associated with the age, the vehicle's maintenance, the driving modes and the lack of an inspection and emission control program. In Mexicali and Tijuana there are no inspection and maintenance program, which is reflected in the significant dispersion of emission values.

The survey conducted in Mexicali and Tijuana, show that 80.7% of on-road vehicles in Mexicali were originally purchased, as new vehicles, in the U.S., while in Tijuana it represents 80.9% [12], which means that for every purchased vehicle in Mexico there are 4 imported used vehicles.

The composition of the vehicular fleet is shown in fig. 4 along with a comparison between Mexicali, Tijuana and MCMA, where there is less penetration of foreign vehicles. It is observed that in Mexicali, 55% of the fleet is older than 10 years, while in Tijuana 63%. In contrast, MCMA 41% of its fleet is older than 10 years. Therefore, MCMA has a higher percentage of vehicles on the last stratum (>=1999).

When the average emissions were compared per city and type of vehicle, in general, in Tijuana CO emissions were higher than in Mexicali, especially for pick-up while HC emissions were higher for SUV/VAN in Mexicali compared to other type of vehicles in both cities. In Mexicali, NO emissions were always higher than in Tijuana.

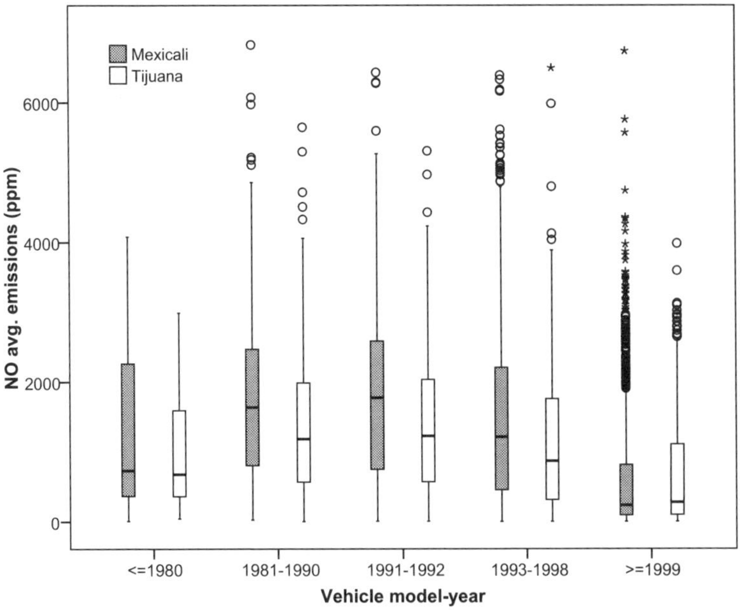

Figure 3: NO emission per vehicular stratum and per city.

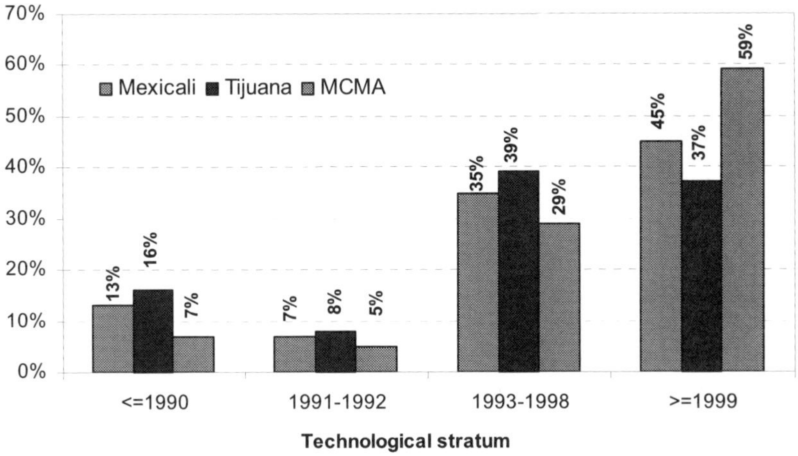

Figure 4: Vehicular stratum classification.

Comparing mean emissions of the three pollutants in the three cities, similar values are observed for CO and HC, while NO is higher in Mexicali, this result match with the ones found in table 3 for NO. However, according to preliminary results of an RSD study carried out in the MCMA in 2006, which remains unpublished [13], it is expected that the data dispersion would be minor in the MCMA compared with Mexicali and Tijuana.

Table 3: Summary of speed, acceleration and emissions per city.

City	Type of vehicle/(n)	CO (vol %)	HC (ppm)	NO (ppm)
Mexicali	Pick up (442)	0.75	132.20	1,320.65
	Sedán (2,509)	0.74	144.33	1,228.70
	SUV/VAN (910)	0.75	174.67	1,473.74
Tijuana	Pick up (384)	1.16	159.86	985.20
	Sedán (1 504)	0.79	135.92	1,041.23
	SUV/VAN (964)	0.92	138.88	1,125.49

Table 4: Mean emissions comparison between MCMA, Mexicali and Tijuana.

Location	CO (vol %)	HC (ppm)	NO (ppm)
MACM, 2005	0.88	156	965
Mexicali, 2007	0.70	132	1,201
Tijuana, 2007	0.83	130	966

4 Conclusions

Results of vehicular emissions and characteristics study show that the importation of used vehicles in Mexicali and Tijuana has a major impact on the composition of its vehicle fleet in terms of age, mechanical condition and vehicle type. This change has caused a great dispersion in the values of emission, reflecting the need to regulate the vehicle emissions. Moreover, based on the results of this study, Mexican environmental authorities have elements to document and implement actions to regulate the importation of used vehicles in Mexico.

References

[1] Peavy H.S. & Rowe, D.R., *Environmental Engineering*, McGraw-Hill: USA, pp. 563-565, 1996.

[2] INE, *Inventario nacional de emisiones de México, 1999,* SEMARNAT: México, pp. xxi-xxxi., 2006.

[3] Selman, P., *Environmental planning*, SAGE publications Ltd: UK, pp 261-262, 2000.

[4] RSD, *Correlation and Remote Sensing Device Application Results. A Summary of the Milestones Reached*, ESP-Environmental Systems Products: USA, full document, 2005.

[5] Shifter I., Díaz L., Mugica V. & López-Salinas E., Fuel-based motor vehicle emissions for inventory metropolitan area of Mexico City, *Atmospheric Environment,* **39**, pp. 931-940, 2005.

[6] Bishop, G.A., Stedman, D.H., De la Graza J. & Dávalos F., On-road remote sensing vehicles emissions in Mexico, *Environmental science & Technology*, **31**, pp. 3505-3510, 1997.

[7] Schifter, I., Díaz, L., Durán, J., Guzmán, E., Chávez, O. & López-Salinas, E., Remote Sensing Study of Emissions from Motor Vehicles in the Metropolitan Area of Mexico City, *Environmental Science & Technology* **37**, pp 395-401, 2003.

[8] Bohren, L., Profile of Vehicles at the Border Crossings between Tijuana, Mexico and San Diego, California. *Proc. of the 93rd Annual Conference & Exhibition*, eds. (CD version): Salk Lake City, pp. full document, 2000.

[9] McClintock, P.M. *The Colorado Enhanced I/M Program 0.5% sample Annual Report*; Prepared for The Colorado Department of Public Health and Environment: Tucson, AZ, January 27, 1998.

[10] EPS-Environmental Systems Products, *Remote Sensing Milestones, "US Experience"* Niranjan Vescio, ESPH August 12, 2004.

[11] SEMARNAT, *Importación definitiva de autos usados. Consecuencias e impactos ambientales*. Documento de trabajo. Subsecretaría de Gestión para la Protección Ambiental, Dirección General de Gestión de la Calidad del Aire y RETC, Dirección de Calidad del Aire, México, D.F. 2007.

[12] TSTES, S.A. de C.V., *Estudio de emisiones y características vehiculares en ciudades mexicanas de la frontera norte*. Documento de trabajo. Elaborando para el Instituto nacional de Ecología bajo contrato INE/ADE-077/2007, México, D.F. 2007.

[13] Personal communication, Gobierno del Distrito Federal, Secretaría del Medio Ambiente, 2008

[14] Appendix 300-A.2 North American Free Trade Agreement [NAFTA] signed the 17th of December of 1992, www.sice.oas.org/trade/nafta/naftatce.asp

Correlations between the exhaust emission of dioxins, furans and PAH in gasohol and ethanol vehicles

R. de Abrantes[1], J. V. de Assunção[2] & C. R. Pesquero[2]
[1]*Vehicular Emission Laboratory, Cetesb, São Paulo, Brazil*
[2]*School of Public Health, University of São Paulo, Brazil*

Abstract

The emissions of seventeen 2,3,7,8 substituted Polychlorinated Dibenzo-p-Dioxins, Polychlorinated Dibenzofurans (PCDD/Fs) and sixteen Polycyclic Aromatic Hydrocarbons (PAH) [14] in the exhaust pipes of spark ignition light duty vehicles considered toxic to human health were investigated. The formations of these compounds were evaluated under the influence of variations of fuels and fuel additives.

Standard tests in a gasohol (gasohol is pure gasoline plus 20% to 25% of anhydrous ethyl alcohol fuel (AEAF)) vehicle and in an ethanol vehicle were performed with variations in the quality of fuels. The sampling of the PCDD/Fs followed the recommendations of a modified 23 method and the analysis basically followed the 8290 method. The recommendations of the TO-13 method were followed for the PAH analysis, with the necessary modifications for a vehicular emission laboratory.

The emission factors of the total PCDD/Fs varied between undetected and 0.157 pg I-TEQ/km. The emission factors of the total PAH varied from 0.01 μg TEQ/km to 4.61 μg TEQ/km.

Significant and positive correlations were observed between the emissions of naphthalene, acenaphthylene, fluorene, phenanthrene, anthracene and fluoranthene and significant and negative correlations were observed between the emissions of CO_2 and fluoranthene in the gasohol vehicle. Significant and positive correlations between carbon monoxide and phenanthrene and between acenaphthylene, fluorene and fluoranthene in the alcohol vehicle were also observed, apart from significant and negative correlations between NOx and phenanthrene. In general way, significant correlations between PAH and PCDD/Fs were not observed, except in the ethanol vehicle considering phenanthrene.

Keywords: vehicular emissions, PCDD/Fs, PAH, air pollution, toxic pollutants, gasohol, ethanol.

WIT Transactions on Ecology and the Environment, Vol 123, © 2009 WIT Press
www.witpress.com, ISSN 1743-3541 (on-line)
doi:10.2495/AIR090191

1 Introduction

Vehicles are responsible for some pollutants that, due to their toxicities, can alter the morbidity and mortality rates of populations. The emissions of some pollutants, such as the PCDD/Fs and PAH from vehicles have the potential to cause damage to human health [9]. Some of these pollutants are carcinogenic to mammalians, even at very low concentrations [5, 16].

PCDD/Fs are formed in combustion process where chlorine atoms are present. The chlorine sources for PCDD/Fs formation can be fuels and fuels additives. Small amounts of chlorine may not be removed from the fuels during refining process. Moreover, chlorine compounds can also be added to premium gasoline to improve engine performance [15].

Information about chemical composition of fuel additives is furnished in just a generic way, but it is possible that organic chlorides can be mixed with fuels additives to improve engine performance [6, 7]; however the Brazilian Federal Administration must be informed about substances that can cause damage to human health [3].

In vehicles, PAH can be formed from incomplete fuel burning and from annealing aromatics rings. PAH emissions can also increase with vehicle aging, due to rising of lubricant oil consumption in the combustion chamber, caused by enlargement of the gaps in the engine's moving parts [10].

Considering the importance of these compounds in relation to human health, the aim of this work was to study the relations between PCDD/Fs and PAH, considered toxic to human health, in the exhaust pipe of spark ignition light duty gasohol and ethanol vehicles.

2 Materials and methods

2.1 Vehicle testing conditions

To carry out this study, two spark ignition light duty vehicles, equipped with catalytic converters and electronic injection systems, have been used, a gasohol vehicle (GV) and a flexible fuel vehicle (EV) fuelled just with hydrated ethyl alcohol fuel (HEAF). They were representative of the vehicle fleet in the State of São Paulo, which corresponds to 36% of the Brazilian fleet [2]. The vehicles characteristics are shown in Table 1.

The assays were carried out in the vehicular emission laboratory of the State Environment Agency (Cetesb) in São Paulo, Brazil. A standardized driving cycle in a chassis dynamometer was performed [1] that simulates the urban conditions. It is identical to the American driving cycle FTP-75 procedure (USEPA).

With the purpose of establishing possible correlations between PCDD/Fs and PAH, factors that could influence the emission of these pollutants, like rate of aromatics in the fuels and fuel additives, were varied under controlled conditions, in nine assays in the GV (from G1 to G9) and six assays in the EV (from A1 to A6). All these variations compose a 2^{3-1} fractional factorial design, which allows determination of the influence of each variable [4].

Table 1: Vehicle characteristics.

	Year	Mass (kg)	Motor (L)	Torque (k/gm at rpm)	Power (kW at rpm)	Odometer reading (km)
GV	1998	1111	1.6	15.1 at 4500	78 at 5500	67 546
EV	2004	1111	1.6	14.4 at 3000	73 at 5750	56 908

2.2 Collection and analysis

Sampling of the exhaust gas was performed during the entire working time of the vehicle. The collection of PCDD/Fs was carried out based on a modified 23 method [13] and in the work developed by RYAN and GULLETT [8].

Raw gas was sampled through a heated line, in order to collect PCDD/Fs. Solid phase was collected in two heated 70 mm diameter quartz fibre filters (120 °C), while the gaseous phase was collected in a cooled at 7 °C 60 mm diameter polyurethane foam (PUF), all of them assembled in series. The condensed material collected upstream the PUF was also sent for analysis. The extraction and analysis were done based in an adapted and validated 8290 method [12]. The PCDD/Fs were identified and quantified by HRGC/HRMS.

Diluted gas was sampled through a simple line, in order to collect PAH. The PAH were identified and quantified by GC/MS according an adapted TO-13 method [14]. The retention of solid phase was done by filtration; two quartz fibre filters with 47 mm in diameter were used, while a 22 mm diameter pre-washed PUF was used for retention of the gaseous phase, all of them assembled in series. Regulated pollutants were also quantified, according Brazilian legislation [1].

3 Results and discussion

The emissions rates of regulated pollutants obtained according Brazilian legislation, and CO_2, are shown in Figure 1. The values obtained are typical of vehicles with these mileages and technology of pollutants control. In the GV, the average emissions rates were 3.3 g/km of carbon monoxide (CO), 0.3 g/km of total hydrocarbons (HC), 0.5 g/km of nitrogen oxides (NOx) and 215.8 g/km of carbon dioxide (CO_2). In the EV, the average emissions rates were 0.7 g/km of CO, 0.3 g/km of HC, 0,3 g/km of NOx and 201.2 g/km of CO_2. In general way, EV emitted less regulated pollutants than GV.

The PAH emissions rates per distance travelled in the GV are shown in Figure 2. The average emissions rates were: 153.1 μg/km of naphthalene, 32.0 μg/km of acenaphthylene, 3.5 μg/km of acenaphthene, 5.3 μg/km of fluorene, 16.3 μg/km of phenanthrene, 3.5 μg/km of anthracene, 5.9 μg/km of

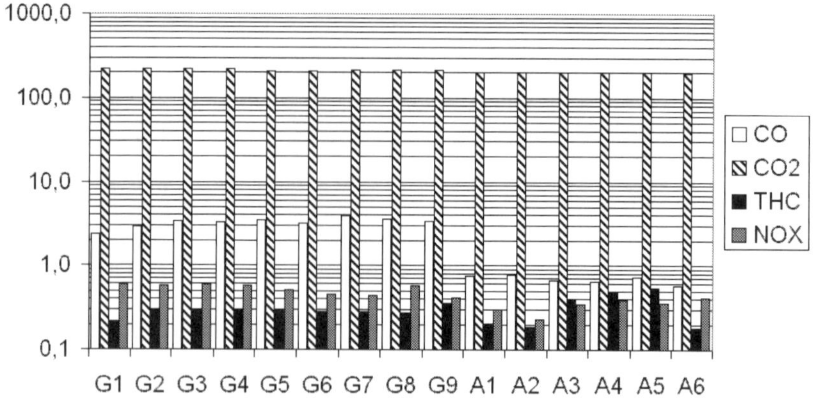

Figure 1: Regulated pollutants emission in the vehicles, in g/km.

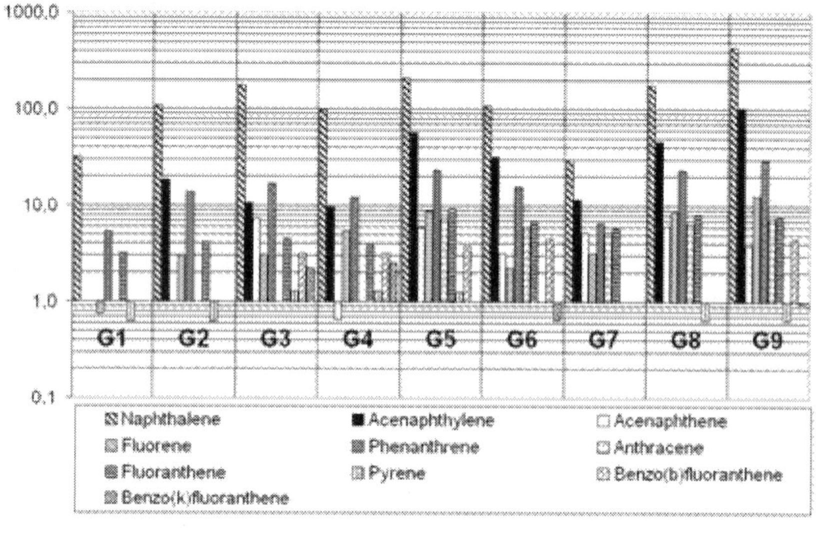

Figure 2: PAH Emission in the GV, in μg/km.

fluoranthene, 0.7μg/km of pyrene, 0.8 μg/km of benzo(a)anthracene, 2.1 μg/km of benzo(b)fluoranthene, 0.7 μg/km of benzo(k)fluoranthene. Benzo(a)pyrene was quantified in just one assay. chrysene, indeno(1,2,3-cd)pyrene, dibenz(a,h)anthracene and benzo(g,h,i)perylene were not detected. In general way, the higher PAH molecular weight represented lower emissions rates.

The PAH emissions rates per distance travelled in the EV are shown in Figure 3. The average emissions rates were 3.0 μg/km of naphthalene, 2.8 μg/km of acenaphthylene, 0.9 μg/km of acenaphthene, 0.6 μg/km of fluorene, 7.7 μg/km of phenanthrene, 4.0 μg/km of fluoranthene. Others PAH were not detected. The emission rates of PAH from EV are far lower than GV, in average 92% lesser. The PCDD/Fs emissions rates per distance travelled are shown in Figure 4. In the GV, the average emissions rates were 2.4 pg/km of 1,2,3,4,6,7,8 Hepta Chlorinated Dibenzo-p-Dioxins (HpCDD), 16.5 pg/km of Octa Chlorinated Dibenzo-p-Dioxins (OCDD). 1,2,3,4,6,7,8 Hepta Polychlorinated Dibenzofurans (HpCDF) was quantified in just one assay. Others PCDD/Fs were not quantified, and none of the 17 PCDD/Fs studied were detected in the G8 and G9 assays. In the EV, average emissions rates were 21.2 pg/km of OCDD. HpCDD was quantified in just one assay. Others PCDD/Fs were not detected. The emissions of PCDD/Fs showed large dispersions and non-regular behaviour.

3.1 Hierarchical cluster analysis

This procedure attempts to identify relatively homogeneous groups of variables based on selected characteristics, using an algorithm that starts with each variable in a separate cluster and combines clusters until only one is left.

The results of samples were submitted to Hierarchical Cluster Analysis (HCA) with the intention of determine the relations between the compounds. The variables were standardized by z-score method, before computing the proximities, because the magnitudes of results are very different. The association levels between the pollutants were defined using the Ward's method (method of minimum variance) as measure of similarity combined with Euclidean distance.

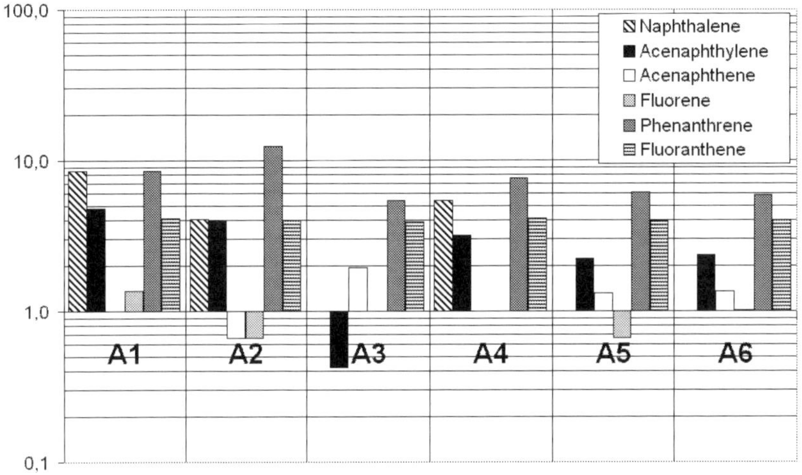

Figure 3: PAH emission in the EV, in μg/km.

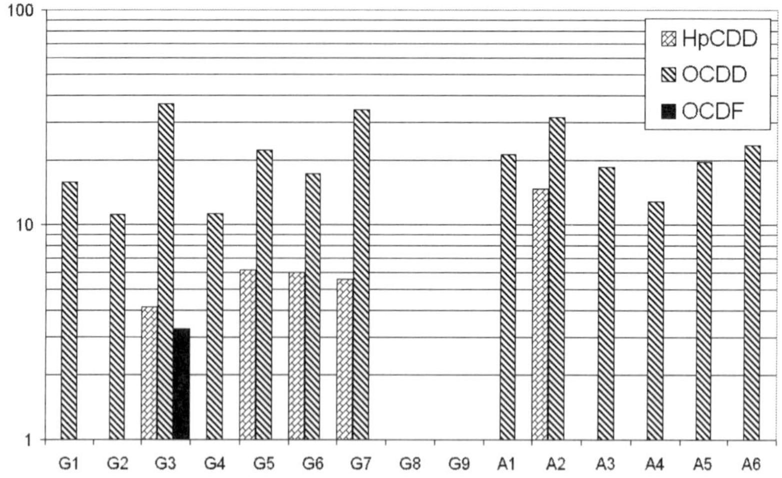

Figure 4: PCDD/Fs emission in the vehicles, in pg/km.

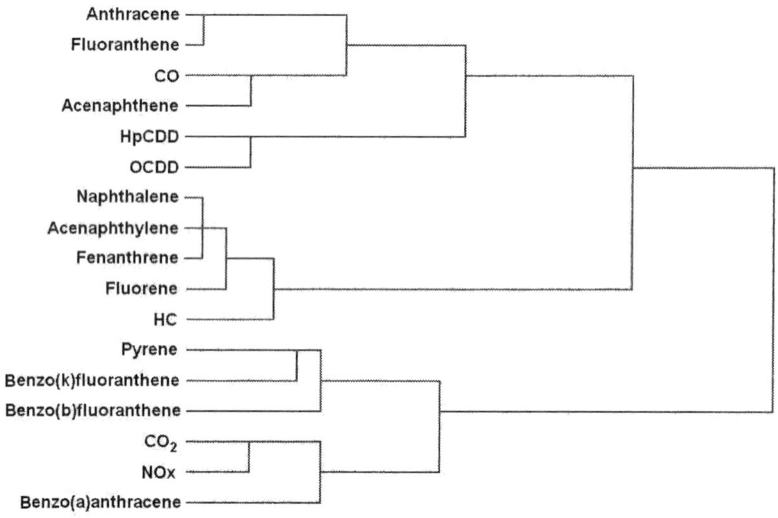

Figure 5: Dendrogram of GV obtained by Ward's method, using Euclidean distance.

The dendrogram in Figure 5 shows the results of HCA and the similarities level between the pollutants from GV. There are five groups of pollutants, the first one characterized by anthracene, fluoranthene, CO and acenaphthene, the correlation coefficient between them is not significant, except between anthracene and fluoranthene, which is 0.94. The second is the PCDD/Fs group that does not show similarities with others compounds.

The third group shows a relationship between HC and naphthalene, acenaphthylene, phenanthrene and fluorene, the correlation coefficients between all of them are very significant, above 0.7. The fourth group shows a relationship between pyrene, benzo(k)fluoranthene and benzo(b)fluoranthene, which were not detected in all assays, and the correlation coefficients are not significant. In the fifth group benzo(a)anthracene are related with CO_2 and NOx, however the correlation coefficients are also not significant. Significant and negative correlation coefficients of –0.87 are observed between the emissions of CO_2 and fluoranthene.

The dendrogram in Figure 6 shows the results of HCA and the similarities level between the pollutants from EV. There are three groups of pollutants, the first one characterized by CO_2, HC and NOx and acenaphthene, the correlation coefficients between them are not significant. In EV, HC not belongs to the light PAH group, because the main part of HC is not burned ethanol, different than occurred with GV, where the main part of HC is not burned gasoline.

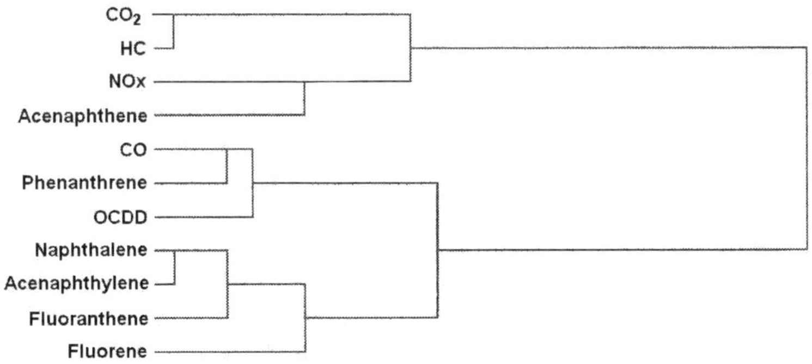

Figure 6: Dendrogram of EV obtained by Ward's method, using Euclidean distance.

Second group relates CO, phenanthrene and OCDD. The correlation coefficients between CO and phenanthrene is 0.71, and the correlation coefficients between phenanthrene and OCDD is 0.68, indicating some relationship between then, and in certain way shows the opposite than proposed by Stanmore [11], that suggests that "de novo" formation process of PCDD/Fs was mainly proportional to the number of phenanthrene skeletons, instead annealing of aromatic rings.

The main part of HC in GV is not burned fuel, what means a phenanthrene skeletons source, however, in GV, phenanthrene and PCDD/Fs correlation is not significant, even belong the same group, according Figure 5. On the other hand, PAH in EV are formed predominantly by annealing process, once there are no aromatics in the fuel, as there is correlation between phenanthrene and OCDD, exists the possibility that PCDD/Fs are also formed mainly by annealing of aromatic rings in ethanol combustion process.

Moreover, in the second group, the correlation coefficient between CO_2 and OCDD is -0.72, and the correlation coefficient between HC and OCDD is -0.72. The correlation coefficient between NOx and phenanthrene is -0.85.

Third group shows a relationship between light and medium molecular weights PAH, the correlation coefficient between acenaphthylene and fluoranthene is 0.73 and the correlation coefficient between fluorene and fluoranthene is 0.98.

The correlation coefficient between acenaphthene and acenaphthylene is -0.98, between acenaphthene and acenaphthylene is -0.98 and between acenaphthene and fluoranthene is -0.87, probably the reason by acenaphthene belongs another group, according Figure 6.

4 Conclusions

The two light duty vehicles used in these experiments were sources of PCDD/Fs, mainly OCDD, and PAH of light and medium molecular weight, high molecular weights PAH were not detected. In general way, with some exceptions, Hierarchical Cluster Analysis grouped the pollutants by their respective families, except in the case of ethanol vehicle where phenanthrene and OCDD have significant correlation coefficients, different than occurred in GV, what suggests that, in vehicles, PCDD/Fs are formed mainly by annealing of aromatic rings.

Acknowledgements

We want to express our gratitude to the Company of Environmental Sanitation Technology - CETESB, for the support given and to the Sao Paulo Foundation for Support to Research – FAPESP for providing the necessary financial support for this project (Grant 2004/02623-6).

References

[1] Associação brasileira de normas técnicas. NBR 6601: Veículos rodoviários automotores leves – Determinação de hidrocarbonetos, monóxido de carbono, óxidos de nitrogênio e dióxido de carbono e material particulado no gás de escapamento. (Regulation: Light self-driven vehicles: determination of hydrocarbons, carbon monoxide, nitrogen oxides, carbon dioxide and particulate matter in the exhaust pipe). Rio de Janeiro, 2005. 44 p.

[2] Associação Nacional dos Fabricantes de Veículos Automotores. Anuário estatístico da indústria automobilística brasileira 2006. (Statistical annual of the Brazilian automobile industry). [Report on line] Available at <URL: http:\\www.anfavea.com.br/Index.html> [accessed in 2007 jan. 26].

[3] Agência Nacional do Petróleo. Portaria n. º 41, de 12 de março de 1999. Establish the selling rules of automotive fuel additives and automotive additivates fuels. Diário Oficial da União. Brasília, DF, 1999 mar.15.

[4] Bruns, R. E.; Barros Neto, B.; Scarminio, I. S. Como fazer experimentos (How to make experiments). São Paulo: Unicamp, 2006. 400 p.

[5] Kogevinas, M. Human health effects of dioxins: cancer, reproductive and endocrine system effects. Human Reproduction Update. v. 7, n. 3, p. 331-9. 2001.

[6] Lubrizol. Ready reference for lubricant and fuel performance lubricant basics – lubricant Additives. Available at <URL: http://www.lubrizol.com/ReadyReference/LubricantBasics?lubeadditives.asp>. [Accessed in 2003 Oct. 14].

[7] Lubrizol. Lubricant properties and the role of additives. Available at <<URL: http://www.lubrizol.com/LubeTheory/prop.asp>. [Accessed in 2007 Jan. 15].

[8] Ryan e Gullett, J. R.; Gullett B. K. On-road emission sampling of heavy-duty diesel vehicle for polychlorinated dibenzo-p-dioxins and polychlorinated dibenzofurans. Environmental Science & Technology. 2000; v. 34: p. 4483–4489.

[9] Saldiva, P. H. et al. Air pollution and child mortality: a time-series study in Sao Paulo, Brazil. Environmental Health Perspectives. 109, p. 347–350, June 2001.

[10] Sher, E. (Ed). Handbook of air pollution from internal combustion engines. London: Academic Press Limited, 1998. 663 p.

[11] Stanmore, B. R. The formation of dioxins in combustion systems. Combustion and Flame. 2004; v. 136: p. 398–427.

[12] United States. Environmental Protection Agency. Method 8290: Analytical procedures and quality assurance for multimedia analysis of polychlorinated dibenzo-p-dioxins and dibenzo-p-furans by high-resolution gas chromatography/high-resolution mass spectrometry. Las Vegas: USEPA, 1994 September.

[13] United States. Environmental Protection Agency. Method 23: Determination of Polychlorinated Dibenzo-p-dioxins and Polychlorinated Dibenzofurans from Municipal Waste Combustors. USEPA, 1997. p.34

[14] United States. Environmental Protection Agency. Compendium method TO-13A - Determination of polycyclic Aromatic hydrocarbons (PAH) in ambient air using gas chromatography/mass spectrometry (CG/MS). Cincinnati: Center for environmental research information, 1999, p.78.

[15] World Health Organization. Chlorine and hydrogen chloride. Finland: 1982. 95 p. (International Programme on Chemical Safety: Environmental Health Criteria, 21).

[16] World Health Organization. Polychlorinated Dibenzo-para-dioxins and Dibenzofurans. Finland: 1989. 409 p. (International Program on Chemical Safety: Environmental Health Criteria, 88).

[17] World Health Organization. Selected non-heterocyclic polycyclic aromatic hydrocarbons. Stuttgart: 1998. 883 p. (International Programme on Chemical Safety: Environmental Health Criteria, 202).

Section 4
Monitoring and measuring

Development of an automated monitoring system for OVOC and nitrile compounds in ambient air

J. Roukos, H. Plaisance & N. Locoge
Département Chimie et Environnement,
Ecole des Mines de Douai, France

Abstract

Few studies have been conducted on Oxygenated Volatile Organic Compounds (OVOC) because of problems encountered during the sampling/analyzing steps induced by water in sampled air. Consequently, there is a lack of knowledge of their spatial and temporal trends and their origins in ambient air. In this study, an analyzer consisted of a thermal desorber (TD) interfaced with a gas chromatography (GC) and a flame ionization detector (FID) has been developed for online measurements of 18 OVOC in ambient air including 4 alcohols, 6 aldehydes, 3 ketones, 3 ethers, 2 esters and 4 nitriles. The main difficulty was to overcome the humidity effect without loss of compounds. Water amount in the sampled air was reduced by the trap composition (two hydrophobic graphitized carbons: Carbopack B/Carbopack X), the trap temperature (held at 12.5°C), by diluting (50:50) the sample before the preconcentration step and a trap purge. Humidity management allowed the use of a polar Lowox column in order to separate the polar compounds from the hydrocarbon/aromatic matrix. The breakthrough volume was found to 405 ml for ethanol by analysing a standard mixture at a relative humidity of 80%. Detection limits ranging from 10 ppt for ETBE to 90 ppt for ethanol were obtained for 18 compounds. Good repeatabilities were obtained at two levels of concentration (Relative Standard Deviation <5%). The calibration (ranging from 0.5 to 10 ppb) was set up at three different levels of relative humidity to test the humidity effect on the response coefficients. Results showed that the response coefficients of all compounds were less affected except for those of ethanol and acetonitrile (decrease respectively of 30% and 20%).

Keywords: OVOC & nitrile analysis, online measurement, analytical validation.

WIT Transactions on Ecology and the Environment, Vol 123, © 2009 WIT Press
www.witpress.com, ISSN 1743-3541 (on-line)
doi:10.2495/AIR090201

1 Introduction

OVOC play an important role in tropospheric chemistry especially in ozone, photooxidant and secondary organic aerosol formation [1]. OVOCs originate from both direct emissions by biogenic and anthropogenic sources, and secondary production from the oxidation of hydrocarbons [1, 2]. Few studies [4–8] have been dedicated to the measurement of OVOC in ambient air due to analytical problems. Consequently, there is a lack of knowledge of OVOC spatial and temporal trends and their origins in ambient air. These problems are mostly caused by water interferences. The studies have reported two distinct ways to overcome the humidity effect: methods that eliminate water from sample and methods that reduce relative humidity in the trap. The use of Nafion membrane dryer, cryogenic water trap or crystalline salt can eliminate water from the sample but lead to losses of OVOC. The warm trap method [9], dilution of sampled air by dry purified air [10] and the reduction of the sampled air volume [11] reduce the humidity from sampled air without OVOC losses.

The online measurement is adapted to study the evolution of OVOC concentration in ambient air thus providing clues on their origins. The most common method for the online measurements is sampling and preconcentration of compounds on solid adsorbents followed by a thermal desorption (TD) and gas chromatographic (GC) analysis. Two different approaches can be used for the separation and the detection of OVOC: (i) a non polar column [3, 4, 6, 7] coupled to mass spectrometry (MS) in order to resolve the co-elution problems (ii) a sufficient separation or a polar column [12] coupled to a non-specific detector, flame ionization detector (FID). The use of MS for field campaign is delicate and need attention for calibration thus we have chosen to use a polar column enabling to isolate the polar compounds from the dominant aliphatic/aromatic matrix coupled to FID [13]. This paper presents the analytical development of a method based on TD-GC-FID suitable for routine, near real time quantitative determination of 18 OVOC in ambient air including 4 alcohols, 6 aldehydes, 3 ketones, 2 ethers, 3 esters and 4 selected nitriles (Table 1). This

Table 1: List of the target compounds and their peak number on chromatograms (MVK: Methyl Vinyl Ketone, ETBE: Ethyl Tert-Butyl Ether).

Familly	compound	Peak number	Familly	compound	Peak number
Alcohols	Ethanol	14	Ketones	Acetone	11
	Isopropanol	15		MVK	12
	Butanol	19		2-Butanone	13
	Isobutanol	17	Ethers	ETBE	5
Aldehydes	Acetaldehyde	4		Furan	1
	Propanal	6		2-methyl furan	2
	Butanal	10	Esters	Ethyl acetate	12
	Benzaldehyde	20		Butyl acetate	18
	Acrolein	7	Nitriles	Acetonitrile	9
	Methacrolein	8		Pentanenitrile	16
Aromatic compound	Toluene	3		Heptanenitrile	21
				Octanenitrile	22

paper focuses on the optimization of the preconcentration and the analytical steps and also on the analytical performances.

2 Material description

2.1 Standard gas generation devices

Tests have been conducted using a standard mixture including the target compounds (Table 1), toluene and methanol in nitrogen (at 3 ppm) contained in a compressed gas cylinder which was prepared by PRAXAIR. A second mean was used to produce continuously various standard gaseous atmospheres from a mixture of pure liquid compounds which is vaporized and diluted by a multistage dynamic dilution. The mixture included the oxygenated compounds and was prepared gravimetrically. Toluene was added to the mixture in order to compare his generated concentration to the one in a compressed gas standard certified by the National Measurement Institute LNE (Laboratoire National d'Essais). A continuous gas generation was ensured by a liquid µ-flow meter (Bronkhorst High-Tech) which injects a known amount of the pressurized liquid mixture into a vaporization chamber. Successive dilutions with purified air were applied. This air is dried and chemically filtered in air purifier (AZ 2020 manufactured by Claind). All flows were controlled by a Mass Flow Controllers (MKS) and measured with a gas flow meter (DryCal DC-Lite) certified by LNE. At the last stage of dilution two controlled airflows reach a 5 L chamber the first is dry and the second comes from a humidifier filled with demineralised water to control the relative humidity of air. The temperature, relative humidity in the chamber are continuously monitored and recorded by means of multifunction probes (Data-logger Testo term 400 and temperature, humidity, wind velocity sensor 0635.1540). The system is able to generate from mg m^{-3} to ng m^{-3}.

2.2 Analytical instrumentation

Continuous air sampling was performed by Air server-Unity I from Markes International. Air samples are collected with a 15 mL min^{-1} flow on a one-stage trap that can be cooled by a Peltier system suitable for field campaign. Samples are then thermally desorbed and analyzed by GC-FID. For separation, a polar CP-LOWOX column (30 m × 0.53 mm × 10 µm) was specially designed by Varian for our application. The extended length of this column enabled us to have a very selective separation. The choice of the trap composition is closely related to the choice of the separation column. Preliminary tests on different adsorbents have shown that the use of a hydrophobic adsorbent is necessary to minimize the water interferences on the analysis with a polar column. Carbopack X, a hydrophobic adsorbent with high specific surface, has been chosen in the trap composition because of its low capacity to trap water [14]. In order to protect this adsorbent from the irreversible adsorption of heavy compounds, a small amount of Carbopack B has been added to the trap composition. The optimized composition of the trap was 5:75 mg Carbopack B: Carbopack X.

3 Optimization of analytical and preconcentration stage

3.1 Optimization of the chromatographic separation

The optimization was performed by analyzing the target compounds and 34 hydrocarbons ranging from 2 to 13 carbon atoms at 10 ppb (dry air) in order to demonstrate the good separation. The oven temperature was programmed from 45°C to 150°C at a rate of 5°C min^{-1} and held for 2 min, then ramped to 250°C at a rate of 3°C min^{-1} and maintained at 250°C for 10 minutes. The analysis lasts for 66 minutes. Figure 1-I shows the separation on the LOWOX column. The hydrocarbons are eluted in the first minutes of the chromatogram, while the polar compounds are eluted after undecane (after the 19th minute). One perfect co-elution persists between ethyl acetate and methyl vinyl ketone (Figure 1(II)).

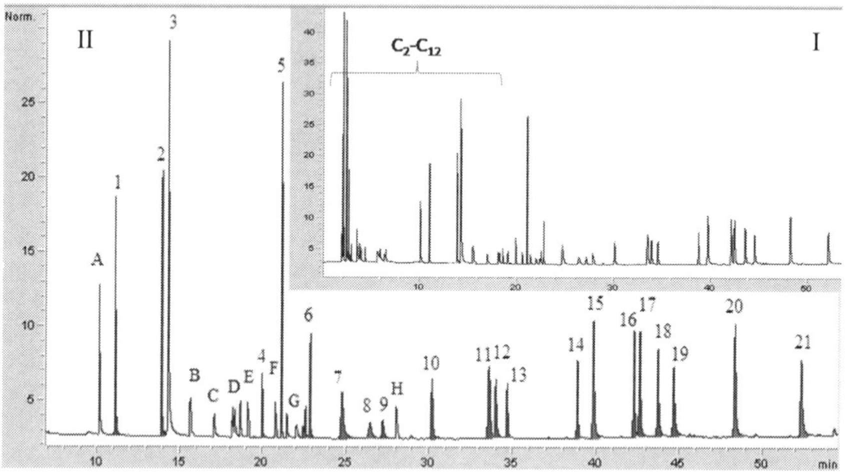

Figure 1: Chromatographic separation of 34 hydrocarbons and the target compounds (I) and a zoom for identification (II). Peak numbers are in table 1. A: benzene, B: decane, C: ethylbenzene, D: xylenes, E: undecane, F: dodecane, G: trimethylbenzene, H: tridecane.

3.2 Optimization of the preconcentration step

3.2.1 Water management

Due to the polar nature of the column, a sharp peak related to the presence of water in the sample appears on the chromatogram and co-elute with propanal and acrolein. These two compounds cannot be quantified with this method. The intensity of this peak is proportional to the amount of water in the sample and lead to a shift in retention time for the compounds eluted close to it. In fact, the content of retained water is affected both by the trap temperature and the initial water vapor concentration in the sampled gas [14]. The aim of this study was to use a one preconcentration stage and to reduce humidity in sample in order to limit the effect of the peak related to the presence of water vapor. The trap temperature optimization, the sample humidity management by dilution (50:50) and the trap purge before the sample analysis are used in order to reduce humidity and are discussed in the following.

3.2.1.1 Warm trap method The temperature of the trap was held above 0°C in order to prevent the ice plug formation and to reduce water condensation in the trap. Several preliminary tests have been conducted by varying the trap temperature from 10 to 30°C and sampling a standard mixture at 3 ppb concentration for all compounds and 50% relative humidity. At 20 and 30°C, the efficiency of trapping has been dramatically reduced especially for the most volatile compound: ethanol. At 10°C, a difficulty for the compounds identification has been encountered (Figure 2-C) and the peak related to water vapor has decreased by 10 minutes and has modified the elution of compounds close to it especially acetaldehyde and ETBE. This result can be explained by the increase of the adsorbed water amount on the trap at 10°C that affects the analysis on the polar column. In contrast a good chromatographic separation was obtained when the trap temperature was at 12.5°C with dry or humid air (Figure 2-A, B). The trap temperature was then optimized at 12.5°C which was a good compromise between an efficient adsorption on the trap and good separation on the polar column.

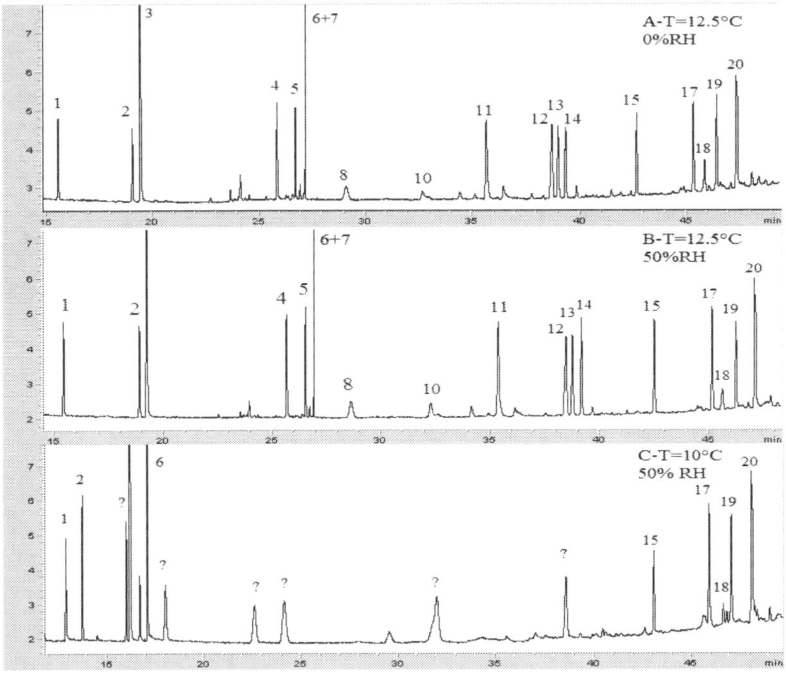

Figure 2: Trap temperature optimization, T_{trap}=12.5°C 0% HR, (top) T_{trap}=12.5°C, 50% HR (middle), T_{trap}=10°C, 50% HR (bottom).

3.2.1.2 Dilution of sampled air by dry purified air The sampled air was diluted with a dry purified air with a ratio of 50:50 in the sampling stream before preconcentration step to guarantee a relative humidity of less than 50% of the sample. The volume of dry air was optimized in order to cover all the meteorological conditions encountered in ambient air. Reducing the relative humidity at the trap enables a good separation and limits retention time shifts on a polar column. Figure 3 shows chromatograms obtained for the target compounds at 3 ppb with RH=0%, 90% with and without dilution at ambient temperature. The two analyses at 0% and 90% RH with dilution show similar chromatograms contrary to the analysis where no dilution with dry air has been applied.

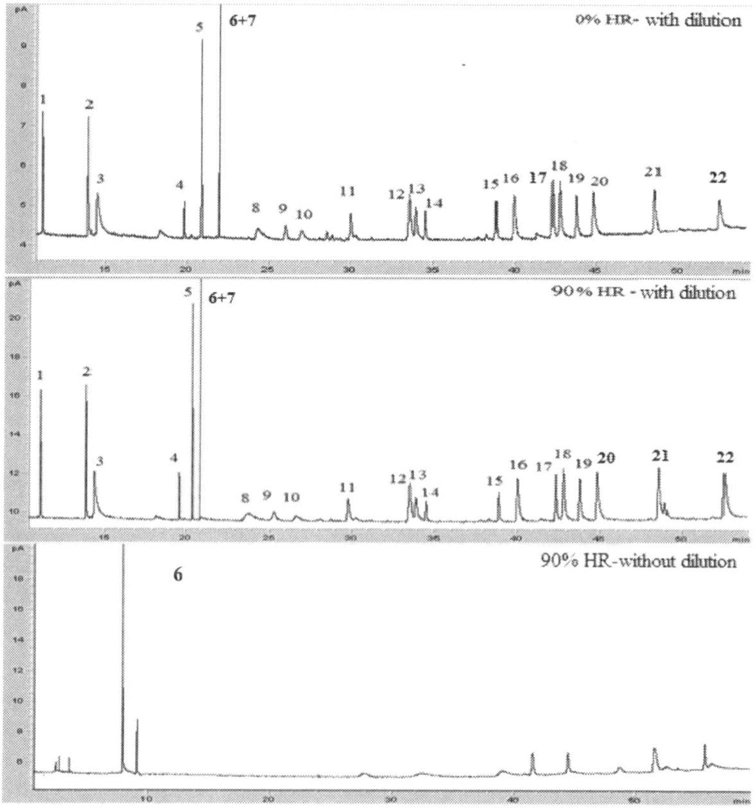

Figure 3: Chromatograms of three analyses of the 22 target compounds at 3 ppb and a trap temperature of 12.5°C dry air (top), 90% RH with dilution (middle), 90% RH without dilution (bottom).

3.2.1.3 Dry purge Dry purging of the trap prior to analysis is accomplished by flowing helium through the trap in the sampling flow direction. The trap purge permits to eliminate the residual water adsorbed on the trap. The trap purge time was optimized, at the same flow (10 ml min^{-1}), in order not to lose the most volatile compound (ethanol). Preliminary tests have permitted to set the trap purge time on 4 minutes with loss limited to 5% for ethanol.

3.3 Breakthrough volume

The determination of the breakthrough volume enables a quantitative analysis. The lightest compound is the first to breakthrough from the trap and the sampling volume will be chosen according to it. In order to determine the breakthrough volume, peak areas versus sampling volume were plotted. The breakthrough volume corresponds to the end of the linear part of the plotted curve. This test was carried out in the trap conditions defined previously (dilution (50:50), T_{trap}=12.5°C and trap purge with 40 mL of helium) at 4.5 ppb and a relative humidity of 80%. Figure 4 displays the plots for the four lightest compounds (ethanol, acetonitrile, acetone and acetaldehyde). Ethanol and acetonitrile show a breakthrough volume at 405 ml and the rest of the compounds had linear response for more than 700 ml.

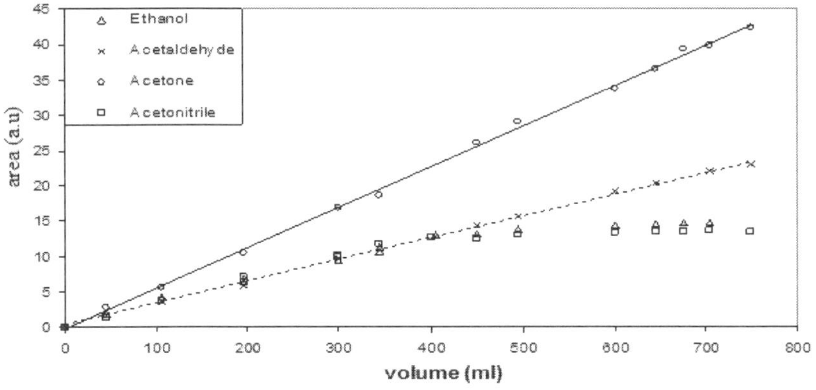

Figure 4: Breakthrough volume of ethanol and acetonitrile at 405 ml and linear plot for acetone and acetaldehyde.

3.4 Temperature of desorption

The optimal desorption temperature should insure a complete desorption for heavy compounds and should not lead to losses of the thermal instable compounds. In order to optimize this temperature, the responses for all of the compounds were tested by varying the desorption temperature. The standard mixture was diluted to 3 ppb level with a relative humidity of 80%. Six analyses were established for each of the five desorption temperatures (200, 250, 280, 300, 350°C). The average and the standard deviation of the six analyses for each compound are calculated. Only the results of 7 compounds are shown on figure 5. For some compounds like furan and toluene, the desorption temperature has no influence on peak area. Nevertheless, for ethanol and acetone, peak areas decrease of 10 and 9% respectively between 280 and 350°C. These two light compounds are sensitive to high temperature. The peak areas of the heaviest compounds (benzaldehyde and heptanenitrile) showed a decrease of 8% between 280°C and 200°C probably due to an incomplete desorption. A temperature of 280°C seems to be the best compromise for the measurement of these compounds.

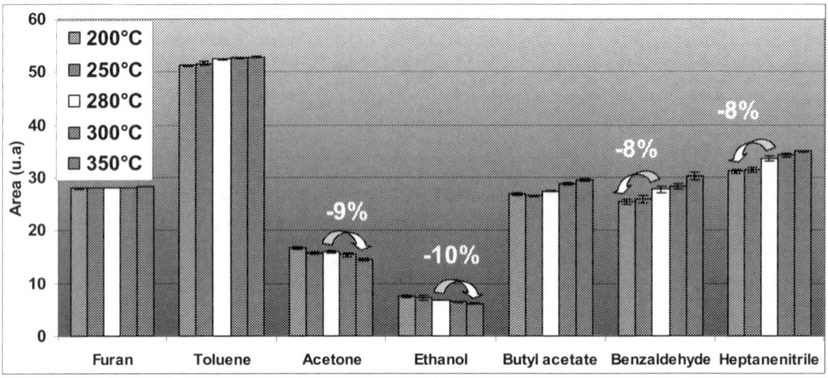

Figure 5: Average and standard deviation of six analyses to demonstrate the influence of the desorption temperature on 7 compounds.

4 Analytical performances

4.1 Calibration

The calibration step is crucial for the analysis of oxygenated and nitrile compounds. The response coefficient k for a compound connects between the area of the compound to the concentration and is determined by the following equation: $k=A/C$ where A is the area for a compound and C is the analyzed concentration. The use of a FID permits to calculate the response coefficient of a compound from the one of a reference compound. Moreover, the response of the FID is connected to the Effective Carbon Number (ECN) in a molecule [15]. The effective carbon number depends on the number of carbons and the chemical function in which the carbons are involved. The theoretical coefficient K_{ti} for each compound is calculated by using this concept as follow: $K_{ti}=(k_r/n_r)\times n_i$ where n_r and n_i are respectively the effective carbon number for the reference compound and the i compound. The laboratory holds a reference toluene standard certified by LNE at ppb concentration (C_r). This standard is analyzed 7 times and the mean value of toluene peak areas (A_r) is determined and thus its k_r. In this section, two calibration methods have been compared from the analysis of a standard gas mixture and a dynamic generation system to the ECN in order to determine the most reliable method. In table 2 are gathered the ECN and the responses coefficient for both of the calibration methods. The response coefficients of toluene, isopropanol, butanol, isobutanol and 2-butanone for the two methods are in agreement with the theory (Table 2). Moreover, six compounds (furan, 2-methyl furan, ETBE, acetone, ethanol, and butyl acetate) have a response coefficient superior to the theory. These compounds respond in a FID better than the theory expects. In contrast, benzaldehyde, methacrolein and especially acetaldehyde show a decrease of their response with dynamic generation. The aldehyde family seems to be not suitable for a dynamic generation. These compounds will be measured by the calibration with the gas mixture. Butanal and acetonitrile show a decrease because of their elution after the peak related to the presence of water in the sample.

Table 2: Comparison of response coefficient of the Dynamic Generation (DG), Gas Mixture (GM) with the theoretical response. ECN (Effective carbon Number), ND: not determined.

Compound	ECN	Theoretical *Kti*	k_i=*Ai/Ci* (DG)	k_i=*Ai/Ci* (GM)	k_i(DG)/*Kti*	k_i(GM)/*Kti*	k_i(GM)/k_i(DG)
Furan	3	7.24	8.6325	9.0585	1.19	1.25	1.05
2-methyl furan	4	9.65	9.8314	10.9940	1.02	1.14	1.12
Toluene	7	16.89	16.8250	17.7970	0.99	1.05	1.06
Acetaldehyde	1	2.41	1.6411	2.2410	0.68	0.93	1.37
ETBE	2	12.06	13.6590	13.2650	1.13	1.10	0.97
Methacrolein	2.9	7.00	2.3636	7.4274	0.34	1.06	3.14
Acetonitrile	2.3	5.55	ND	3.8905	ND	0.70	ND
Butanal	3	7.24	3.5722	3.4210	0.49	0.47	0.96
Acetone	2	4.83	5.8334	5.8916	1.21	1.22	1.01
2-Butanone	3	7.24	6.9996	6.9935	0.97	0.97	1.00
Ethanol	1.4	3.38	3.5922	4.0200	1.06	1.19	1.12
Isopropanol	2.25	5.43	5.1879	5.2080	0.96	0.96	1.00
Pentanenitrile	5.3	12.79	ND	12.0240	ND	0.94	ND
Isobutanol	3.4	8.20	7.9536	8.7293	0.97	1.06	1.10
Butyl acetate	4	9.65	10.5340	11.5760	1.09	1.20	1.10
Butanol	3.4	8.20	8.0097	8.4337	0.98	1.03	1.05
Benzaldehyde	6	14.48	1.3583	10.7550	0.09	0.74	7.92
Heptanenitrile	7.3	17.61	ND	13.8100	ND	0.78	ND
Octanenitrile	8.3	20.02	ND	13.0050	ND	0.65	ND

4.2 Linearity

Linearity is defined as the range of sample concentration where the peak area (A) is proportional to concentration (C). Seven dilutions of the ppm gas standard mixture have permitted to set up the graph A= f(C) for a VOC concentration ranging from 0.5 to 10 ppb. The linearity of responses was evaluated by means of the linear regression square coefficient R^2. Beside, the humidity effect has been also tested on the calibration curves with three different RH (0%; 20% and 80%). Table 3 gathers the plot slopes, their ratios and R^2.

The R^2 is higher than 0.99 for all of the compounds (Table 3). The humidity has no significant effect on 9 compounds (Table 3). Although, at 20% RH, ethanol and acetonitrile have shown a decrease in their response of about 30% and 20% respectively. This fact can be related to their water solubility. Moreover, isopropanol and isobutanol have also shown a decrease of 10% with a 90% RH. These two compounds are less soluble than ethanol but still sensitive to humidity at a high RH. In contrast heavy compounds (with 4 or more atoms of carbon) have shown an increase of 10% in their response which is related to the solubilisation of the adsorbed compound on the path of the sample leading to an increase of their concentration. The calibration should than be realised with humid air for a reliable quantification and in order to be in agreement with ambient air.

Table 3: Calibration slope equation, their ratios and square coefficient for the target compounds at 3 RH levels.

	0% RH		20%RH		80%RH		Slope Ratio	
slope	slope	R^2	slope	R^2	slope	R^2	20/0% RH	80/0% RH
Furan	9.06	0.9996	9.23	0.9999	9.13	0.9997	1.0	1.0
2-methyl furan	11.00	0.9998	11.18	0.9996	11.01	0.9998	1.0	1.0
Toluene	17.80	0.9984	17.66	0.9961	17.74	0.9989	1.0	1.0
Acetaldehyde	2.25	0.9935	2.56	0.9992	2.77	0.9975	1.1	1.2
ETBE	13.27	0.9992	13.26	0.9996	13.19	0.9996	1.0	1.0
Methacrolein	3.68	0.9975	3.67	0.9969	3.77	0.9991	1.0	1.0
Acetonitrile	3.90	0.9998	3.30	0.9997	3.13	0.9976	0.8	0.8
Butanal	3.43	0.9969	3.61	0.9971	3.63	0.9959	1.1	1.1
Acetone	5.90	0.9988	6.01	0.9967	5.64	0.9986	1.0	1.0
Ethyl acetate+MVK	5.96	0.9998	6.15	0.9996	5.88	0.9998	1.0	1.0
2-Butanone	7.00	0.9998	7.36	0.9995	6.86	0.9974	1.1	1.0
Ethanol	4.02	0.9995	2.81	0.9994	2.63	0.9970	0.7	0.7
Isopropanol	5.21	0.9983	5.18	0.9992	4.85	0.9991	1.0	0.9
Pentanenitrile	12.03	0.9992	12.39	0.9997	12.06	0.9998	1.0	1.0
Isobutanol	8.73	0.9979	8.45	0.9998	7.91	0.9991	1.0	0.9
Butyl acetate	11.58	0.9980	12.19	0.9995	11.64	0.9992	1.1	1.0
Butanol	8.44	0.9983	8.86	0.9998	8.35	0.9995	1.1	1.0
Benzaldehyde	10.76	0.9970	12.22	0.9995	11.71	0.9992	1.1	1.1
Heptanenitrile	13.81	0.9981	14.37	0.9985	13.58	0.9996	1.1	1.0
Octanenitrile	13.01	0.9985	15.03	0.9964	13.69	0.9989	1.2	1.1

4.3 Repeatability

In order to evaluate this parameter, two standard mixtures of the target compounds (at 0.5 and 5 ppb and 80% RH) have been successively analysed seven times. The Relative Standard Deviation (RSD) of each compound peak area for the two concentration levels was evaluated. The RSD results for the target compounds for the two levels of concentration are gathered in table 4. For the two levels, RSD were lower than 5% for all the compounds.

4.4 Detection limit

The detection limit (DL) was evaluated by signal to noise ratio of 3. Table 4 gathers the detection limits obtained for each compound. For 19 compounds, the detection limits were acceptable and vary between 0.01 ppb for ETBE to 0.09 ppb for ethanol. Methacrolein, acetonitrile and butanal have higher detection limits varying between 0.2 and 0.6 ppb. These compounds are eluted after the peak related to the presence of water in the sample affecting their quantification. These detection limits for the 19 compounds are similar to those obtained by other studies [4, 8] and are reasonable for quantification of the compound in ambient air.

Table 4: Henry's constant, detection limits (DL) and the relative standard deviation (RSD) for the target compounds.

Compound	DL (ppb)	RSD (0.5 ppb)	RSD (5 ppb)	Compound	DL (ppb)	RSD (0.5 ppb)	RSD (5 ppb)
Ethanol	0.09	4.0	4.9	2-Butanone	0.07	4.1	2.4
Isopropanol	0.06	3.7	2.2	ETBE	0.01	0.5	1.1
Butanol	0.04	3.8	4.2	Furan	0.02	1.6	0.8
Isobutanol	0.04	1.8	0.9	2-methyl furan	0.02	1.4	0.8
Acetaldehyde	0.04	2.6	4.6	Ethyl acetate	0.07	2.3	1.3
Propanal	0.07	*	*	Butyl acetate	0.02	3.5	1.7
Butanal	0.44	4.7	3.4	Acetonitrile	0.29	3.7	4.9
Benzaldehyde	0.02	2.6	3.4	Pentanenitrile	0.06	2.7	3.6
Methacrolein	0.64	2.9	4.5	Heptanenitrile	0.06	3.6	3.7
Acetone	0.04	4.0	1.2	Octanenitrile	0.05	3.6	3.5
MVK	0.07	2.3	1.3	Toluene	0.04	2.2	2.5

*Not determined.

5 Conclusion

The development of an automated method based on thermal desorption coupled to a GC-FID for the online measurement of OVOC and nitrile compound was successfully achieved. This method required a combination of three methods of reducing water amount from sampled air: warm trap temperature held at 12.5°C, diluting (50:50) of the sample before the preconcentration step and a trap purge. The advantage of this method is the analysis of the integrality of the sample without any losses of compound by a drying device and the limitation of shifts of retention time for compounds on the polar column. The analytical performances (DL and repeatability) were satisfied. Calibration should be realized by analyzing humidified standards. This method is ready for a routine continuous monitoring in the field work.

References

[1] Atkinson, R. & Arey, J., Atmospheric degradation of Volatile Organic Compounds. *Chemical Reviews*, **103(12)**, pp. 4605–4638, 2003.

[2] Ciccioli, P., Cecinato, A., Brancaleoni, E., Brachetti, A. & Frattoni M. Polar volatile organic compounds (VOC) of natural origin as precursors of ozone. *Environmental Monitoring and Assessment*, **31**, pp. 211–217, 1994.

[3] Riemer, D., Pos, W., Milne, P., Farmer, C., Zika, R., Apel, E., Olszyna, K., Kliendienst, T., Lonneman, W., Bertman, S., Shepson, P. & Starn, T., Observations of nonmethane hydrocarbons and oxygenated volatile organic compounds at a rural site in the southeastern United states. *Journal of Geophysical Research,* **103(D21)**, pp. 28111–28128, 1998.

[4] Apel, E.C., Hills, A.J., Lueb, R., Zindel, S., Eisele, S. & Riemer, D.D., A fast-GC/MS system to measure C_2 to C_4 carbonyls and methanol aboard aircraft. *Journal of Geophysical Research,* **108(D20)**, pp. 8794, 2003.

[5] Singh, H.B., Salas, L.J., Chatfield, R.B., Czech, E., Fried, A., Walega, J., Evans, M.J., Field, B.D., Jacob, D.J., Blake, D., Heikes, B., Talbot, R., Sachse, G., Crawford, J.H., Avery, M.A., Sandholm, S. & Fuelberg, H., Analysis of the atmospheric distribution, sources, and sinks of oxygenated volatile organic chemicals based on measurements over the Pacific during TRACE-P. *Journal of Geophysical Research,* **109(D15S07)**, doi:10.1029/2003JD003883, 2004.

[6] Legreid, G., Balzani Lööv, J., Staehelin, J., Hueglin, C., Hill, M., Buchmann, B., Prevot, S.H.A. & Reimann, S., Oxygenated volatile organic compounds (OVOCs) at an urban background site in Zürich (Europe): Seasonal variation and source allocation. *Atmospheric Environment*, **41**, pp. 8409–8423, 2007.

[7] Monod, A., Bonnefoy, N., Kaluzny, P., Denis, I., Fostern, P. & Carlier, P., Methods for sampling and analysis of tropospheric ethanol in gaseous and aqueous phases. *Chemosphere*, **52,** pp. 1307–1319, 2003.

[8] Hopkins, J.R., Boddy, R.K., Hamilton, J.F., Lee, J.D., Lewis, A.C., Purvis, R.M. & Watson N.J., An observational case study of ozone and precursors inflow to South East England during an anticyclone. *Journal of Environmental Monitoring,* **8**, pp. 1195–1202, 2006.

[9] Gawryś, M., Fastyn, P., Gawlowski, J., Gierczak, T. & Niedzielski, J., Prevention of water vapour adsorption by carbon molecular sieves in sampling humid gases. *Journal of Chromatography*, **933**, pp. 107–116, 2001.

[10] Karbiwnyk, C.M., Mills, C.S., Helmig, D., Birks, J.W., Journal of Chromatography. A 958 (2002) 219.

[11] Chang, C.C., Lo, S.J., Lo, J.G., Wang, J.L., Analysis of methyl tert-butyl ether in the atmosphere and implications as an exclusive indicator of automobile exhaust. *Atmospheric Environment*, **37**, pp. 4747–4755, 2003.

[12] Hopkins, J.R., Lewis, C.A. & Read, A. K., A two-column method for long-term monitoring of non-methane hydrocarbons (NMHCs) and oxygenated volatile organic compounds (o-VOCs). *Journal of Environmental Monitoring*, **5**, pp. 8–13, 2003.

[13] Dettmer, K., Felix, U., Engewald, W. & Mohnke, M., Application of a unique selective PLOT Capillary Column for the Analysis of Oxygenated Compounds in Ambient air. *Chromatographia Supplement*, **51**, pp. S-221–S-227, 2000.

[14] Dettmer, K. & Engewald, W., Adsorbent materials commonly used in air analysis for adsorptive enrichment and thermal desorption of volatile organic compounds. *Analytical and Bioanalytical Chemistry*, **373**, pp. 490-500, 2002.

[15] Tranchant, J., Gardais, J.F., Gorin, P., Prévôt, A., Serpinet. J., & Untz, J., *Manuel pratique de chromatographie en phase gazeuse*, Masson, edition 3, pp. 504, 1982.

Multispectral gas detection method

M. Kastek, T. Sosnowski, T. Orżanowski, K. Kopczyński
& M. Kwaśny
Institute of Optoelectronics, Military University of Technology, Poland

Abstract

The article presents the problem of methane detection with a multispectral infrared camera. It also presents some commercially available devices and method used for the detection of chemical substances (gases). The project of a multispectral infrared camera and theoretical calculation regarding the possibility of methane detection is also reported in this paper. The calculations included the properties of optical path: camera-cloud of methane background. Verification of theoretical results was made during laboratory measurement. Some initial results of methane detection are also presented.
Keywords: gas detection, passive detection of gases, multispectral detection.

1 Introduction

Thermal cameras (fitted with a set of appropriate filters and special software) and IR imaging spectroradiometers make it possible to obtain the image of observed scenery with marked areas, where specific gaseous chemical compounds can be found. The detection of gases is possible, because absorption and emission bands of many chemical compounds (molecular vapours and aerosols) are located in the IR range. Two spectral bands are usually analysed: 3 – 5 μm and 8 – 12 μm, which are the typical working bands of IR devices due to atmospheric transmission. The development of imaging devices for gas detection was greatly influenced by the progress in focal plane array IR detectors. The modern devices fitted with highly sensitive IR detectors can detect the presence of chemical compounds from the distance of 1 – 2 kilometres therefore they must have high temperature and spectral resolution. Two main groups of such devices can be distinguished: thermal cameras (additionally equipped with spectral filters and signal processing units) and devices built around IR spectroradiometers.

WIT Transactions on Ecology and the Environment, Vol 123, © 2009 WIT Press
www.witpress.com, ISSN 1743-3541 (on-line)
doi:10.2495/AIR090211

Gas detection by thermal cameras consists in the comparison of two thermal images: one taken in the absorption (or emission) band of sought after gas and the other (so called reference image) taken slightly outside spectral area where given gas exhibits absorptive (or emissive) properties.

2 Methods and devices used for the detection of gases

Thermal camera used for detection of volatile chemical substances has to be equipped with the systems ensuring adequate spectral resolution. It is connected with the necessity of application of the filters (usually tunable ones, increasing universality of a device), the systems of signal analysis, and often special optical system.

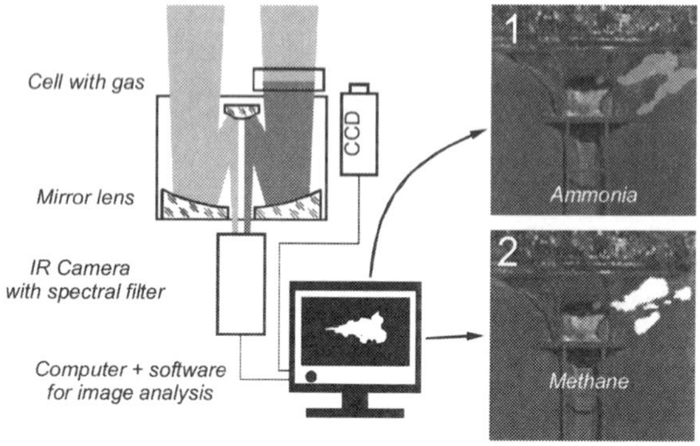

Figure 1. Gas detection system with a thermal camera [1].

Sample application of a thermal camera equipped with an optical filter for gas detection is schematically presented in Fig. 1. The system consists of: an Agema THV900LW camera equipped with the bandpass filters having the characteristics matched to the absorption bands of gases, a Cassegrainian telescope, frame grabber, CCD camera, and PC computer with special software for image analysis. The IR camera is equipped with suitable interference filters to isolate narrow spectral region containing absorption features of the gas. The image processing is performed according to the following scheme: A and B images are captured at the same time using the IR camera and frame grabber (A – image of the infrared scene from one of telescope openings, B – image of the same scene from the other telescope opening, with gas cell in front of it), next an error normalization image E is created to handle imperfections such as symmetrical vignette and stray-light from the two openings ($E=A_0/B_0$ where A_0, B_0 are images registered with no gas in the scene). Finally images are digitally overlapped and gas correlation image G is calculated as $G=A/B/E$. In this way the appropriate offset is subtracted from the images. Resulting image, which presents the region of gas presence in the observed scenery, is superposed on a visual image from CCD camera [1]. The result of these operations is shown in Fig. 1. Image (1) presents the total leakage of ammonia. Due to filling up the gas chamber with methane, detection of this substance and its visualization is also possible – image (2).

Another approach of a thermal camera application for the searching of gas leakages is a thermal camera system equipped additionally with a Fabry-Perot interferometer. An interferometer in IR system plays the role of a tunable optical filter (Fig. 2). It selects wavelength of IR radiation illuminating, at the given moment, the pixels of FPA of a thermal camera. An operation range of the interferometer is 3 – 5 μm (MWIR version) or 8 – 12 μm (LWIR) and tuning velocity is of about 10 – 20 ms. An interferometer module, containing a detector matrix, has been developed by the Physical Sciences, Inc. and it is known as AIRIS (Adaptive InfraRed Imaging Spectroradiometer) [2].

Figure 2. The scheme of thermal camera and Fabry-Perot interferometer [2].

By applying the aforementioned tunable optical filters it is possible to create a gas detection system around a thermal camera. The filter is tuned to the absorption band of a chosen compound and the interpretation of thermal image is performed by a human operator. The gas cloud appears as a "black smoke" on the display of a thermal camera. It should be mentioned that only detection of certain gas can be achieved without the possibility of quantitative analysis [2].
The Telops has developed an innovative instrument that can not only provide an early warning for chemical agents and toxic chemicals, but also one that provides a "Chemical Map" of the field of view. To provide to best field imaging spectroscopy instrument, Telops has developed the FIRST, Field-portable Imaging Radiometric Spectrometer Technology (Fig. 3). This instrument is based on a modular design that includes: a high performance infrared FPA and data acquisition electronics, onboard data processing electronics, a high performance Fourier transform modulator, dual integrated radiometric calibration targets and a visible boresighted camera. These modules, assembled together in an environmentally robust structure, used in combination with Telops' proven radiometric and spectral calibration algorithms make this instrument a world-class passive standoff detection system for chemical imaging [3].

There are several systems and devices for gas detection in the infrared range that use thermal camera or IR-sensitive FPA, like GasFindIR produced by FLIR Systems. The analysis showing the estimated gas detection capabilities as

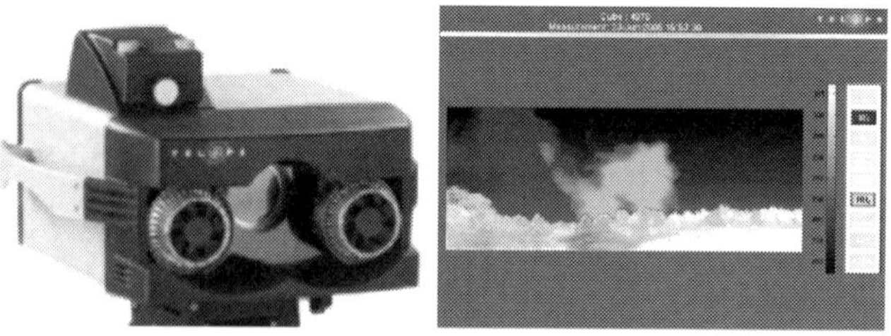

Figure 3. The Field-portable Imaging Radiometric Spectrometer Technology FIRST [3].

a function of gas concentration and NETD of a thermal camera will be presented in the following part of this paper.

3 Thermal camera capability for gas detection

The computer simulation of a methane gas detection process was performed in order to estimate the detection capabilities of a thermal camera. The variation of a signal reaching the camera caused by the presence of a gas was calculated and compared with the reference signal obtained without the presence of a gas in the camera's field of view.

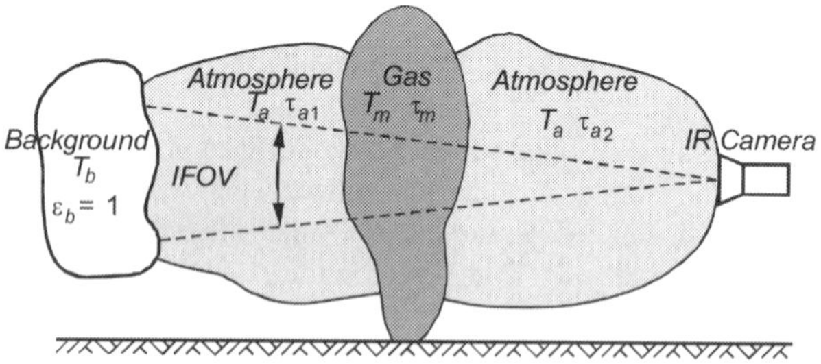

Figure 4. Scheme of a measuring process and notations accepted for gas analysis in atmosphere.

When there is a methane gas in a field of view of the thermal camera (fig. 4), total infrared radiance reaching the sensor at given wavelength N_{MSC} is the sum of the contributions from each layer and given by:

$$N_{MSC}(\lambda) = \tau_{MSC}\left[\tau_a \tau_m N_b(\lambda,T_b) + \tau_a(1-\tau_m)N_m(\lambda,T_m) + (1-\tau_a)N_a(\lambda,T_a)\right]. \quad (1)$$

where N_b is the Planck radiance of the background, N_m is radiance of the methane cloud and N_a is the atmospheric radiance. The quantities τ_m and τ_a are the spectral transmission of the methane cloud and the transmission of the atmosphere between the methane cloud and IR camera. The τ_{MSC} represents total spectral transmission of the objective, spectral filters and detector window.

The radiance from the background, attenuated by the gas cloud and intervening atmosphere, is the first term in (1). The second term is the radiance of the methane cloud, also attenuated by the atmosphere between the cloud and the camera. The third term in (1) is the radiance of the atmosphere between the gas cloud and the camera.

The transmission of the methane cloud τ_m is computed from the spectral properties of the chemical species using Beers' Law:

$$\tau_m(\lambda) = exp\left[-\sum k_i(\lambda)C_i d\right], \tag{2}$$

where C_i is the average concentration of the chemical compound over the optical path length d and $k_i(\lambda)$ is the wavelength-dependent absorption coefficient [2]. The sum over index i in (2) is calculated over all spectrally relevant chemical species.

PC MODWIN 3 v.1.0 computer program was used for the calculation of the transmission of the methane cloud. The results of this simulation are presented in Fig. 5.

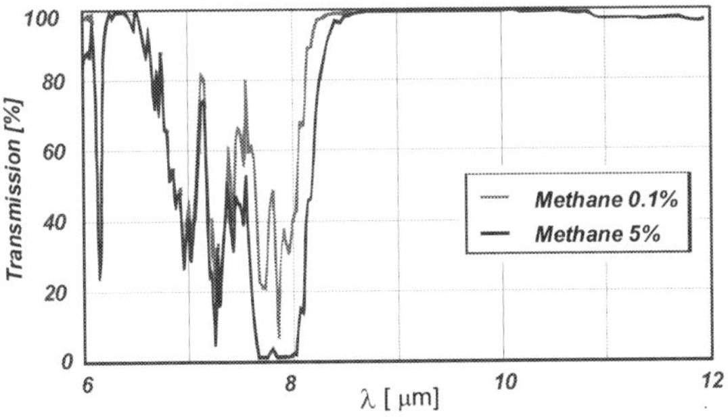

Figure 5. The transmission of the methane cloud for various gas concentrations.

The differential radiance observed by the camera, resulting from the presence of the gas cloud can be approximated as:

$$\Delta N_{MSC}(\lambda) = k_i(\lambda)\Psi\left[\frac{dN}{dT}\right]\Delta T, \tag{3}$$

where ΔT is the temperature difference between background and methane cloud (T_m - T_b), and Ψ is the column density of the methane.

This approach has been commonly applied to the analysis of detection of chemical agents and their simulants by all passive sensors.

NETD is s an important figure of merit used to characterize the performance of thermal imaging systems. Some analyses were performed of the NETD values required for a successful detection of a gas cloud. The relation between camera NETD and temperature difference between background (T_b=293 K) and gas cloud that allows for gas detection (with gas concentration as a parameter) is presented in Fig. 6 [4].

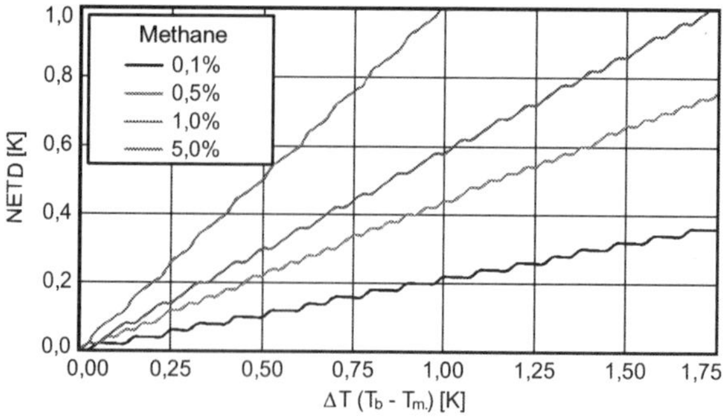

Figure 6. Required camera NEDT as a function of temperatures difference between background (object) and gas for various values of methane concentration.

It can be stated, on the basis of the presented analyses, that for typical measuring tasks it is possible to detect chemical compounds using a camera with NETD lower than 0.2 K. Thermal detection of chemical substances is more and more frequently used method for the monitoring of industrial installations. Having in view a serious progress in IR array detectors, it should be supposed that such solutions will be commonly applied.

4 Project of multispectral IR camera

A multispectral IR camera, the scheme of which is shown in Fig. 7, consists of the following basic blocks: objective, set of optical filters, microbolometer FPA, electronic read-out system and image analysis system.

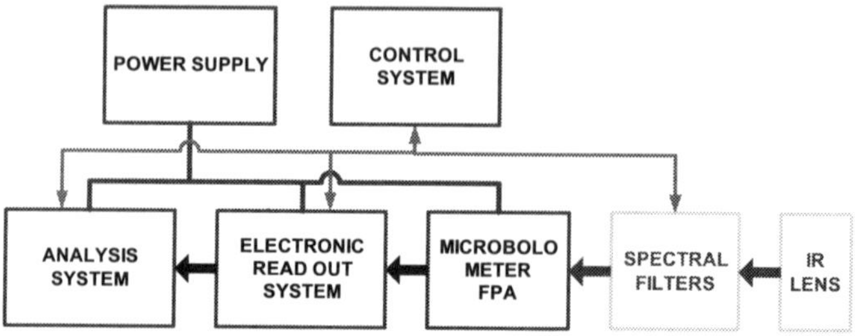

Figure 7. Block diagram of multispectral IR camera.

The Umicore Gasir Standard Lens 60 mm F/1.1 was used as the objective of the presented multispectral IR camera. It has the field of view of 18.5° and the transmission better than 92% in LWIR range. The UL 03 08 1 (ULIS, France) microbolometer array detector working in the long wavelength infrared (LWIR) range of 8-14 μm was used as a detector of the camera. It consists of 384×288 microbolometers (pixels) with the 35 μm pixel pitch. The chosen FPA is made in the photolithographic technology based on amorphous silicon and it features high performance: responsivity of 7.2 mV/K, thermal time constant less than 7 ms, and NETD equal to 40 mK at the F/1 optical aperture, 303K FPA temperature, and 60 Hz frame rate.

For the tests, a laboratory circuit for signal readout from the FPA has been designed, as well as a complete path of digital processing, data collecting, and image displaying on VGA monitor. A clock signals generator to control the readout integrated circuit (ROIC) in the FPA and all modules for digital signal processing have been made by means of Altera DSP Development Kit Stratix II Edition with the field programmable gate array (FPGA) EP2S60-1020C4 device. The raw image data were received from a FPA board with the 14-bit resolution [5]. The whole system that analyses the image from a microbolometric FPA has been designed in a FPGA chip [6]. This system was also used to control the optical filters between the detectors array and objective lens. A set of optical filters consists of two filters: methane filter 1 and methane filter 2, the transmission band of both filters covers the methane absorption area in a long IR wavelength range. These filters have been purchased from Spectrogon. Methane filter 1 consists of two filters: SP-8855 and LP-6500 and methane filter 2 consists of the SP-8100 filter and LP-6500 filter. As a reference filter, the BP 8350-12660 filter was used, also made by Spectrogon. Its transmission band is, in turn, outside the area of methane absorption. Transmission characteristics of those filters are shown in Fig. 8. The characteristics have been measured by means of Backman FAR IR 720 spectrophotometer.

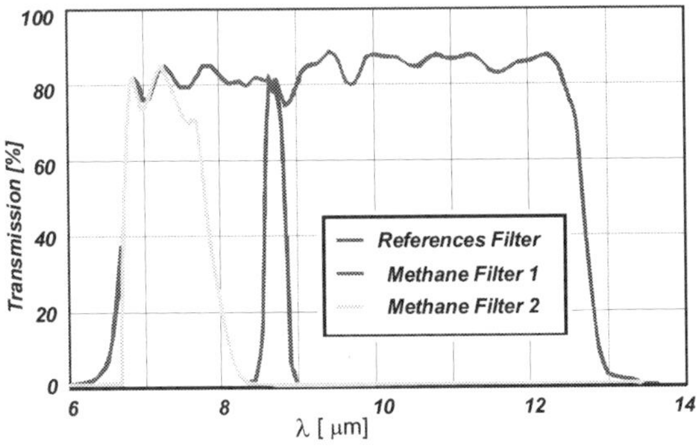

Figure 8. The transmission of optical filters used in multispectral IR camera.

The model of radiation transmission has been developed to calculate the amount of radiation reaching the input aperture of the optical system and camera's detector. The calculations were performed using PC MODWIN 3 v.1.0 software. A horizontal transmission path was assumed and Midlatitude Summer atmosphere model was chosen, for which the spectral characteristics of the atmosphere visibility coefficient τ_a have been determined. A series of additional calculations of a transmission coefficient for various parameters have been made to obtain a complete analysis of the influence of an atmosphere transmission coefficient on the possibility of gas detection.

The spectral characteristic of the objective and optical filters used in the multispectral IR camera can be described by the following relation:

$$\tau_{MSC} = \tau_o \tau_F , \quad (4)$$

where τ_o is the spectral characteristic of the objective and τ_F is the spectral characteristic of the actually used filter.

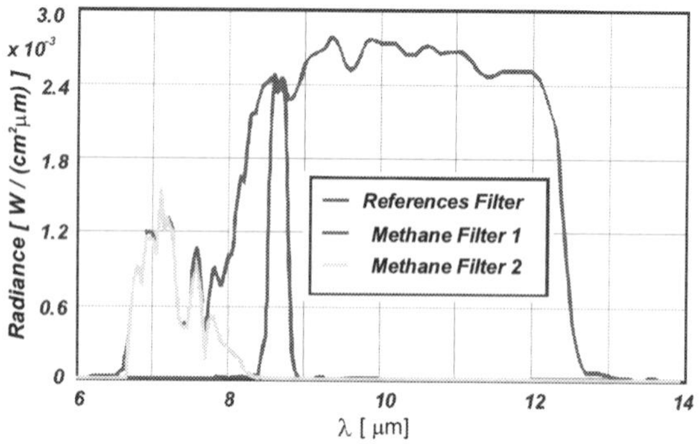

Figure 9. Spectral distribution of radiance reaching the detector of multispectral IR camera (background temperature T_b=293 K, atmosphere temperature T_a=295 K, methane concentration 5%).

On the basis of (1) and spectral characteristic of the camera detection path described by (4), the spectral distribution of luminance causing the output electric signal of the detector during methane cloud detection can be calculated. Assuming the background temperature $T_b = 293\ K$ and the atmosphere temperature $T_a = 295\ K$, the spectral distribution of radiance has been calculated and the results are presented in Fig. 9 [7].

Spectral distribution of luminance causing the output electric signal of the detector (when reference filter was used) can be calculated according to the relation:

$$N_{BG}(\lambda) = \tau_{MSC} \tau_a N_b(\lambda, T_b) . \quad (5)$$

The simulations were carried out for several methane concentrations and various temperatures of a background and a gas cloud to determine the possibility of gas cloud detection. To estimate the possibility of methane cloud detection using the multispectral camera, the analysis of the signal ratio, expressed by (6), was made.

$$S = \frac{N_{BG}(\lambda)}{N_{MSC}(\lambda)} . \quad (6)$$

On the basis of the results of computer simulations, it can be stated that the multispectral IR camera with aforementioned parameters will provide detection of methane if its concentration is above 5% and for temperature difference between the background and atmosphere of, at least, 2.3 K. The measurements on laboratory stand have been performed to verify theoretical results.

5 Laboratory tests of multispectral IR camera

In order to verify theoretical calculations, laboratory tests of the designed multispectral IR camera have been performed. Figure 10 illustrates the model of multispectral IR camera used during the tests.

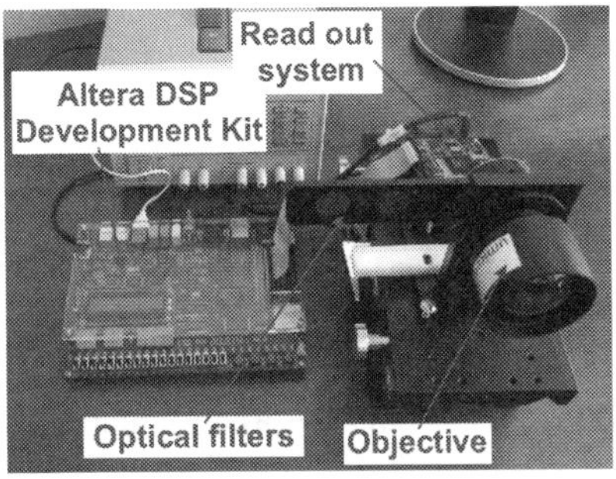

Figure 10. The model of multispectral IR camera.

The measurements were basically made to determine the possibility of methane detection. For two various methane concentrations (5% and 1%) a series of thermal images were registered, as presented in Fig. 11. During the measurements, the temperature difference between methane and background (T_m - T_b) was 2.6 K. To distinguish an area of methane leakage, the first thermogram was registered with the reference filter applied. This was the reference image. The following images were recorded with the appropriate methane filters. Next, due to digital image processing, the data obtained inside and outside methane absorption range were superimposed and clearly visible methane cloud has been obtained.

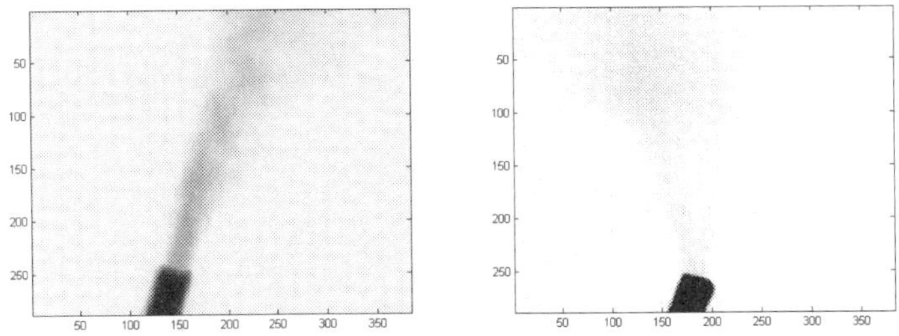

Figure 11. The thermograms registered during the tests: a) thermogram with methane concentrations 5%, b) thermogram with methane concentrations 1%.

Summing up, it can be stated that a thermal camera equipped with the adequately chosen set of optical filters and the software for image processing can be used as an imaging detector for gas detection.

During theoretical analysis and laboratory verification it was proven that the microbolometer FPA has sufficient sensitivity to be applied for methane detection. It can reduce the costs of devices used for methane detection and enable real-time observations of gas leakages.

References

[1] Sandsten J., Weibring P., Edner H. and Svanberg S., Real-time gas-correlation imaging employing thermal background radiation, OPTICS EXPRESS 92, vol. 6, No. 4, (2000).

[2] Gittins M., Marinelli W. J. and Jensen J. O., Remote Sensing and Selective Detection of Chemical Vapor Plumes by LWIR Imaging Fabry-Perot Spectrometry, Proc. SPIE 4574, pp. 63-71, (2001).

[3] Vallières A., Chamberland M., Farley V., Belhumeur L., Villemaire A., Giroux J. and Legault J. F., High-performance field-portable imaging radiometric spectrometer technology for chemical agent detection, SPIE 5590, (2005).

[4] Kastek M., Madura H., Sosnowski T. and Firmanty K., Thermovision method of gas detection in far infrared range, Proc. of the 9 Advanced Infrared Technology and Applications 2007, Leon Mexico, pp. 387-394, (2008).

[5] Orżanowski T., Madura H., Powiada E. and Pasierbiński J., Analysis of a readout circuit for microbolometer detectors array, Measurements, Automation, and Checking, No. 9, pp. 16-20, Poland, (2006).

[6] Sosnowski, T., Orżanowski, T., Kastek, M., Chmielewski, K., Digital image processing system for thermal cameras, Proc. of the 9 Advanced Infrared Technology and Applications 2007, Leon Mexico, pp. 252-262, (2008).

[7] Kastek M., Piątkowski T., Polakowski H. and Sosnowski T., Methane detection in far infrared using multispectral IR camera, 9th International Conference on Quantitative InfraRed Thermography, pp 347-350, Poland, (2008).

Application of advanced optical methods for classification of air contaminants

M. Wlodarski, K. Kopczyński, M. Kaliszewski, M. Kwaśny,
M. Mularczyk-Oliwa & M. Kastek
Institute of Optoelectronics, Military University of Technology, Warsaw, Poland

Abstract

Biochemical, highly sensitive methods of pollutant identification have common drawbacks – relatively long analysis time and the necessity of on-site sample collection. Optical methods are less sensitive and less selective, but they allow real time analysis. These optical methods include Fourier Transform Infrared Spectroscopy (FTIR), and Laser Induced Fluorescence (LIF). Both methods give the possibility of stand-off detection. The aim of the presented work is to study the possibilities of detection and classification of air contaminants, based on their optical properties. In this work we present measurement results of UV-VIS fluorescence characteristics and IR optical absorption of vegetable pollens, bacterial spores, and non-biological air contaminants (diesel fuel, dust, syloid). Results of the principal components analysis show that both LIF and FTIR methods allow distinguishing between groups of materials, i.e. pollens and non-biological materials, thus making application of these methods for detection and preliminary classification possible.

Keywords: biological contaminants, LIF, FTIR, stand-off detection, UV-VIS fluorescence, pollution monitoring.

1 Introduction

Air pollutants have strong impacts on human health. Inhaled particulate matter is often responsible for severe side effects like respiratory tract infections caused by bacterial or viral agents and allergic reactions triggered by pollens.

Modern biochemical assays, although highly sensitive and specific, are time consuming. The fast analysis is important in case of potential warfare attack (Pan

WIT Transactions on Ecology and the Environment, Vol 123, © 2009 WIT Press
www.witpress.com, ISSN 1743-3541 (on-line)
doi:10.2495/AIR090221

et al [1]). For this reason, methods based on optical or spectral properties of contaminants, for detection and analysis have become more important in the last decade (Hairston *et al* [2], Pan *et al* [3]).

In our project we are currently developing the application of two optical methods of biological pollution monitoring: passive FTIR spectroscopy, and dual wavelength fluorescence LIDAR for fast stand-off detection and classification.

Biological materials excited with UV light show strong fluorescence in two bands. The first, more intensive band has excitation maximum at about 280 nm, and emission maximum at about 330 nm, and comes mainly from fluorescence of Tryptophan. Second band is result of NADH fluorescence, with excitation and emission maxima at about 340 nm and at about 450 nm respectively (Wlodarski *et al* [4]).

2 Materials and methods

2.1 Excitation-emission matrices

Fluorescence spectra of the samples were measured with FS 900 spectrofluorimeter (Edinburgh Instruments) provided with a front-surface holder that eliminates inner filter effect. A sample solution was placed in 1 cm length quartz cuvette. The solutions were mixed during the experiment with a magnetic stirrer. The excitation wavelength ranged from 250 to 400 nm in 5 nm increments, and emission was scanned from 260 to 550 nm with step 2 nm. The slits on both monochromators were set for 2-nm bandwidth. The Raman signal and the background were subtracted from the sample spectra.

Table 1: Alphabetical list of measured materials.

Sample name	**Group**	**Abbrev.**
AC Fine Test Dust	Non-biological	DUST
Bacillus atrophaeus spores (Edgewood)	Bacterial spores	BGst
Bacillus thuringiensis spores (Turex)	Bacterial spores	BTst
Bermuda grass pollen	Vegetable pollens	BER
Bermuda grass smut spores	Vegetable smut spores	BERs
Black walnut pollen	Vegetable pollens	WAL
Corn pollen	Vegetable pollens	COR
Diesel fuel	Non-biological	OIL
Johnsons grass smut spores	Vegetable smut spores	JON
Paper mulberry pollen	Vegetable pollens	PAP
Pecan pollen	Vegetable pollens	PEC
Ragweed pollen	Vegetable pollens	RAG
Syloid	Non biological	SYL

2.1.1 Chemicals

Technical preparation of Bacillus turingiensis spores (Turex) and Bacillus atrophaeus spores were gently provided from PvTT, Tampere, Finland. Pollens and AC Fine Test Dust standard were obtained from Duke Scientific.Corp, Faber Place, Palo Alto, CA. L-Tryptophan was obtained from Sima-Aldrich, Poland. Syloid 244 FP was from Grace GmbH&Co. KG. Diesel fuel was commercially available from petrol station. All chemicals were of the highest purity commercially available.

2.1.2 Stock solutions

Solutions were prepared from powders. Distillated water was added to obtain the final concentration 1 mg/ml. In some cases it was necessary to dilute prepared samples due to a strong fluorescence signal that caused detector saturation.

2.2 Laser Induced Fluorescence

LIF spectra were measured with the ICDD camera (Princeton Instruments PI-MAX 2) connected to monochromator (Acton Research Corporation SpectraPro 2150i). The exposure time of the camera was set to 100 ms. High signal-to-noise ratio, was obtained by averaging 100 single measurements of spectrum. The spectrometer resolution was set to 5 nm. Two lasers were used for excitation: 266 nm (Intelite Inc. quadrupled Nd:YAG, 20 mW average power), and 375 nm (Power Technology Inc. laser diode, 15 mW power). Laser wavelengths were chosen in consideration with fluorescence LIDAR requirements, therefore they are not the same as optimal excitation wavelengths for biological materials. Lasers and monochromator were connected to a measurement chamber via optical fibers. Laser signal was rejected from emission spectra using two Semrock long-pass filters: FF-300LP, and FF-409LP for 266 nm and 375 nm excitation respectively. Fluorescence was measured in the range of 300 – 550 nm for 266 nm excitation, and 390 – 640 nm for 375 nm excitation. Solid samples were fixed between two thin quartz plates. The diesel fuel sample was measured in quartz 1 cm cuvette. Fluorescence signal was collected from the surface of the sample to avoid inner filter effect.

2.3 Medium infrared absorption

IR absorption spectra were collected using Perkin-Elmer Spectrum GX Optica FTIR spectrometer. Measurement range was 4000 cm^{-1} – 650 cm^{-1} (2.5 μm – 12.5 μm) with 4 cm^{-1} resolution. Two methods of measurement were applied: Diffuse Reflectance Infrared Fourier Transform (DRIFT), and Horizontal Attenuated Total Reflectance (HATR).

DRIFT technique allows for measurement of solid, highly absorbing samples ([5], Brimmer *et al* [6]). HATR technique allows for direct measurement of liquid samples [7].

Solid samples for DRIFT measurements were mixed with KBr powder in 5% proportion. Diesel fuel was measured in liquid form using the HATR technique.

Both techniques allow measurements from a small amount of samples. Those methods were chosen as promising for real-time monitoring applications.

2.4 Principal Components Analysis

PCA was conducted using SIMCA-P 11.5 program from Umetrics AB. Results were presented in the form of score plots.

3 Results and discussion

3.1 Excitation-emission matrices

Fluorescence spectra recorded in map mode provide full information on excitation and emission maxima. Presented matrices are normalized to the maximum of emission intensity.

Representative EX-EM maps are shown in Figure 1. The shapes and intensities vary between different materials.

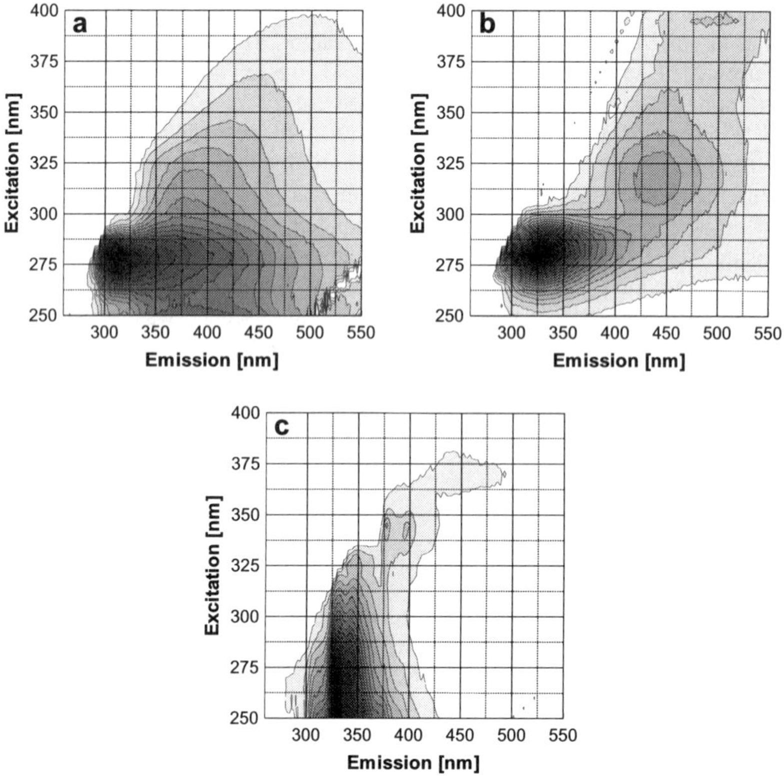

Figure 1: EX-EM matrices of: a) Turex, b) Paper Mulberry Pollen, c) Diesel fuel.

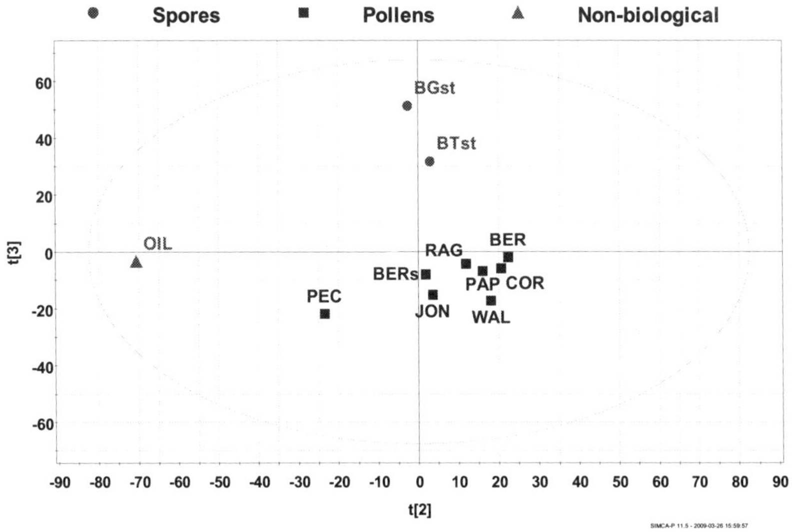

Figure 2: PCA results of Excitation-Emission matrices.

Classification with PCA method shows that particular groups are well separated (Figure 2). Syloid and dust were not analyzed due to lack of fluorescence. Oil, as a non-biological material, is placed in a distant position compared to spores and pollens.

3.2 Laser Induced Fluorescence spectra

Laser Induced Fluorescence is promising technique for stand-off detection of biological contamination. Analysis of EX-EM matrices allows for optimal choice

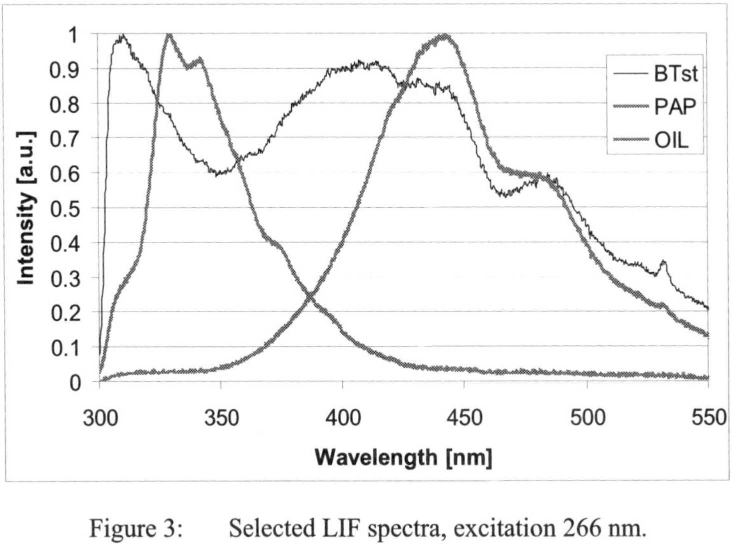

Figure 3: Selected LIF spectra, excitation 266 nm.

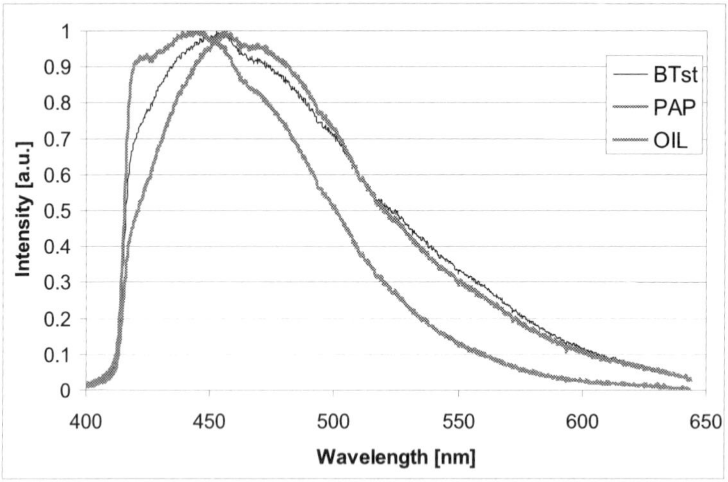

Figure 4: Selected LIF spectra, excitation 375 nm.

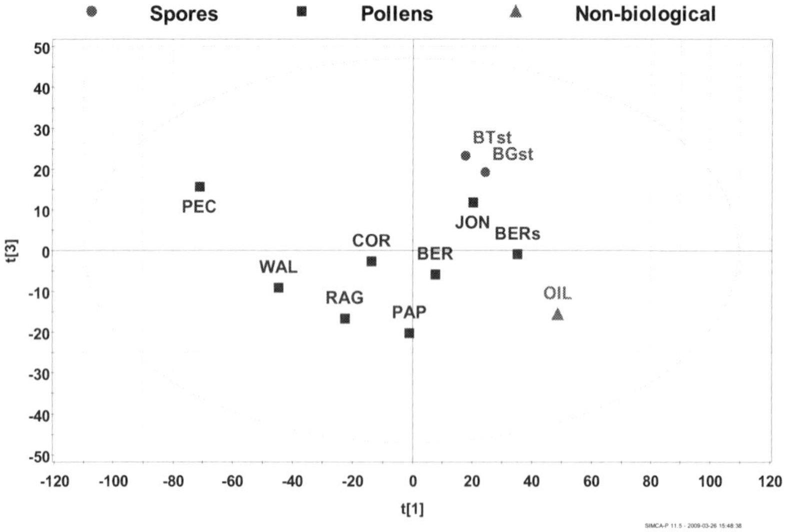

Figure 5: PCA results of LIF spectra.

of light source. LIF spectra recorded at excitation 266 nm differ significantly (Figure 3), while those from excitation at 375 nm are very similar (Figure 4).

PCA of LIF spectra presented in Figure 5 show that bacterial spores are forming separate group. Discrimination of pollens is unequivocal because of considerable distribution over the plot area.

PCA graphs of whole EX-EM maps (Figure 2) show more precise discrimination comparing to LIF (Figure 5). Differences can be explained by

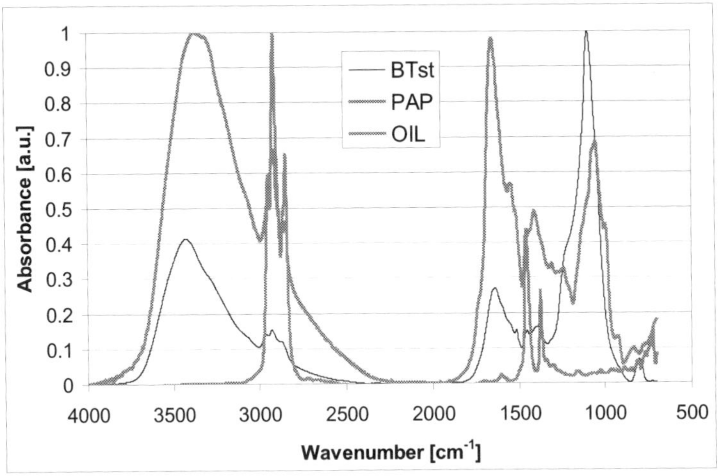

Figure 6: Selected FTIR spectra.

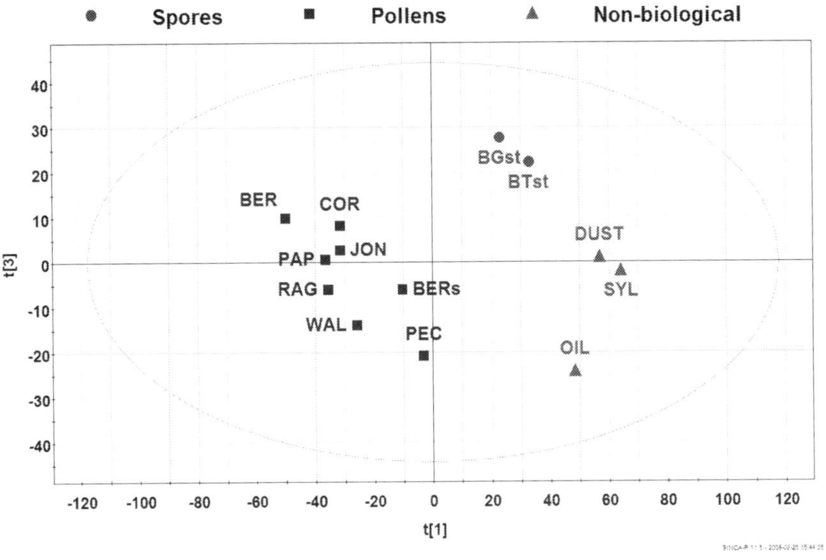

Figure 7: PCA results of FTIR spectra.

30 excitation wavelengths included in map while LIF applies only two. Despite reduction of data it is still possible to differentiate between biological and non-biological material.

3.3 Medium Infrared absorption spectra

Infrared absorption spectroscopy has been widely used for identification of specific groups and chemical bondings. The great advantage of this method is identification of non-fluorescent substances.

Despite very complex composition, biological substances contain similar groups of proteins, lipids, organic and inorganic elements. Their presence is manifested by absorption peaks at specific region of the spectrum.

Normalized spectra of three different substances are presented in Figure 6. The plots exhibit considerable differences mainly due to intensities of the absorption peaks. That feature is common for biological samples like PAP and BTst. Generally, Diesel fuel absorbs IR light in different spectral regions than biological samples. The only similarity is apparent in region 3000-2800 cm^{-1} that is characteristic to C-H bounds.

Results of PCA analysis of IR spectra are shown in Figure 7. Points representing particular groups are well separated from each other.

4 Conclusions

Spectroscopy in infrared region showed better discrimination than fluorescence methods. IR spectra provide more information due to higher number of different peaks. Simultaneous application of passive FTIR and fluorescence LIDAR provides better independence from environmental conditions during monitoring (i.e. FTIR for daytime and LIDAR for nighttime application). Identification of biological air pollutants with advanced optical systems can be applied for both civil and military purposes. Unquestionable advantages are continuous, real-time, and stand-off monitoring. Today's FTIR and LIDAR technologies allow stand-off detection of chemical aerosol contaminants (Mierczyk *et al* [8], Killinger [9]). Creation of biological material database is unavoidable and very important element for development of monitoring systems. Our results show that preliminary classification of biological air contaminants using advanced optical methods is possible.

References

[1] Y-L. Pan, V. Boutou, J. Bottiger, S. Zhang, J-P. Wolf, R. Chang. *A Puff of Air Sorts Bioaerosols for Pathogen Identification*, Aeros. Sci. and Techn, 38, 2004, 598-602.

[2] P.P. Hairston, J. Ho, F.R. Quant. *Design of an instrument for real-time detection of bioaerosols using simultaneous measurement of particle aerodynamic size and intrinsic fluorescence*, J. Aerosol Sci., 28, 3, 1997, 471-482.

[3] Y-L Pan, J. Hartings, R.G. Pinnick, S.C. Hill, J. Halverson, R.K. Chang. *Single-Particle fluorescence spectrometer for ambient aerosols*, Aerosol Sci. Technol., 37, 8, 2003, 627-638.

[4] M. Wlodarski, M. Kaliszewski, M. Kwasny, Z. Zawadzki, K. Kopczynski Z. Mierczyk, J. Mlynczak E. Trafny, and M. Szpakowska, *Fluorescence excitation-emission maps database of biological agents*, Proc. SPIE, 6398, 06-1-12 (2006).

[5] *Diffuse Reflectance Accessory; User's Guide*, Copyright © 1998 The Perkin-Elmer Corporation United Kingdom

[6] Paul J. Brimmer, Peter R. Griffiths, and N. J. Harrick, *Angular Dependence of Diffuse Reflectance Infrared Spectra. Part I: FT-IR Spectrogoniophotometer*, Appl. Spectrosc. **40**, 258-265 (1986)
[7] *Horizontal ATR Accessory; User's Guide*, Copyright © 1998 The Perkin-Elmer Corporation United Kingdom
[8] Z. Mierczyk, M. Zygmunt, A. Gawlikowski, A. Gietka, M. Kaszczuk, P. Knysak, A. Młodzianko, M. Muzal, W. Piotrowski, J. Wojtanowski, *Two-wavelength backscattering lidar for stand-off detection of aerosols,* Proceedings of SPIE, Vol. 7III, 7III0R, SPIE Europe Remote Sensing, Cardiff (Wales), (2008).
[9] Dennis Killinger, *Detecting chemical, biological, and explosive agents*, SPIE Newsroom, archived invited paper, August 2006.

Electronic application to evaluate the driver's activity on the polluting emissions of road traffic

D. Pérez, F. Espinosa, M. Mazo, J. A. Jiménez, E. Santiso,
A. Gardel & A. M. Wefky
Department of Electronics, University of Alcalá, Spain

Abstract

The polluting emissions (gases and particles) produced by the traffic of automobiles are directly related to the activity of vehicle, but they are also affected by the route conditions and moreover, by the driver's behavior. However, PAMS commercial systems do not usually include elements to register this last component. This article presents an electronic system, specially designed to evaluate, ad-hoc, the effect of driver's activity on polluting emissions. This electronic application integrates a hardware and a software component, both designed concerning MIVECO research project. From the hardware component the sensorial part stands out, formed by potentiometers connected to the pedals that control the vehicle and the inertial device, which allows one to evaluate the instantaneous accelerations in the x-y-z-axes as well as the turns with respect to these axes. Once the signals are conditioned and acquired, the software component processes them for on-line monitoring in a GUI and stores them in a database to facilitate its evaluation off-line. This electronic application has two important properties: it can be incorporated in any vehicle of the market (light or heavy, diesel or gasoline, pre or post-eobd) and it allows the capture and registration of information about the driver's activity synchronously with PEMS (gases and/or particles) systems. The work includes experimental results obtained in an urban circuit in the city of Madrid.

Keywords: driver's behavior, PAMS, eco-driving, multi-sensorial system, onboard electronic system.

WIT Transactions on Ecology and the Environment, Vol 123, © 2009 WIT Press
www.witpress.com, ISSN 1743-3541 (on-line)
doi:10.2495/AIR090231

1 Introduction

According to various studies realized by distinct investigation groups [1–5] and official organizations from different countries [6–12], it was concluded that both the surrounding environment and the driving form affect both the fuel consumption and the gaseous as well as particles pollutant emissions of the vehicle.

Other studies show that relationship using only the speed measurement to determine if the driving is more efficient in terms of the emission of the pollutant gases (CO, HC, NO_x, CO_2) [13, 14].

In addition to savings in pollutant emissions, ecological driving produces a multitude of enhancements such as an average saving of 15% of fuel, less noise pollution, reduction in the risk of accidents of between 10 and 25%, reduction in maintenance costs of the vehicle: brakes, clutch, gearbox and engine, and increased comfort for the driver and the passengers [14].

In this context the challenge is to have onboard equipment that records information from the driver's activity in synchronisation with complementary systems for measuring emission (PEMS).

A solution that integrates two types of measuring, direct and indirect, the activity of the driver is proposed in this paper. On one hand, the direct action on the pedals to control the vehicle is measured. On the other hand, motion variables, as linear acceleration, speed and inclination of the vehicle in 3 axes, are recorded using an inertial sensor.

2 Methodology

An in-vehicle electronic system was designed to measure driver activity. The designed system consists of three basic subsystems, as shown in Fig. 1.

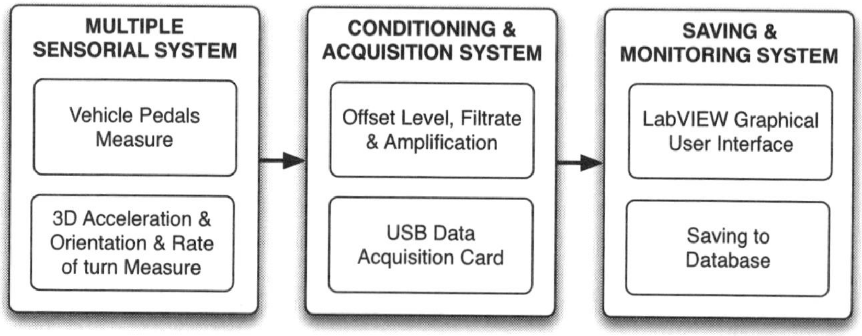

Figure 1: Block diagram of the global system.

The first subsystem contains the study of selection and placement of the sensory system and its measurement characteristics. The second subsystem contains the development and acquisition of signals from the sensors. And the third subsystem is responsible for monitoring and recording information from sensors in a database on a laptop computer.

2.1 Sensorial system

The sensorial system consists of two parts that enable the measurement of the driver behavior. On one hand, three potentiometers are used to provide a signal proportional to the action of the driver on the three pedals of the vehicle. On the other hand, an inertial measurement unit is used, to provide signals proportional to acceleration, direction, and speed of the vehicle on the XYZ-axes.

2.1.1 Potentiometers measurement characteristics

Three ASM WS42C displacement potentiometers were used to measure the activity of the pedals [15]. Such potentiometers are based on the variation of the resistance in proportion with the displacement. That displacement is produced by stretching the wire rolled around a pulley and controlled by a spring. When the force on the cable is released, it returns to its original position automatically.

The three potentiometers are fixed by a metal plate with screws to the internal part of the vehicle under the dashboard, specifically to the left of the pedals as shown in the left side of Fig. 2. In order not to bother the driver, each end of the three potentiometers is tied to a nylon thread which is attached to the upper part of each pedal at its other end as shown in the right side of Fig. 2.

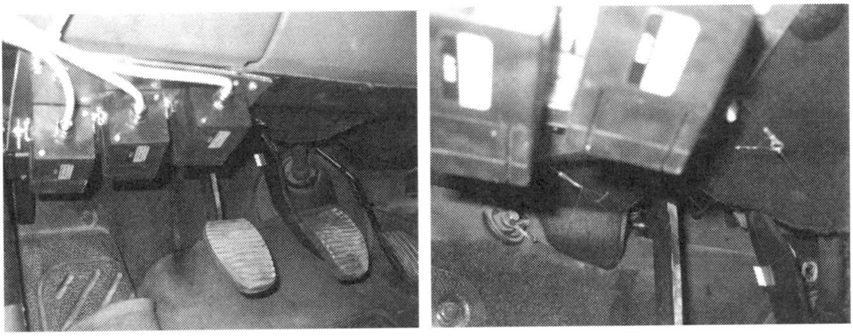

Figure 2: Potentiometers measurement system.

Once the system is set up, it is necessary to make an initial calibration to establish the dynamic range of the pedal displacement. That dynamic range depends on each vehicle model. In order to standardize and draw valid conclusions using different vehicles, the displacement values were normalized between 0 and 100% of the pedal path. Eqn. (1) uses the voltage resulting from each potentiometer which is proportional to the displacement to accomplish the normalization process.

$$V_{pedals}(\%) = \frac{V_{measured} - V_{\min}}{V_{\max} - V_{\min}} \cdot 100 \qquad (1)$$

2.1.2 Inertial sensor measurement characteristics

An inertial measurement sensor MTi of the company Xsens was used to indirectly measure the driver behaviour [16].

This device integrates gyroscope and accelerometers in the three axes and internally makes digital signal processing through a sensorial fusion process. It is able to provide the inclination (tilt), the angular speed and linear acceleration in the three axes. One of the advantages of such type of sensors is that due to its small and compact size, it can be easily installed in the vehicle. It is necessary to take into account making an initial reset of the system to compensate for the possible initial angle between the surface where the sensor is placed and the reference system of the vehicle. In order not to be affected by any possible unwanted vibrations, this sensor has been placed at the bottom of the vehicle as shown in the right side of Fig. 3.

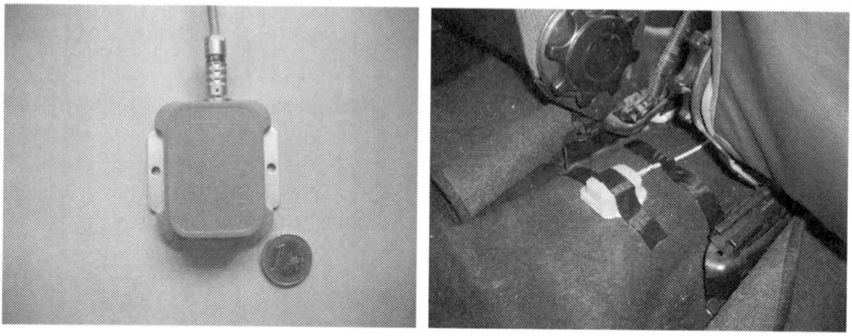

Figure 3: Inertial sensor measurement system.

2.2 Conditioning and acquisition system

The two sensory sources have different electrical properties. Consequently, two stages of data acquisition and conditioning were designed as shown below.

2.2.1 Conditioning and acquisition characteristics of potentiometers

It is necessary to make some signal conditioning operations on the output voltage of the potentiometers such that it occupies the entire dynamic range of the data acquisition card. Therefore, the signal conditioning module DAQP-BRIDGE-B, shown in Fig. 4, was used [17]. This module allows a gain adjustment by a differential amplifier. Moreover, it adjusts the offset of the signal. Finally, it contains a low pass filter to eliminate the possible high-frequency noise.

Once the signal was conditioned, a Wheatstone bridge configuration was used to measure small increments of voltage before proceeding to one of the channels of the data acquisition card.

The portable and high-speed data acquisition card USB NI-6211 of the company National Instruments was used to make the acquisition of the previously conditioned signals from the potentiometers. This card performs data communication with a laptop computer via a USB port and is fed through the

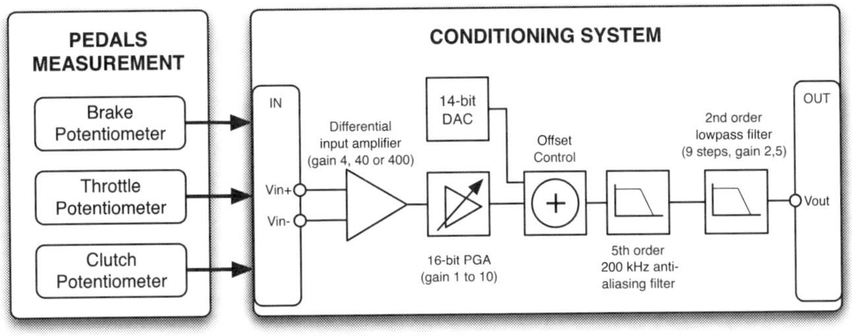

Figure 4: Potentiometers conditioning system.

same bus. A sampling period of 20 ms, which is sufficient to represent the dynamics of the driver's action, was chosen to capture any quick action of the driver on any of the pedals.

2.2.2 Inertial sensor conditioning and acquisition characteristics

The inertial sensor is connected directly to the laptop that is responsible for storing the data acquired during a tour through another USB ports.

This sensor provides a built-in system of conditioning and acquisition. Therefore, it is not necessary to use one channel of the data acquisition card as in the case of potentiometers. This system, shown in Fig. 5, is software configurable and provides an output synchronized by a trigger. This trigger activates the acquisition and delivery of the desired data after a process of sensorial fusion. The sampling period of the system was also set to 20 ms to compare with the measurement system of the potentiometers.

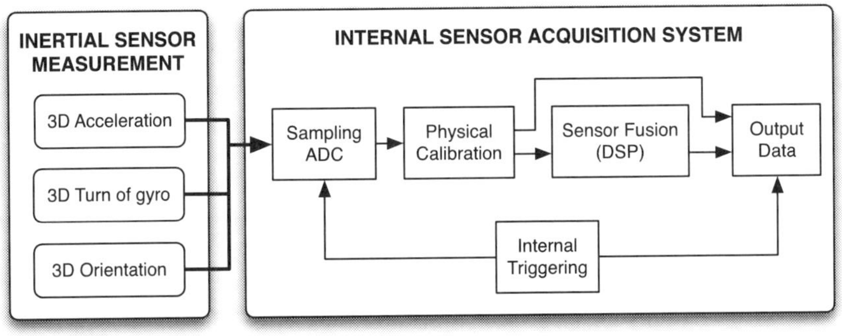

Figure 5: Inertial sensor acquisition system.

2.3 Saving and monitoring system

This system allows visualize and record the activity level of the driver. On one hand, the system shows the behavior of the variables measured by the sensorial system in real time. It was programmed with LabVIEW, an application that has a GUI as shown in Fig. 6 [19]. This application allows monitoring in a laptop onboard the vehicle.

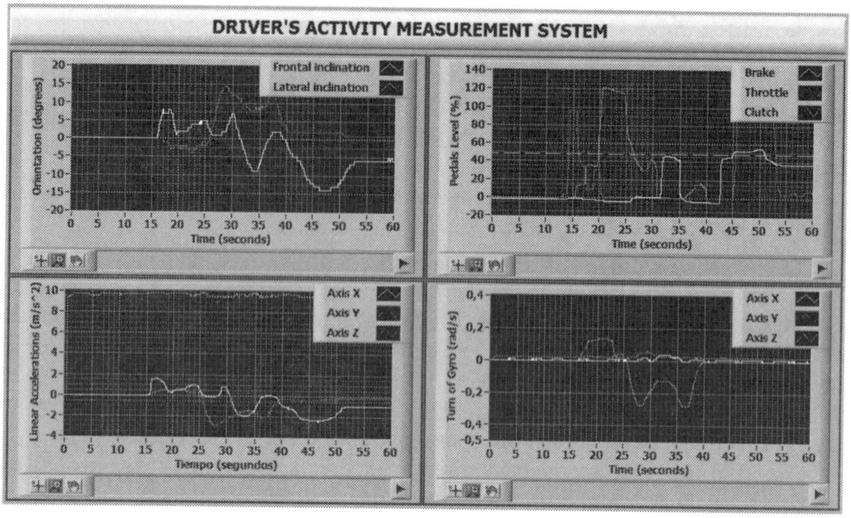

Figure 6: GUI for monitoring of the driver's activity.

Besides visualization, the designed software application allows recording of all signals captured by sensors in a database for off-line visualization and analysis. The database was designed using MySQL [20].

3 Results and discussion

The system has been designed to characterize the behavior of the driver by two complementary techniques whose results are detailed below.

3.1 Driver's activity measurement with potentiometers

This method allows determine the degree of drivers' activity on each of the three pedals in the car, i.e., accelerator, brake, and clutch, relative to the full scale of the pedal path.

The measurement of the accelerator and brake pedals helps decide whether the driver has a more or less aggressive driving behavior. In other words, he may drive with sudden and repeated accelerations and brakes. Or on the contrary, he may drive with soft and maintained accelerations and brakes.

The measurements of the three pedals for the driver realized during a tour in the center of Madrid are shown in Fig. 7. There are constant accelerations and brakes as well as gear changes are reflected in the variation of the clutch pedal.

Due to the complexity of analyzing the signals in urban environments where there are continuous changes in the actions of the pedals, it has employed.

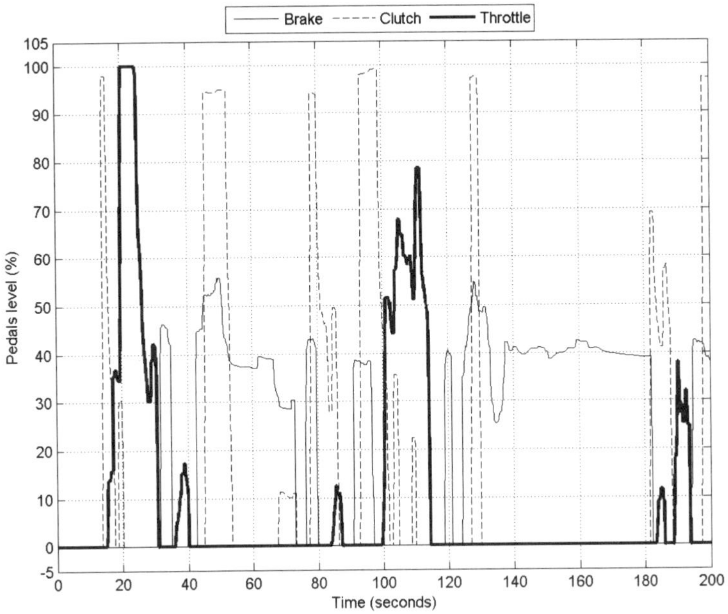

Figure 7: Vehicle pedals measurement in urban route.

A simple statistical analysis to evaluate the driver's behavior was made. A tour of the center of Madrid with an aggressive driving and a more relaxed one were conducted. As an example, the result obtained for the signal of the accelerator pedal position in the case of an aggressive driving with continuous and sudden accelerations as compared to a relaxed driving can be seen in the upper part of Fig. 8.

A statistical analysis was performed to compare the two signals of the accelerator pedals of Fig. 8. Therefore, the mean and standard deviation of the signal were calculated, which are plotted in the lower part of Fig. 8, and then analytically in Table 1.

A higher value of the mean indicates that the values of the accelerator pedal while driving are of higher amplitude. On the other hand, a higher value of the standard deviation indicates the dispersion around the mean is higher, and consequently, there have been more pronounced changes in the accelerator pedal for the case study of aggressive driving.

On the other hand, the measurement of the clutch pedal should be analyzed in a different way from the brake and accelerator pedals. This signal helps determining whether the driver is travelling over low gear levels during a long time period or on the contrary, he uses high gear levels [14]. This criterion is also decisive to save fuel and extend engine life.

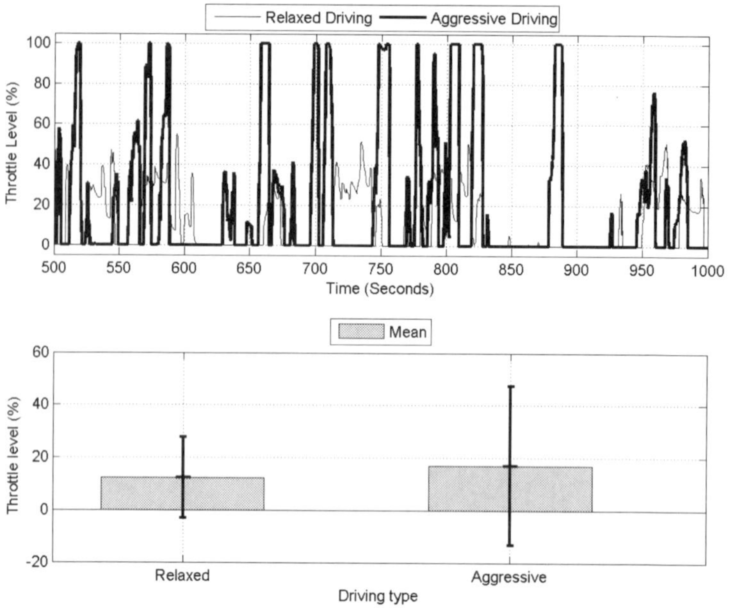

Figure 8: Accelerator level study in urban route.

Table 1: Throttle dispersion measure.

Driving	Mean	Standard deviation
Relaxed	12.2768	15.4476
Aggressive	17.0336	30.3452

3.2 Driver's activity measure with inertial sensor

This method tells us in an indirect way when the vehicle slows down and speeds up paying attention to the measurement of acceleration in the X-axis. This axis was chosen to represent the longitudinal movement of the vehicle. Increments in acceleration caused by sharp turns in curves, roundabouts or changing direction can be observed paying attention to the transverse axis and the Y-axis.

The measurements of the pedals during the route stated previously for measuring the driver's activity by means of the pedals are shown in Fig. 9. Constant accelerations and decelerations in the X- and Y-axes can also be observed here. It seems logical that the acceleration in the Z-axis remains approximately constant around the value of gravity.

The measurement of the linear accelerations in the X- and Y-axes principally allows deciding whether the driver has a more or less aggressive driving behavior, i.e., whether he drives with sharp accelerations in the forward direction of the vehicle, or he passes roundabouts in a more or less smooth manner.

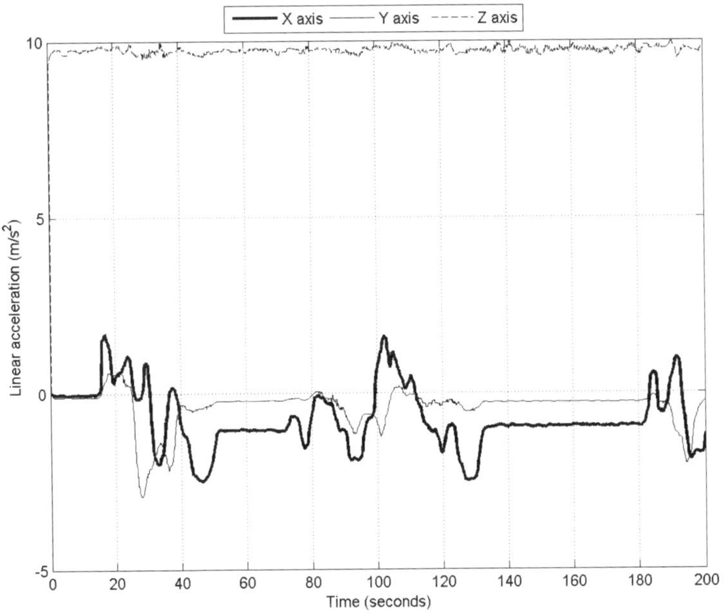

Figure 9: X-axis linear acceleration measurement in urban route.

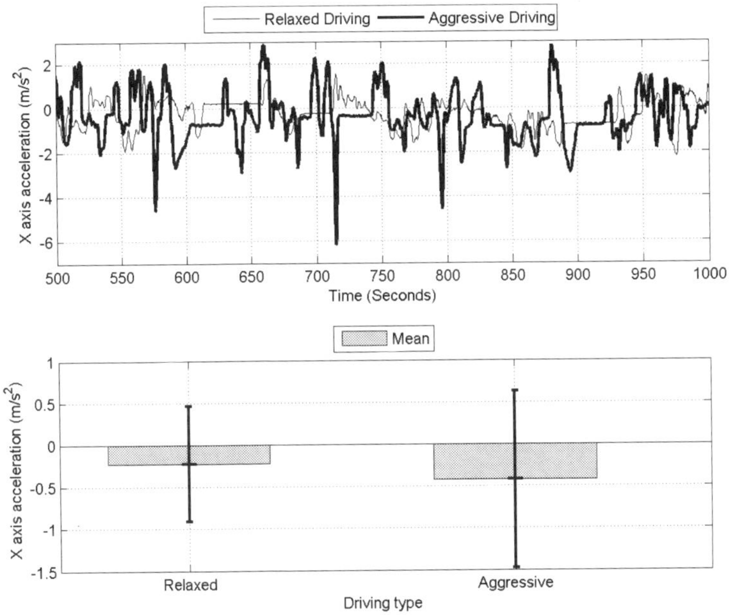

Figure 10: Axis X acceleration study in urban route.

A simple statistical analysis to evaluate the driver's behavior was used due to the complexity of analyzing the signals in urban environments where there are continuous changes in the frontal and lateral accelerations of the vehicle. A tour of the center of Madrid with an aggressive driving and a more relaxed one was realized. For example, the result for the signal from the vehicle's linear acceleration in the X-axis in case of aggressive driving as compared to a more relaxed driving can be seen in the upper part of Fig. 10.

A statistical analysis was performed to compare between the two signals of acceleration in the X-axis in Fig. 10. We calculated the mean and standard deviation for both signals. The results are plotted in the lower part of Fig. 10 and illustrated numerically in Table 2.

Table 2: Axis X acceleration dispersion.

Driving	Mean	Standard deviation
Relaxed	-0.2237	0.6877
Aggressive	-0.4251	1.0504

In this case, the ideal value of the mean for the measurement of acceleration in the X-axis should be zero. The measured values are close to the ideal value. The standard deviation over the mean value in the ideal case of an aggressive driving is more than that of a relaxed driving. Because in the relaxed driving, there were minor amplitude changes in the X-axis acceleration signal.

4 Conclusions

The activity of the driver is a key parameter in efficient driving in terms of fuel economy and pollution levels. Therefore, it is very useful to have an electronic system that records the variables associated with driving behavior synchronously with the measurement of gases and / or particles.

The designed and implemented system allows measuring the activity of the driver directly (using accelerator, clutch, and brake pedals activity) and indirectly (using instantaneous changes in the linear and angular movement of the vehicle). Simultaneously, the proposed system provides information on variables associated with more or less aggressive behavior of the driver.

The resulting electronic system in both its hardware and software components is autonomous, universal and easily integrated with other measurement applications in the car.

References

[1] Wada, T., Doi, S., Imai, K., Tsuru, N., Isasi, K., Kaneko, H., "*Analysis of Drivers' Behaviors in Car Following Based on A Performance Index for Approach and Alienation*", SAE Technical Papers, Document Number: 2007-01-0440, 2007.

[2] Gordon, D., *"Steering A New Course: Transportation, Energy, and the Environment"* Union of Concerned Scientists, 26 Church Street Cambridge MA 022238, 1991.

[3] Li, H., Andrews, G.E., Daham, B., Bell, M.C., Tate, J.E., Ropkins, K., "Impact of Traffic Conditions and Road Geometry on Real World Urban Emissions using a SI Car", SAE Technical Papers, Document Number: 2007-01-0308, 2007.

[4] Li, H., Andrews, G.E., Khan, A. A., Savvidis, D., Daham, B., Bell, M., Tate, J., Ropkins, K., *"Analysis of Driving Parameters and Emissions for Real-World Urban Driving Cycles Using an On-Board Measurement Method for a EURO 2 SI Car"*, SAE Technical Papers, Document Number: 2007-01-2066, 2007.

[5] Mierlo, J.V, G. Maggetto, E. Burgwal, and R. Gense. *Driving Style and Traffic Measures-Influence on Vehicle Emissions and Fuel Consumption. J of Automobile Engineering*, v.218 part D:46-50, 2004.

[6] Treatise UK. *Treatise Ecodriving manual: Smart, efficient driving techniques*, 2007. http://www.treatise.eu.com/downloads-uk.html

[7] EcoDriven. *European Campaign On improving DRIVing behaviour, ENergy-efficiency and traffic behaviour (ECODRIVEN)* – Benefits of EcoDriving, 2006. http://www.ecodrive.org/Benefits-of-ecodriving.277.0.html

[8] United Nations Framework Convention on Climate Change. The First Ten Years. ISBN 92-9219-010-5, 2004.

[9] Kyoto Protocol to the United Nations Framework Convention on Climate Change, 1998.

[10] Green Paper: Towards a new culture for urban mobility, 2007.

[11] European Federation for Transport and Environment, 2007. http://transportenvironment.org

[12] Frey, H.C., A. Unal, and J. Chen, *Recommended Strategy for On-Board Emission Data Analysis and Collection for the New Generation Model*, Prepared by North Carolina State University for the Office of Transportation and Air Quality, U.S. Environmental Protection Agency, Ann Arbor, MI. February 2002.

[13] Vermuelen, R. J., TNO Report 2006: *The effects of a range of measures to reduce the tail pipe emissions and/ort he fuel consumption of modern passenger cars on petrol and diesel.* http://www.ecodrive.org/Downloads.203.0.html

[14] Wilbers, P., *Eco-Driving in Netherlands: The smart driving style. Energy Forum*, Pamplona, Spain, 2005.

[15] WS31 / WS42 Position Sensors http://www.asm-sensor.com/asm/homepage.php

[16] MTi-Miniature Attitude and Heading sensor. http://www.xsens.com/en/products/machine_motion/mti.php

[17] DAQP-BRIDGE-B Module (revision 2). Technical reference manual. http://www.dewetron.com/products/

[18] USB Bus-Powered M Series Multifunction DAQ. National Instruments. http://sine.ni.com/
[19] LabVIEW Development Systems (2008). http://www.ni.com/labview
[20] MySQL 5.1 Reference Manual. MySQL AB, 2008 Sun Microsystems, Inc. http://www.mysql.com

The importance of atmospheric particle monitoring in the protection of cultural heritage

I. Ozga[1,2], N. Ghedini[1,2], A. Bonazza[1], L. Morselli[2] & C. Sabbioni[1]
[1]*Institute of Atmospheric Sciences and Climate, CNR, Bologna, Italy*
[2]*University of Bologna, Italy*

Abstract

It is now well known that air pollution is responsible for the accelerated damage encountered on cultural heritage located outdoors. Although several works on atmospheric pollutants have been performed, studies of atmospheric pollutant monitoring close to monuments remain rare. In addition, the few cases reported in the literature mostly regard indoor environments. As the protection and conservation of monuments and historic buildings constitutes a priority for each country, knowledge of particle composition near monuments over time is an important issue in conservation strategies. For this reason, the atmosphere in proximity of the Florence Baptistery, located in the city centre, was continuously monitored during 2003 and 2004 by means of aerosol sampling performed close to two of the three doors of the monument. In particular, the monitoring was performed close to the North Door, realized by Lorenzo Ghiberti (1403-1424), currently utilized as the entrance to the monument, and the South Door, a masterpiece of Andrea Pisano (1330), employed as the exit for visitors. The sampling sites were characterized by different expositions to road traffic emissions. The non-carbonate carbon and soluble ionic components of the total suspended matter were measured. The data obtained is presented and discussed with the goal of contributing to the formulation of guidelines for a suitable safeguard of the built cultural heritage.

Keywords: monitoring, urban pollutants, non-carbonate carbon, ions, conservation, cultural heritage.

WIT Transactions on Ecology and the Environment, Vol 123, © 2009 WIT Press
www.witpress.com, ISSN 1743-3541 (on-line)
doi:10.2495/AIR090241

1 Introduction

Since human history is documented by the artistic-cultural heritage, its preservation in time is a priority for all countries. Moreover, every year movable and immoveable heritage attracts millions of visitors, incrementing tourism and constituting an important source of income. Thus, the safeguard, conservation and maintenance of cultural heritage constitute a common challenge of all nations, and the least onerous strategy, but the one most able to protect the artworks, is to avoid damage by reducing the potential causes. Today, anthropogenic air pollution represents an element of serious risk to the preservation of works of art, particularly cultural heritage located outdoors [1]. Such historic monuments and buildings are the most susceptible to deterioration risks because as they are mainly located in city centres, directly exposed to atmospheric pollutants, which are the main cause of damage in urban areas [2].

The conservation of monumental heritage is directly correlated with the atmospheric pollution and its future depends on the reduction of air pollutants. Thus the monitoring pollution and surface deposition must be among the main goals in cultural heritage preservation [3]. In urban areas, outdoor stone and metal artworks are particularly vulnerable to pollutant action [4, 5], particularly to gases and particles from combustion processes, the main cause of the aesthetic and material damage encountered today on the monumental and built heritage [6, 7].

To protect cultural heritage effectively it is necessary to have a detailed knowledge of the characteristics of the environment in which single assets are situated. Such information is a fundamental pre-requisite for identifying and quantifying multi-pollutants, singling out emission sources, calculating the entity of deposition on the various materials of interest, forecasting and estimating the damage potential, setting tolerance thresholds, and establishing mitigation and adaptation strategies.

In the past, the attention of researchers was focused mainly on gaseous pollutants, especially sulphur dioxide, considered to be one of the most important causes of stone deterioration [8, 9]. However, in recent years a change has come about: in many areas of Europe, levels of SO_2 have been reduced, while the increase in automobile traffic has brought in its wake a rise in levels of ozone concentrations and total suspended particulate (TSP), and, among the constituents of the latter, there has been an increase in carbonaceous and nitrogenous fractions. This has generated an entirely new air pollution scenario [10, 11].

Ever-increasing human activities have produced a continual enrichment of the carbonate fraction in suspended particulate, which is drawing increasing attention of researchers. Carbonaceous particles are not only the cause of the black colour of the patinas that mar the appearance of monuments and buildings [12], but it has also been demonstrated that they play an active role in calcite sulphation, because of their sulphur content and the presence of heavy metals, which trigger the said process [13]. Moreover, due to their high specific surface

(10–100 m^2/g), they themselves act as a catalytic support in deterioration reactions [14].

The preservation of cultural heritage and its protection against possible damage due to atmospheric air pollution has only recently become the focus of well-deserved scientific interest. So far, very little coherent work has been performed to assess and monitor the impact of air pollution on the most important historic buildings and monuments, with the aim of protecting them from damage. Most of the monitoring undertaken in urban areas has been with a view to studying the health-related effects of air pollution. Thus, the assessment of the impact of multi-pollutants is based on monitoring that is generally performed according to the air quality directives adopted for the protection of human health. The resulting data often regards samples collected far from monuments of interest, and obviously do not allow the evaluation of the spatial and temporal variations of multi-pollutants in proximity to the monuments to be protected.

In order to investigate variation trends in particle quantity and composition in the air immediately surrounding monuments of historic-artistic importance, during 2003 and 2004 atmospheric monitoring was carried out in the area of the Baptistery in Florence, and particularly close to the two famous bronze doors realized by Lorenzo Ghiberti (Northern Door) and Andrea Pisano (Southern Door).

The aim of this work is to characterize the local atmosphere around the Baptistery. The results of the study will be of use to cultural heritage managers, and especially the authorities responsible for the preservation of the Florence Baptistery.

2 Materials and methods

The Baptistery of Florence (fig. 1) is situated in the historic city centre, in an area mainly reserved for pedestrians, with the exception of one road bearing intense traffic of heavy diesel-run vehicles (more than 2000 buses a day – ARPAT 2003), accessible also to motorcycles and taxis, which run a few meters' distance from the north-facing Door.

Figure 1: The Baptistery in the center of Florence.

The specific location of the monument means that its sides have a different exposition to pollutant agents, in particular the two bronze doors of the northern and southern sides: the North Door (ND), the work of Lorenzo Ghiberti (1403-1424), is currently used as the entrance to the monument, and the more ancient South Door (SD), a splendid example of Gothic sculpture, masterpiece of Andrea Pisano (1330), is used as the visitor exit.

Sampling of total suspended particulate (TSP), i.e. without granulometric selection, was performed by the Agenzia Regionale per la Protezione Ambientale della Toscana (ARPAT) using two TECORA samplers, equipped with membrane filter systems (Ø = 47 mm), positioned in proximity to each of the two doors. Each system consists of several sampling lines: quartz fibre filters were positioned on two of them, for the determination of organic substances and carbon, while the other two were equipped with polycarbonate membranes, used to determine the ionic fraction and for morphological analyses.

TSP sampling took place from February 2003 to November 2004 in the form of nine three-day seasonal campaigns: winter (December, January, February), spring (March, April, May), summer (June, July, August), and autumn (September, October, November).

On the TSP, measurements were performed of the soluble ionic, anionic and cationic components, as well as non-carbonate carbon (NCC), i.e. the fraction not linked to soil dust.

For the measurement of the soluble saline fraction carried out by ion chromatography (IC), a Dionex Chromatograph model 4500i with conductivity detector (Dionex CD II) was used. During the analyses the addition of fructose prevents the oxidation of the sulphites and sulphates [15]. The cations measured were: lithium, sodium, potassium, ammonium, magnesium, calcium, strontium, manganese, and the following anions: sulphate, bisulphite, nitrate, nitrite, fluoride, chloride, bromide, hydrogenophosphate, formate, acetate, oxalate.

The analysis of the carbonaceous fraction was determined by combustion/gas chromatography, using a specific analytical methodology designed, developed and validated for carbon measurements in complex, minimal mass samples, such as those consisting of aerodisperse particulate collected on a filter. For the analysis of NCC, portions of filter of known area (fig. 2) are weighed and introduced into the instrument's silver micro-crucible.

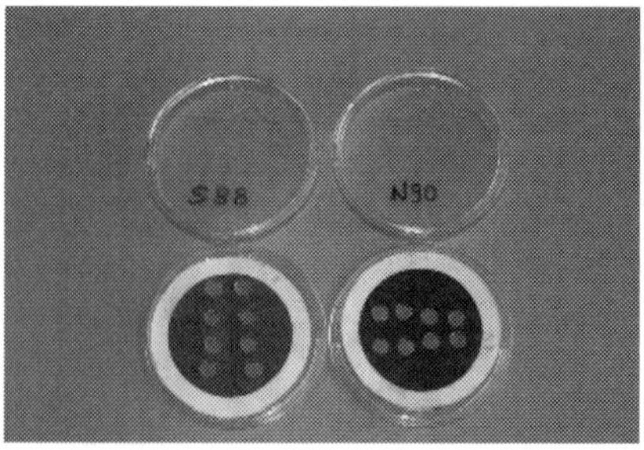

Figure 2: Quartz fibre filters without the portions used for NCC analysis (sampled from SD on the left and ND on the right).

Subsequently, 5-10 μl of 50% HF solution is added in order to eliminate the quartz through the formation of soluble, highly volatile, and strongly acid SiF_4. The crucible is then maintained for 24 hours in a dryer containing potassium hydroxide, in order to eliminate SiF_4, excess HF and water. The dry residue remaining in the crucible is analysed employing a CHNSO EA 1108 FISONS Instrument. The same pre-treatment and the same measurement are performed also on a blank filter and the measured quantity of non-carbonate carbon (NCC) is subtracted from those obtained on particulate samples deposited on the filter.

3 Results and discussion

IC data of all samples collected near both northern door (ND) and southern door (SD) during the two year period indicates that the major anions are $SO_4^{=}$, NO_3^- and Cl^- and the major cations are Ca^{++}, Na^+ and Mg^{++}.

In figures 3 and 4 the concentrations of the anions sulphate and nitrate show, at the two sampling sites, rather similar trends during 2003 and 2004, but with mean seasonal concentrations higher at ND than SD. In general, while $SO_4^{=}$ concentrations undergo an appreciable decrease during the two years period those of NO_3^- are on the rise. This increase in mean concentrations of NO_3^-, from 2003 to 2004 is particularly evident in winter and spring, when fuel consumption increases due to the greater demand for heating. In all seasons, both in 2003 and 2004, the quantities of NO_3^- measured at ND are higher than those recorded at SD.

These results reflect the gradual reduction of the sulphur content in fuels and, at the same time, their increased consumption.

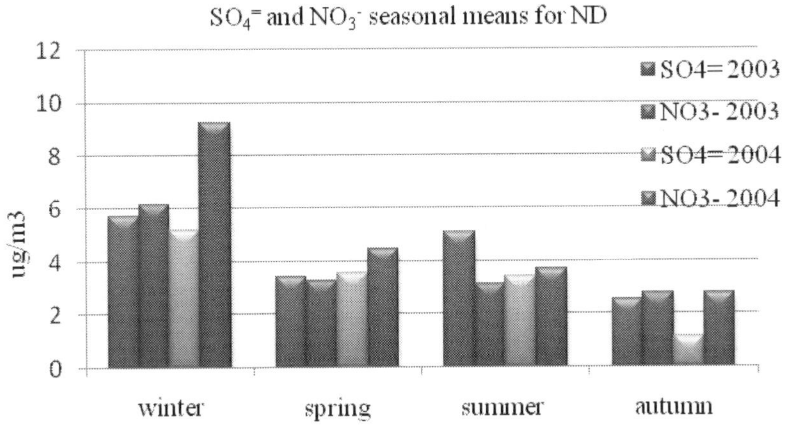

Figure 3: Seasonal means of sulphate and nitrate anions at ND in the monitoring period.

Chlorides are always present in considerable concentrations in all the aerosol samples analysed. In winter, spring and autumn at ND, mean total values of Cl^- are much higher than those measured at SD, while in summer the mean concentration measured at SD are about double the one detected at ND (0.50 e 0,99 $\mu g/m^3$, respectively).

In the two-year period in question, measurements of organic anions show acetates to have the highest mean concentrations at both sampling sites (0.81 $\mu g/m^3$ at ND and 0.87 $\mu g/m^3$ at SD), followed by oxalates and formates. With regard to organic anions, no major differences between the concentrations measured at ND and SD are observed throughout the period under consideration.

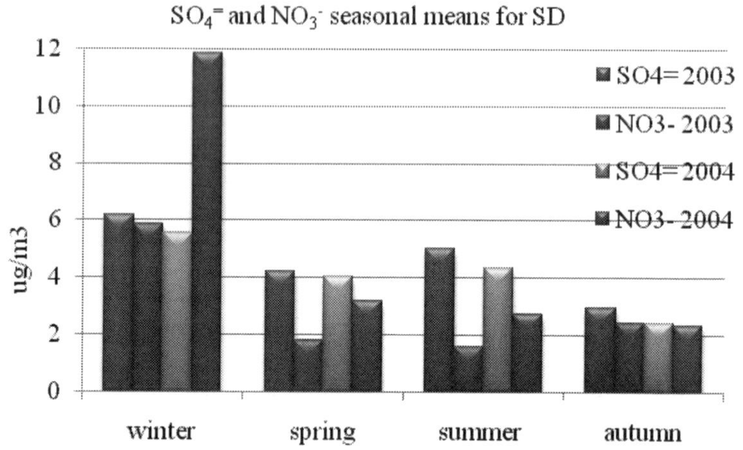

Figure 4: Seasonal means of sulphate and nitrate anions at SD in the monitoring period.

Since organic anions have a number of natural and anthropogenic origins [16], their yearly mean concentrations do not reveal particular variations. However, at ND the highest concentration is observed for $C_2H_3O_2^-$ in winter and for $C_2O_4^=$ and CHO_2^- in spring, while at SD the highest concentrations are measured for $C_2H_3O_2^-$ and $C_2O_4^=$ in winter and CHO_2^- in autumn.

The cations monitored at the two doors are found to have very similar seasonal mean concentrations during both years (fig. 5, 6).

In 2003 and 2004, Ca^{++} and Mg^{++} ions are the cations that are always present in the highest concentrations in the sampled aerosol, both at ND and SD. The origin is natural, being among the dominant macro-constituents of the earth's crust. The Ca^{++} presents the same seasonal trend at ND and SD, reaching a maximum concentration in spring (5.20 and 7.04 $\mu g/m^3$, respectively) and a minimum in autumn (2.57 and 2.04 $\mu g/m^3$). With regard to Mg^{++}, the maximum is reached in spring at ND (0.89 $\mu g/m^3$) and in winter at SD (0.48 $\mu g/m^3$), while the minimum occurs in summer at both doors. On average over the two-year period, at both ND and SD, Na^+ shows maximum concentrations in winter and autumn, and minimum in spring and summer. The K^+ ion is present in higher quantities in winter and summer at ND and SD, while lower ones are encountered in autumn and spring at both doors. In winter and summer, the NH_4^+ ion has higher mean seasonal concentrations at both ND and SD, while the minimum is found at ND in spring and at SD in autumn.

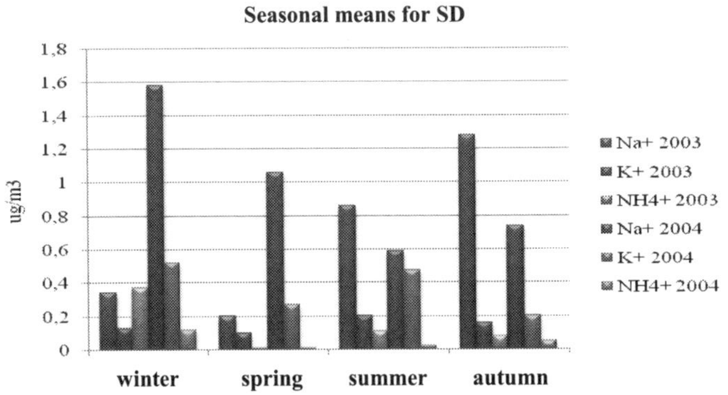

Figure 5: Seasonal mean concentrations of Na^+, K^+ and NH_4^+ ions at ND in the monitoring period.

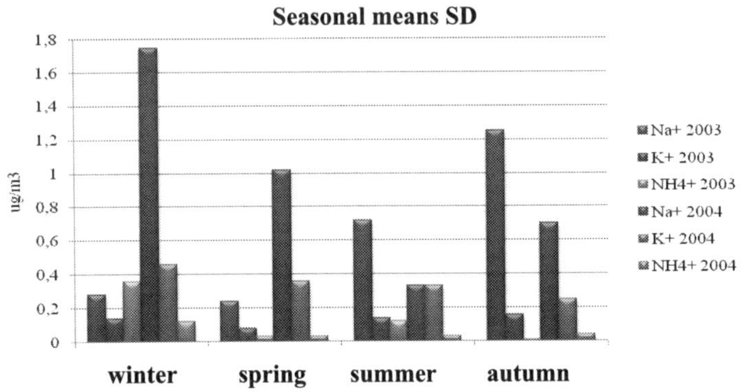

Figure 6: Seasonal mean concentration for Na^+, K^+ and NH_4^+ ions at SD in the monitoring period.

Regarding NCC, the concentrations measured at ND are always higher than those measured at SD in the same period (fig. 7, 8).

In fact the overall mean of NCC shows higher values at ND compared to SD (18.55 and 14.84 μg/m^3, respectively) and, at both doors, the annual mean of NCC is higher in 2004 than in 2003. On a seasonal basis, again in terms of mean values of NCC relative to the period February 2003–October 2004, at ND and SD, respectively, the quantities measured are: 20.15 and 15.40 μg/m^3 in winter, 18.12 and 17.90 μg/m^3 in spring, 17.10 and 13.67 μg/m^3 in summer, and 17.48 and 12.03 μg/m^3 in autumn.

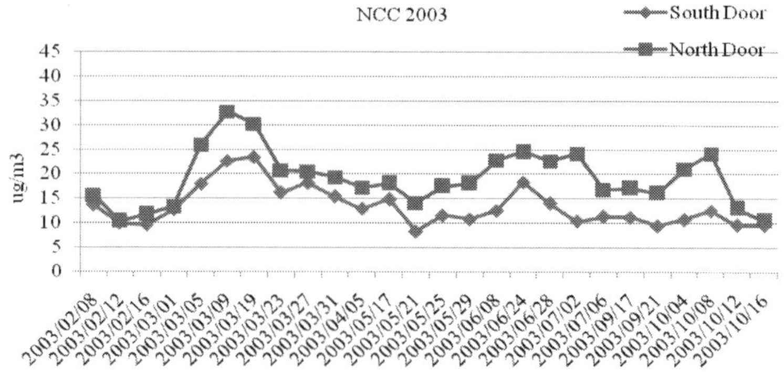

Figure 7: Non-carbonate carbon measured in 2003 at both Baptistery doors.

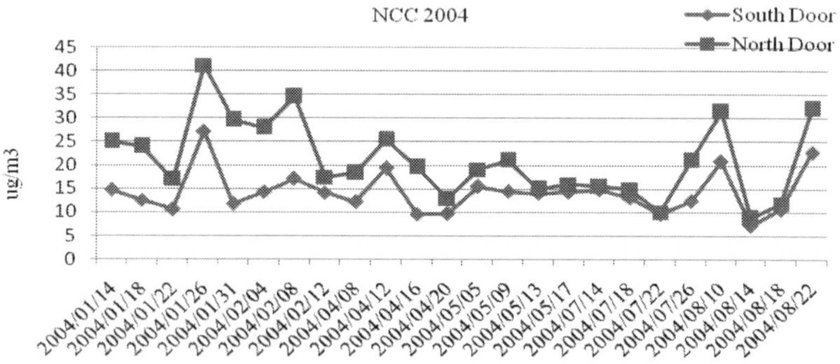

Figure 8: Non-carbonate carbon measured in 2004 at both the Baptistery doors.

The comparison among the seasonal mean values of NCC (fig. 9) shows that, in 2003, the highest concentration was in summer at ND and in spring at SD, while in 2004, the highest concentrations are those found in winter at ND and in spring at SD. In 2003 the lowest seasonal values of NCC are those relating to the winter at ND and to autumn at SD, while in 2004 the minima are recorded in summer at both ND and SD. The different seasonal trends for mean NNC concentrations during 2003 and 2004, can be explained by the high temperature and exceptional lack of precipitations recorded in the summer of 2003. During summer 2003 due to high temperatures and prolonged drought, the increase of TSP concentration was affected by road and soil dust re-suspension.

Such re-suspension was favoured by the flow of tourists and, at ND where visitors queue at length to purchase entrance tickets, there is an excellent correlation between concentrations of NCC and TSP.

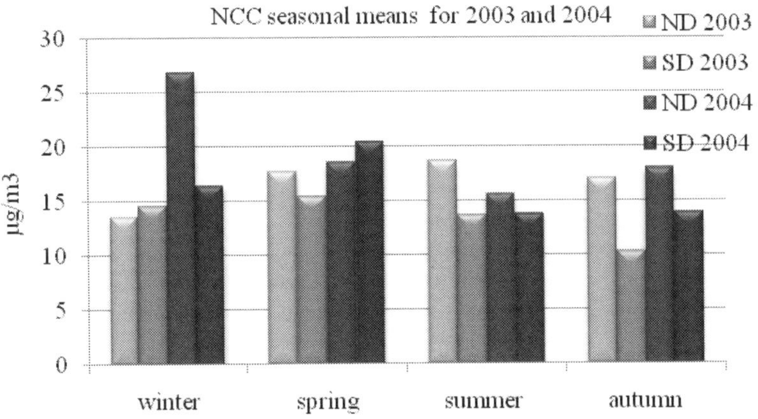

Figure 9: NCC seasonal mean concentration, for 2003 and 2004, at the two doors.

The mass balance based on the results obtained show that, at ND, the aerosol consists on average of about 27% NCC, 9% cations, 15% anions and the rest, approximately 49%, constitutes the non determinate fraction (soil dust); at SD it is composed on average of about 25% NCC, 12% cations, 18% anions and about 45% non determinate fraction.

4 Conclusions

Since the degradation of historic monuments and buildings situated outdoors is known to depend on their exposure to airborne chemical, physical and biological agents, continuous monitoring was performed of TSP in proximity to two of the three doors of the Florence Baptistery, the north-facing and south-facing ones. The study aimed to assess the impact of multi-pollutants by measuring their concentrations in the atmosphere close to the monument. The results obtained give evidence of pollution due to fossil fuel (diesel, petrol) combustion processes, typical of urban centres, the source being emissions by both mobile and fixed combustion sources such as vehicles, domestic heating systems and industrial activities. Because of its location, northern door (ND) turns out to be more exposed to the action of atmospheric multi-pollutants than southern door (SD), nearly always showing higher concentrations of TSP, NCC and NO_3^-.

While the mean concentration of $SO_4^=$ gradually decreases over time, that of NO_3^- undergoes an increase from 2003 to 2004, especially in winter and spring. In all seasons, concentrations of the nitrate anion at ND are higher than those measured at SD, in both 2003 and 2004, and this difference can be attributed to the emissions of the intense vehicular traffic characterizing the road adjacent to ND.

The mean concentration of NCC also rises from 2003 to 2004, with a trend over time similar to that of TSP. The quantity of NCC measured at ND (on average about 27% of TSP) always turns out higher than that at SD (about 25% of TSP). Here again the higher values can reasonably be attributed to the emissions of the heavy traffic (cars, heavy diesel-run vehicles, taxis, motorcycles) running along the road at about three meters distance from ND, which is therefore directly exposed, while SD is only indirectly exposed. At both ND and SD the ions NH_4^+ and $SO_4^=$ show the same mean seasonal trend, which seems to point to a common origin.

Most aerosol samples showed significant concentrations of acetates and oxalates, near both ND and SD. The origin, particularly that of the oxalate ion, remains an unresolved problem, since neither an anthropogenic nor biological origin can be excluded with certainty. In this case too, it is likely that both combine to determine the quantities measured during the two years monitored. The constant absence of sulphites underscores the oxidant character of the local atmosphere.

The results obtained show how the concentration and composition of particles, especially those relative to ND, appears to be influenced by the presence of visitors in the area of the Baptistery, emphasizing that tourism can have a strong impact on the re-suspension of particulate (road-soil dust), especially in the dry period. The monitoring carried out ascertains that every monument located outdoors is exposed to chemical, physical and biological agents characterizing the area in which it is situated. Thus, only extensive local monitoring can produce the data required to identify and quantify them. The study also shows that, even in outdoor environments, the flow of visitors in proximity to cultural heritage (fruition) can influence the surrounding atmospheric aerosol.

Because of the crucial role of atmospheric pollutants in the damage of cultural heritage, especially in urban areas, atmospheric pollutant monitoring close to single monuments and historic buildings is necessary to establish correct mitigation and adaptation strategies, specific for each site of interest, to protect it from degradation and preserve it for future generations.

Acknowledgements

The authors wish to thank the Agenzia Regionale per la Protezione Ambientale della Toscana (ARPAT) for performing the atmospheric particles sampling within the "Progetto Battistero: studio dell'impatto ambientale sui monumenti nel centro urbano di Firenze".

This work was performed within the “European Ph.D. in Science for Conservation (EPISCON)” funded by the European Commission with 6FP EU-Marie Curie EST Action and within the EC Project “Technologies and Tools to prioritize Assessment and diagnosis of air pollution impact on immovable and movable Cultural Heritage (TeACH)”(contract number: 212458).

References

[1] Rossval, J., *Air Pollution and Conservation*, Elsevier, Amsterdam, 1988.

[2] Brimblecombe, P., *The effects of air pollution on the built environment*, Air Pollution Reviews, Vol. 2, Imperial College Press, London, 2003.

[3] Morselli, L., Bernardi, E., Passarini, F., The City of Tomorrow: Towards a Definition of "Limit Values" for the Pollutants in Relation to Decay of Cultural Heritage. A Proposal. *Environmental Science and Pollution Research - International*, **9**, pp. 287-288, 2002.

[4] Bernardi, E., Chiavari, C., Lenza, B., Martini, C., Morselli, L., Ospitali, F., Robbiola, L., The atmospheric corrosion of quaternary bronzes: the leaching action of acid rain. *Corrosion Science*, **51,** pp. 159-170, 2009

[5] Camaiti, M., Bugani, S., Bernardi, E., Morselli, L., Matteini, M., Effects of atmospheric NO_x on biocalcarenite coated with different conservation products. *Applied Geochemistry*, **22**, pp. 1248-1254, 2007.

[6] Sabbioni, C., & Zappia, G., Decay of sandstone in urban areas correlated with atmospheric aerosol. *Water, Air and Soil Pollution,* **63**, pp. 305-316, 1992.

[7] Sabbioni, C., Zappia, G., Gobbi, G., Carbonaceous particles and stone damage in a laboratory exposure system. *Journal of Geophysical Research,* **101**, pp. 19621-19627, 1996.

[8] Sabbioni, C., Contribution of atmospheric deposition to the formation of damage layers. *Science of the Total Environment*, **167**, pp. 49-56, 1995.

[9] Gobbi, G., Zappia, G., Sabbioni, C., Anion determination in damage layers of stone monuments. *Atmospheric Environment*, **29A**, pp. 703-707, 1995.

[10] Jacob, D.J., *Introduction to atmospheric chemistry*, Princeton University Press, Princeton, pp. 144-159, 2000.

[11] Bonazza, A., Sabbioni, C., & Ghedini, N., Quantitative data on carbon fractions in interpreting black crusts and soiling on European built heritage. *Atmospheric Environment*, **39**, pp. 2607-2618, 2005.

[12] Grossi, C. M., Esbert, R. M., Diaz-Pache, F., Alonso F. J., Soiling of building stones in urban environments. *Building and Environment,* **38**, pp. 147-159, 2003.

[13] Rodríguez-Navarro, C., & Sebastián, E., Role of particulate matter from vehicle exhausts on porous building stone (limestone) sulphation. *The Science of the Total Environment,* **187**, pp. 79-91, 1996.

[14] Benner, W.H., Brodzinsky, R., Novakov, T., Oxidation of SO_2 in droplets which contain soot particles. *Atmospheric Environment*, **16**, pp. 1333-1339, 1982.

[15] Gobbi, G., Zappia, G., Sabbioni, C., Sulphite quantification on damaged stones and mortars. *Atmospheric Environment*, **32**, pp. 783-789, 1998.

[16] Sabbioni, C., Ghedini, N., & Bonazza, A., Organic anions in damage layers on monuments and buildings. Atmospheric Environment, **37**, pp. 1261-1269, 2003.

Section 5
Aerosols and particles

CFD modelling of radioactive pollutants in a radiological laboratory

G. de With
Department of Radiation and Environment,
Nuclear Research & Consultancy Group, Arnhem, The Netherlands

Abstract

An important aspect of indoor air quality is the presence of radioactive pollutants. These pollutants can be present in the form of gas or particles, and are typically found in nuclear installations and radiological laboratories. In this work the dispersion of radioactive pollutants in an indoor environment is studied using Computational Fluid Dynamics (CFD). The aim of this work is to evaluate the exposure to radioactive particles during an accidental release, and to evaluate suitable ventilation design to minimise exposure. These CFD findings are used towards improvement of the Dutch assessment procedures for evaluating the risk of radioactive exposure to radiological workers. For the purpose of this work a CFD model is developed to simulate the dispersion and nuclear decay of gas and aerosols, and the attachment and deposition of radioactive aerosols.
Keywords: CFD, radioactive pollution, particle modelling.

1 Introduction

Dispersion of radioactive pollutants in nuclear installations and radiological laboratories can form a potential health hazard to those people working nearby. The dispersion can come from an accidental release of radioactive gasses, or alternatively from a burst of radioactive dust particles. After release these pollutants are dispersed in the enclosed environment through the existing air recirculation. Consequently, this leads to an increase in radioactive concentrations in the humans' inhalation region, resulting in increased health risks.

The primary role of the laboratory ventilation system is to mitigate those health risks and minimise the exposure. Nevertheless, achieving an efficient

WIT Transactions on Ecology and the Environment, Vol 123, © 2009 WIT Press
www.witpress.com, ISSN 1743-3541 (on-line)
doi:10.2495/AIR090251

ventilation system is a challenging task. At present most assessment procedures for air quality in nuclear laboratories are primarily based around some design values for the air exchange rate (AER). Nevertheless, varies studies have demonstrated that increasing the AER only has little effect on environmental conditions and exposure to pollutants [1]. More important is the air diffusion and air movement, which have considerably more impact on the ventilation performance. Consequently, the role of ventilation and its 3 dimensional flow characteristics must be an essential aspect in the air quality assessment. However, these 3 dimensional features are controlled through good ventilation design. To develop and optimise a good ventilation design faces design engineers with considerable challenges.

The purpose of this study is to investigate the dispersion of radioactive pollution using (Computational Fluid Dynamics) CFD computation. The computations take account of the dispersion and nuclear decay of gas and particles. In addition attachment of nuclear particles with aerosols and deposition of particles and aerosols are included. The algorithms for dispersion and nuclear decay of particles and gas are based on the work by Zhuo *et al.* [2]. The attachment of particles with aerosols and its deposition on the walls is based on the work by Porstendörfer [3] and Lai and Nazaroff [4].

This paper is organized in the following manner. In Section 2 a description of the CFD model is provided, followed by a validation (Section 3). Section 4 provides an overview of the CFD results, and the paper is finished with a summary of the conclusions (Section 5).

2 Mathematical model equations

As part of this investigation the CFD software FLUENT© is used. This section will provide a brief overview of the models that are used. Some of those models are already available in the CFD software and are used in the CFD computation where possible. However, algorithms to predict the deposition and dispersion of aerosols as well as the attachment of radio nuclides with the surrounding aerosols are developed in the framework of this study.

2.1 Airflow modelling

The basis for the CFD calculation is a set of two conservation equations. The first conservation equation refers to the conservation of mass and is defined in the following manner:

$$\rho(\nabla \cdot u_k) = 0 \tag{1}$$

Here u is the velocity and k is the index for the 3 velocity components. The second conservation equation is generally known as the Navier-Stokes equation. This equation describes the momentum conservation and is an implementation of Newton's second law applied to gas and liquid.

$$\rho\left(\frac{\partial(u_k)}{\partial t} + \nabla \cdot (u_k u_l)\right) = -\nabla P + \nabla \cdot (\mu_{eff} \nabla u_k) + S_{u,k} \tag{2}$$

In this equation u represents the velocity vector (m/s), P is pressure, μ_{eff} is the effective viscosity (Ns/m^2), S_u is the source term and ρ is the air density. The indices k and l are used to indicate the three velocity components. It is important to stress that most turbulent flow movements are not computed explicitly. Instead those flow movements are incorporated through an additional viscosity μ_t. This concept is developed by Prantl around 1940 and is still adopted in most CFD studies. Therefore, the effective viscosity μ_{eff} is the sum of both dynamic viscosity μ_l and turbulent viscosity μ_t.

$$\mu_{eff} = \mu_l + \mu_t \tag{3}$$

For calculation of the turbulent viscosity the well-known $k-\varepsilon$ turbulence model is used.

2.2 Modelling of gas dispersion

For the dispersion of hazardous gas in the laboratory environment an additional conservation equation is applied. This conservation equation is shown below:

$$\frac{\partial C_m}{\partial t} + \nabla \cdot u C_m = \nabla \cdot \left(\Gamma_m \nabla C_m\right) + S_{C,m} \tag{4}$$

In this equation C is the concentration of activity expressed in Bq/m^3 and Γ is the diffusion coefficient (m^2/s) of the hazardous gas. The two terms on the left hand side represent the convective transport of activity. On the right hand side the dispersion from diffusion is shown followed by the source term S_C. Where necessary, radioactive decay of the hazardous gas is incorporated in the source term S_C. Radioactive decay represents a sink to the activity of the gas and is therefore defined as $S_C = -\lambda C$. The diffusion coefficient Γ in the dispersion equation is identical to the effective viscosity in the Navier-Stokes equation (μ_{eff}/ρ) [5].

2.3 Modelling of particle dispersion

For the dispersion of particles a Eulerian based conservation equation is applied. The approach is based on the drift-flux method described by Lai and Nazaroff [4]. In literature the drift-flux method is described extensively and the method is also specifically developed for the dispersion of particles in an indoor environment [4–9].

In the drift-flux method the dispersion of particles is described by means of a continuity equation. This equation is comparable with the continuity equation for hazardous gas.

$$\frac{\partial C_m}{\partial t} + \nabla \cdot \left[\left(u + v_{s,m}\right) C_i\right] = \nabla \cdot \left[\left(\Gamma_m + D_m\right) \nabla C_m\right] + S_{C,m}. \tag{5}$$

In this equation v_s is de terminal velocity (m/s) and D is the Brownian diffusion coefficient of particles. The equation is comparable with eqn (4);

however, there are some subtle differences. The convective transport is based on a modified flow field. In this flow field the terminal velocity of the particles is incorporated. In addition the Brownian diffusion coefficient is added to the effective diffusion. In this approach it is assumed that the particles do not influence the flow field significantly. For particles that normally do not exceed 10 μm this is a very acceptable assumption.

2.4 Deposition of particles

The deposition of particles J_d (Bq/m^2/s) on the surface is the product of particle concentration near the surface and the speed with which the particles are deposited. In mathematical terms this can be described as follows:

$$J_d = v_d \cdot C_b \,. \tag{6}$$

Here v_d is the deposition velocity (m/s) of the particles and C_b is the particle concentration in the vicinity of the surface (Bq/m^3). Calculation of the deposition velocity is a difficult task. The deposition depends on a large number of factors including, gravitation, turbulence, thermal forces. In addition particles can be re-entrained as a result of resuspention and rebound. It is therefore important to choose a deposition model that is suitable for indoor flow conditions. For this reason the deposition model of Lai and Nazaroff [4] is chosen. Their model is based on the experiments from Zhang *et al.* [10] and is developed specifically for deposition in the indoor environment.

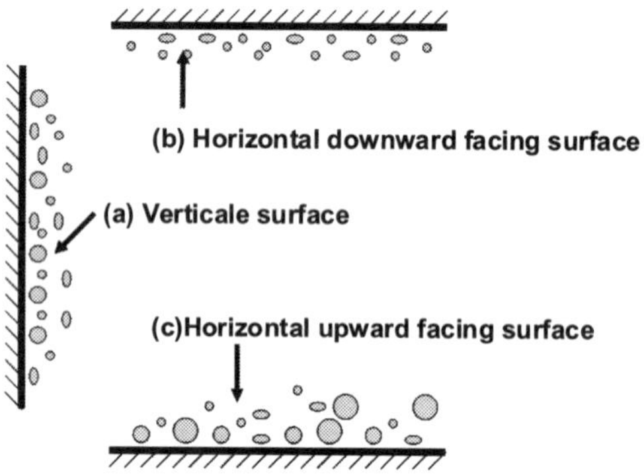

Figure 1: Orientation of the deposition surfaces.

The model equations from Lai and Nazaroff [4] describe the following three scenarios as shown in Figure 1:

- Deposition on vertical surfaces; these include the side walls.
- Deposition on horizontal surfaces facing downwards; these include the ceiling.
- Deposition on horizontal surfaces facing upwards; these include the floor.

Subject to orientation of the surface the smaller particles will deposit on the horizontal surfaces facing downwards and the vertical surfaces. The larger particles will deposit primarily on the horizontal surfaces facing upwards (Figure 1).

A summary of the model equations is presented in Table 1. In the equations the integral parameters *I, a* and *b* are used. Those parameters are based on experimental data and algebraic relations for idealized turbulent flows in the vicinity of the wall.

Table 1: Summary of the drift-flux deposition model.

Deposition speed	Correlations
-Vertical surface	$v_{dv} = \frac{u^*}{I}$
- Horizontal upward facing surface	$v_{dd} = \frac{v_s}{1-\exp\left(-\frac{v_s I}{u^*}\right)}$
- Horizontal downward facing surface	$v_{dd} = \frac{v_s}{\exp\left(\frac{v_s I}{u^*}\right)-1}$

$$I = \left[3{,}64 Sc^{2/3}(a-b)+39\right]$$

$$a = \frac{1}{2}\ln\left[\frac{\left(10{,}92\cdot Sc^{-1/3}+4{,}3\right)^3}{Sc^{-1}+0{,}0609}\right]+\sqrt{3}\cdot\tan^{-1}\left[\frac{8.6-10{,}92\cdot Sc^{-1/3}}{\sqrt{3}\cdot 10{,}92\cdot Sc^{-1/3}}\right]$$

$$b = \frac{1}{2}\ln\left[\frac{\left(10{,}92\cdot Sc^{-1/3}+r^+\right)^3}{Sc^{-1}+7{,}669\cdot 10^{-4}\left(r^+\right)^3}\right]+\sqrt{3}\cdot\tan^{-1}\left[\frac{2r^+-10{,}92\cdot Sc^{-1/3}}{\sqrt{3}\cdot 10{,}92\cdot Sc^{-1/3}}\right]$$

* *Nomenclature:* $Sc = v_l D^{-1}$, *Sc is the Schmidt number,* v_l *de kinematic viscosity of air and D is the diffusion coefficient of the particle;* $r^+ = d_p u^*(2v_l)^{-1}$, d_p *is the diameter;* u^* *is the friction velocity;* v_s *is de terminal velocity of the particle.*

The deposition velocity is calculated on the basis of the equations described in Table 1. The results are shown in Figure 2. The deposition velocity is presented for three different friction velocities u*. A higher friction velocity is associated with higher wind speed and turbulent intensity in the near wall region.

The model assumes that the air velocities are sufficiently small that no resuspention or rebound of particles occurs [4]. Lai and Nazaroff [4] have indicated that modelling of resuspention and rebound is not essential due to low air velocities and turbulent intensities in the indoor environment. It is however important to note that implementation of a resuspention model in de drift-flux model is feasible [6].

3 Model verification

To validate the numerical model for simulating the indoor particle distribution, the measured data by Chen *et al.* [7] is adopted. Chen *et al.* [7] performed laboratory experiments for a model room with a geometry of $L_{ength}\times W_{idth}\times H_{eight}$ = 0.8m×0.4m×0.4m. The inlet and outlet are of the same size (0.04m×0.04m) and

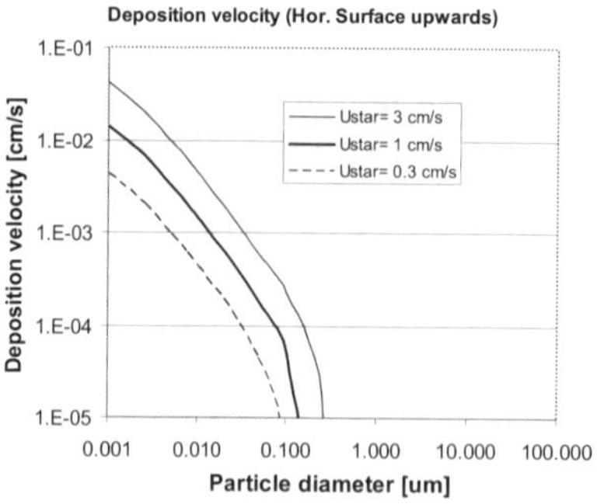

(a) Hor. downward facing surface

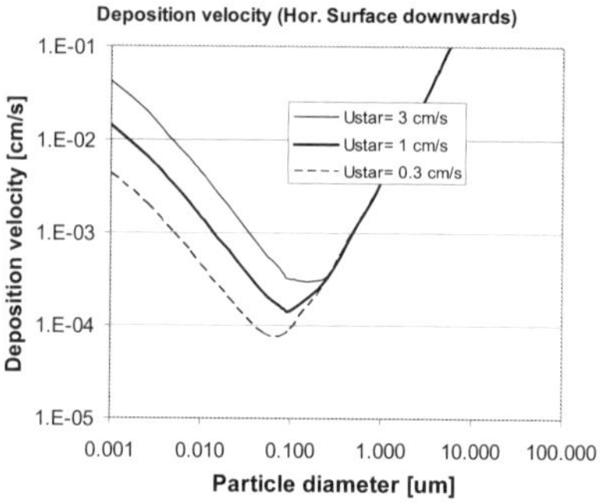

(b) Hor. upward facing surface

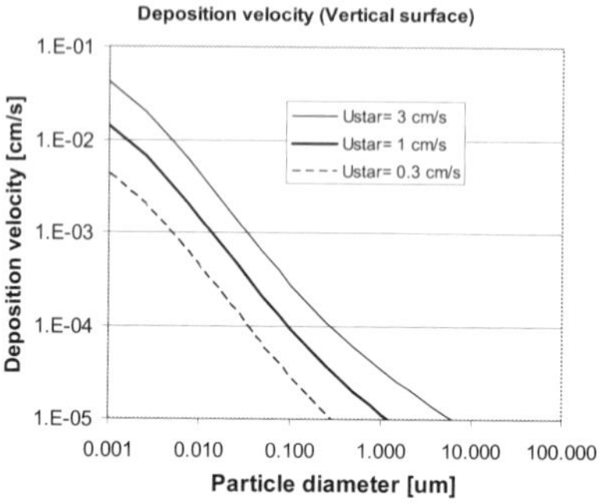

(c) Vertical surface

Figure 2: The deposition velocity as function of the diameter and the friction velocity. The calculations are based on an atmospheric pressure of 1 bar, temperature is 293 K and the density is 1000 kg/m^3.

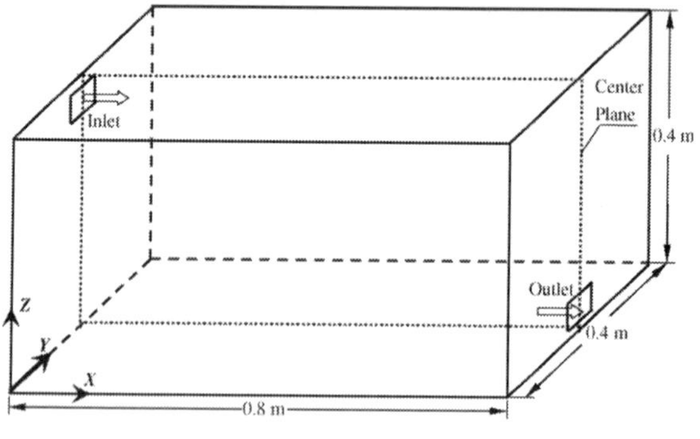

Figure 3: Schematic overview of the ventilation chamber for validation of the numerical model.

both are symmetrical with the center plane at $y = 0.2$ m. A schematic overview of the model is shown in Figure 3.

The imposed velocity at the inlet boundary is 0.225 m/s. At the same inlet boundary particles are injected and the particle concentration in the room is normalized with the concentration at the inlet. The particle density is 1400 kg/m^3 and the size of the particles is 10 μm. This particle size is suitable to validate the drift flux model as the drift flux is dominant for particles of this size. Chen *et al.* measured the airflow velocity and particle concentration with a Phase Doppler Anemometry (PDA) system. Picture (a) in Figure 4 shows a comparison of the simulated x-velocity component with measured data. Comparison of the particle concentration is shown in picture (b) in Figure 4. The comparison is shown along the vertical z-axis at x=0.4 m and y=0.2 m. The results show that both the airflow and particle concentration distribution is simulated accurately by the drift flux model.

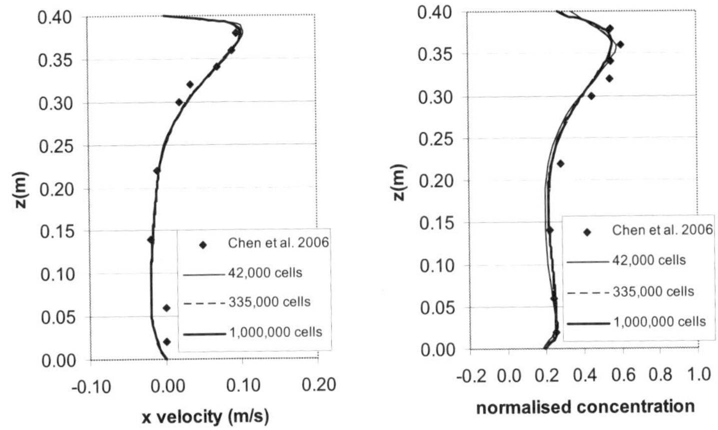

(a) (b)

Figure 4: Comparison of the simulated airflow velocity and particle concentration distribution with the measured data of Chen *et al.* [7].

4 Simulation results

Simulations are performed for a medium size laboratory room of $L_{ength}\times W_{idth}\times H_{eight}$ = 6m×6m×3m. A workbench is located in the center of the laboratory and the pollution source is located directly above the workbench. The workbench is located in the center of the laboratory and is $L_{ength}\times W_{idth}\times H_{eight}$ = 1m×1m×1m. In this work a release of radioactive particles is studied. The particles' radioactivity is long-lived providing a direct relation between the concentration of particles and the exposure to radioactivity. Varies particle diameters are evaluated in combination with different ventilation scenarios. Two flow features are assessed in each simulation. They include the mean concentration of particles at 1.5 m above ground and the particle concentration directly above the release at 1.5 m above ground. Both features highlight the exposure to humans and the ability of the ventilation system to minimize particle concentration in the area of interest. The height above ground of 1.5 m is based on the average height of human's inhalation, and provides a best indicator for the human's internal exposure to aerosols.

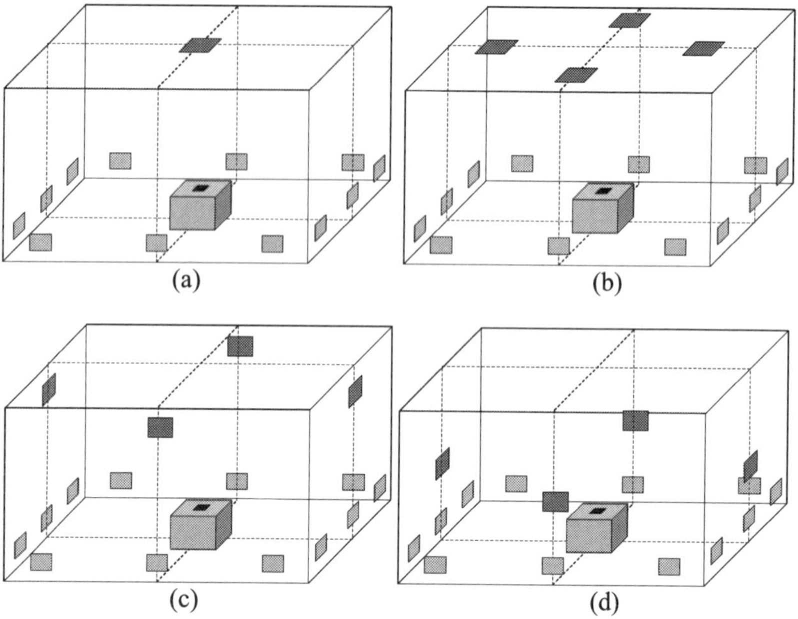

Figure 5: Laboratory ventilation scenarios selected for simulation. The 12 inlet ventilation slots, highlighted in light grey, are located at the bottom of the side walls. The pollution source is above the table and indicated by a small black surface. The ventilation extractors are highlighted in dark grey and vary for each scenario.

The concentration levels are expressed relative to the concentration at the release. In addition it is important to note that coagulation of the radioactive particles with the surrounding aerosols is not selected. The four ventilation scenarios that are selected are shown in Figure 5. All ventilation scenarios are equipped with 12 ventilation slots (0.1m×0.1m) at the bottom of the four side walls. Location of the ventilation extractors varies for each ventilation scenario. In addition the effect from air exchange is studied and the air exchange varies from 2 hr^{-1} to 10 hr^{-1}.

The simulation results are presented in Figure 6 and Figure 7. The first figure contains two pictures and shows the particle concentration directly above the release. The second figure contains two pictures as well and shows the mean particle concentration in the laboratory at 1.5m above ground.

Picture (a) of Figure 6 shows a number of interesting phenomena. For all three ventilation cases, the particle concentration reduces as the particle diameter increases. However, for smaller particles up to 5 μm an increase in the ventilation's air exchange rate reduces the particle concentration, while for particles larger then 5 μm there is an opposite effect. For larger particles the concentration above the release reduces rapidly. This is partly due to the large gravitational forces that result in considerable deposition. However, when ventilation is increased the larger particles are less prone to deposition and are instead entrained into the main flow stream. As a result there is an increase in particle concentration. A further phenomenon is the diminishing effect from additional ventilation above 5 AER for particles up to 5 μm. This phenomenon has been reported in varies studies [1] and is also confirmed in this work.

The simulation results from scenario B, C and D are shown in picture (b) of Figure 6. The three ventilation cases suggest that the location of the ventilation extractors have limited effect on the particle concentration near the release. In contrast with the findings from scenario A, scenario B to D have no negative adverse effects when the AER is increased.

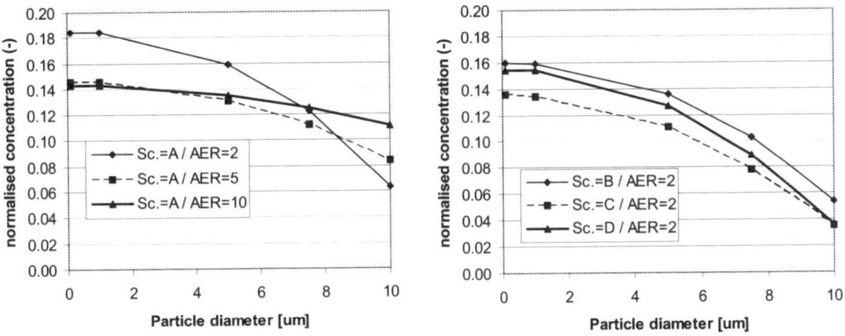

(a) (b)

Figure 6: Particle concentration above the release at 1.5m above ground level.

Figure 7 shows the mean particle concentration in the laboratory at 1.5 m above ground. In this figure a distinct difference between scenario A (picture a) and the other scenarios B to D (picture b) is shown. Where ventilation is applied directly above the release (scenario A) concentration levels are more than ten times smaller when compared against the other three ventilation scenarios. Similar to Figure 6 the results from scenario A also show some adverse effects from increased ventilation for particles of around 10 μm. The results clearly suggest that the scenarios B to D have comparable ventilation features. In contrast scenario A shows features that are typically found in a fume hood.

For particles smaller than 1 μm concentration levels in the laboratory are unaffected by the particle diameter. For those types of particles the terminal velocity becomes insignificant in a well ventilated room with more then 2 air exchanges per hour.

At this stage more simulations are required for different ventilation arrangements and laboratory setup before general conclusions can be drawn. However, the above findings provide some first data to review existing assessment procedures for air quality in radiological laboratories.

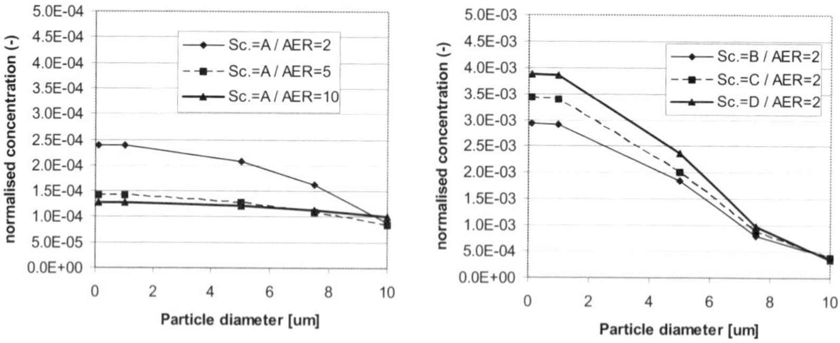

(a) (b)

Figure 7: Mean particle concentration in the room at 1.5m above ground level.

5 Conclusions

The conclusions of this work are as follows:

- A CFD model is developed to predict the dispersion and decay of radioactive gasses and particles. Particle deposition on the surfaces and effects from gravitational settling are accounted for.
- The prediction of particle dispersion is validated against experimental data of Chen *et al.* [7].
- A total of four ventilation scenarios are investigated and in addition the effects from increased air exchange are studied. The simulation results suggest the following:
  - o In scenario A an increase in ventilation has a negative adverse effect on the particle concentration for particles large then 10 μm.
  - o The dispersion of particles in scenario B to D is comparable. In contrast the particle dispersion in scenario A is more similar with the dispersion found in fume hoods.
  - o The results suggest concentration levels for particles smaller then 1 μm are unaffected by the particle diameter.
- Suggestions for further work are to take account of induced turbulence from laboratory workers and thermal driven flows. Both flow features enhance mixing and may affect concentration levels in the laboratory.

References

[1] Crane J., Ellis I., Siebers R., Grimmet D., Lewis S., Fitzharris P., A pilot study of the effect of mechanical ventilation and heat exchange on house-dust mites and Der p 1 in New Zealand homes. *Allergy*, Vol. 53, No 8, 755-762, 1994.

[2] Zhuo W., Iida T., Moriizumi J., Aoyagi T., Takahashi I., Simulation of the concentrations and distributions of indoor radon and thoron. *Rad. Prot. Dosim.*, Vol. 93 No 4, pp 357-368, 2001.

[3] Porstendörfer J., Properties and behaviour of radon and thoron and their decay products in the air. Proc. Fifth Int. Symp. on the Natural Radiation Environment. EU commission, Luxembourg, 2003.

[4] Lai A.C.K., Nazaroff W.W., Modelling indoor particle deposition from turbulent flow onto smooth surfaces. *J. Aerosol Sci.,* Vol. 31, No 4, pp. 463-476, 2000.

[5] Hinze J.O., Turbulence. 2nd Edition, McGraw-Hill, New York, 1975.

[6] Schneider T., Kildesø J., Breum N.O., A two compartment model for determining the contribution of sources, surface deposition and resuspention to air and surface dust concentration levels in occupied rooms. *Build. Environ.,* Vol. 34, pp. 583-595, 1999.

[7] Chen F., Yu S.C.M., Lai A.C.K., Modelling particle distribution and deposition in indoor environments with a new drift-flux model. *Atmospheric Environ.,* Vol. 40, pp. 357-367, 2006.

[8] Gao N.P., Niu J.L., Modelling particle dispersion and deposition in indoor environments. *Atmospheric Environ.,* Vol.41, pp. 3862-3876, 2007.

[9] Zhao B., Wu J., Particle deposition in indoor environments: Analysis of influencing factors. *J. Hazard. Mat.,* Vol. 147, pp. 439-448, 2007.

[10] Zhang J.S., Shaw C.Y., Nguyen-Thi L.C., MacDonald R.A., Kerr G., Field measurements of boundary layer flows in ventilated rooms. *ASHRAE Trans.,* Vol. 101, Part2, pp. 116-124, 1995.

Indoor aerosol transport and deposition for various types of space heating

P. Podoliak, J. Katolicky & M. Jicha
Brno University of Technology, Brno, Czech Republic

Abstract

Computational modelling of aerosol transport under various heating systems in a room was conducted with the goal to understand and evaluate a regional deposition and to assess an optimum position for air cleaning device. The room is equipped with a ventilation inlet integrated into one of the windows. The computational model room simulates a real room in an experimental house. Three heating systems are taken into account, namely radiators, floor heating and ceiling radiation panels. In total four cases are modelled for each heating system. Thermal diversity is simulated by varying the ventilation air flow and outdoor temperatures. Aerosol entering the room has the size of particles 1, 2.5 and 10 μm diameters. The room is divided into several regions and sub-regions where the deposition is evaluated for individual aerosol sizes and outdoor and incoming air temperatures. Also the air velocity and temperature fields in the room are depicted to support the analysis of deposition patterns. The void fraction of the aerosol phase also shows the space distribution inside the room and provides us with an overall view of the aerosol transport. Based on the results of aerosol transport and deposition, an optimal location of an air cleaner can be selected.
Keywords: indoor aerosol, deposition, heating systems, CFD.

1 Introduction

Suspended particles belong to the most dangerous pollutants in both outdoor and indoor environments. As people spend the most of their lives in the indoor environment where they are exposed to suspended particles by breathing the ambient air, the issue of transport and deposition of particles indoors gets more and more attention [8]. Whilst this does not mean that indoor exposures produce

WIT Transactions on Ecology and the Environment, Vol 123, © 2009 WIT Press
www.witpress.com, ISSN 1743-3541 (on-line)
doi:10.2495/AIR090261

more harmful health effects, the evidence is that indoor concentrations of many pollutants are often higher than those typically encountered outside [4].

The motion of particles in ventilated rooms is governed by the flow field, which determines their local and temporal concentrations and deposition. Numerous factors affect indoor particle concentrations including outdoor particles that penetrate indoors, indoor activities that generate particles, deposition of particles, and air exchange [2]. The influence of the geometry with furnishing of the room and thermal conditions are matter of course in order to generate the specific flow field. In the prediction of indoor particle concentration, it is necessary to take into consideration the particle size, wall texture and orientation [1].

The transport and deposition of aerosol particles was simulated by CFD in a model of a room using the Lagrangian multiphase flow in a steady state. Three types of space heating systems were simulated, namely: radiators, floor and ceiling heating. Different airflow patterns in the room were reached by changing the temperature of ventilation air and outdoor temperature. Four different temperatures of ventilation air in combination with outdoor temperature were simulated for each kind of heating. A heat load of thermal system was changed to ensure the thermal comfort in the room. The source of aerosol was situated to the inlet of the ventilation air. Three sizes of particles in diameters 1, 2.5 and 10 μm were used with the characteristic concentrations of outdoor aerosol for the part of the city, where the experimental house is located. The effects of different heating systems on aerosol transport and deposition were investigated. The deposition was affected by gravitational settling, buoyancy caused by temperature gradient in the room, and also by the interaction between the discrete and continuous phase. The electrostatic or mechanical aspects of surfaces and re-suspension of particles were neglected. Particles were assumed to deposit when they reach the wall in the distance shorter than their radius in the near-wall region. Collected data were analysed to observe the interaction between the type of heating system and the amount of particles settled down on several surfaces of the room. The results of particles settled on surfaces were compared relatively in percentage.

2 Methods

The computational model for simulation of particles deposition was based on the real room situated in an experimental house in the Czech Republic on the campus of the Brno University of Technology in Brno. Main dimensions of model were 4.2 x 6.3 m for the area of the floor and the room height was 2.7 m. The main parts of the model are shown in figure 1. The model of the room consists of nearly 450000 hexahedral cells. The size of the cells was changing down from 10 cm to 0.5 cm. These fine cells were needed to achieve more accurate solution in the places where the gradients of values were high. It was primarily the area of the inlet, outlet, and radiators. Finer cells were also used in the near wall area.

The task was simulated in a steady state. The Lagrangian approach was used to describe the two-phase flow of particles in the indoor air. For describing the turbulence behaviour of air, the k-ε model with standard wall functions was used. Radiative heat transfer between internal surfaces was solved using the discrete ordinate method. For this transfer, the surface was divided into 1110 smaller surfaces, if possible, primarily as a square with the side length of 0.4 m.

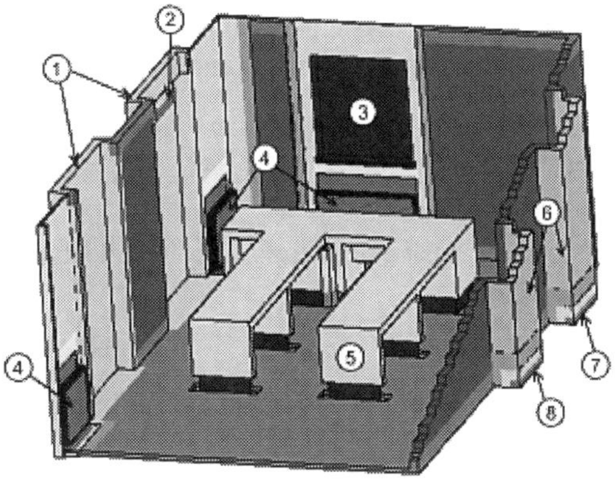

Figure 1: Model of the room with main parts: 1 - outer windows (French windows), 2 – inlet, 3 – window, 4 – radiators, 5 – furniture, 6- inside doors, 7 and 8 – outlets.

Several boundary conditions, as needed, were used in the model. Wall boundary conditions were used for the floor, ceiling, surrounding walls, and windows. It was necessary to determine the thermal resistance of structures and also their radiative properties. Baffle boundary conditions were used for the radiators and furniture situated inside of the domain. The specific values used for each boundary are shown in table 1.

Instead of thermal resistance parameters of material, the heat flux was set for radiators in case the radiator heating was turned on. In different cases of used heating systems it was necessary to set up the heat flux generated by the floor or ceiling as needed. The heat flux depended on the heat balance of the room and the ambient environment to reach the values of indoor air temperatures according to thermal comfort.

The change of temperatures resulted in a consequent change of heat loads in order to simulate several thermal situations. Owing to different temperature distribution, the air distribution in the room is also changed. As the characteristic values of the outdoor temperatures for the heating season the following values were assumed: the minimum (-12°C) and the average (3.6°C) outdoor temperatures given by national standard for the location of the house.

The ventilation of the room was realised by the inlet of the fresh outdoor air, which was built in the upper part of one of the windows. The fresh air was entering the direction perpendicular to the plane of the inlet with the velocity

www.witpress.com, ISSN 1743-3541 (on-line)

0.126 m/s. This value corresponded to the air exchange rate 0.35 1/h for the volume of the simulated room. These ventilation conditions were chosen in terms of minimum hygienic requirements for an administrative kind of activity in the room. The temperature of the incoming air was the same as the outdoor temperature or the air was heated to the temperature 10°C and 20°C using preheating or recuperation in order to get several conditions indoor. The outlet was realised by the pressure boundary conditions at the bottom of both doors opposite the inlet. Four different thermal situations were simulated for three different types of space heating systems. Values of the specific heat flux for simulated temperature conditions and the heating area for each simulation are in table 2.

Table 1: The values of thermal parameters and radiation properties.

	thermal resistance [m^2K/W]	emissivity	reflectance	transmittance
wall to indoor	3.484	0.91	0.09	0.00
wall to outdoor	6.71	0.91	0.09	0.00
indoor doors	0.48	0.91	0.09	0.00
french windows	0.953	0.84	0.08	0.08
ceiling	3.47	0.91	0.09	0.00
floor	7.186	0.91	0.09	0.00
window	0.953	0.84	0.08	0.08
radiators	0.0005	0.91	0.09	0.00
furniture	0.02	0.91	0.09	0.00

As for aerosol, the only source was the outdoor air coming into the room through ventilation. Under normal ventilation ratios of indoor to outdoor, particle number and mass are very closely to one [5]. Concentration of aerosol was taken from previous measurements of several concentrations of aerosol size classes in the streets of the city close the experimental house. We have chosen the average values of dust concentration in wintertime from the spectrum of size classes for three particle sizes with diameter 10, 2.5 and 1 μm with concentrations 12, 3 and 15 μg/m^3 respectively. Also the average density of dust particles was taken 1500 kg/m^3.

The deposition of particles was reached using the subroutine for the behaviour of particles near the boundary condition. The point was that for the particle reaching the wall cell, zero velocity was assumed. After particles had settled down, they were removed numerically from the domain to avoid re-suspension. This rule then enabled us to simplify and realize the deposition. Hence the effect of aerosol transport and deposition is very complex this simulation took into account only some aspects of this complex phenomenon.

The interaction between discreet and continuous phase and the action of gravitation and buoyancy were taken into consideration but an electrostatic interaction or mechanical properties of surfaces were neglected.

Table 2: Values of heat flux with appropriate heating area for several simulations. t_{out} - outdoor temperature, t_{vent} - ventilation air temperature.

used type of heating	heat flux [W/m^2] needed in the case of t_{out}				area of heating [m^2]
	3.6°C; 3.6°C	3.6°C; 10°C	3.6°C; 20°C	-12°C; -12°C	
radiator	105.3	92.8	82.7	156.7	9
floor	31.7	26.1	26.2	44.6	27.58
ceiling	34.6	31.5	28.1	53.8	27.58

3 Results and discussion

To carry out the comparison of different simulation cases, five planes of interest were defined. Because of the action of airflow, momentum is transmitted to aerosol particles, and the location of planes was chosen to capture the main aspects of forming the fields of air distribution. The cold air coming from the outside fell down below the inlet in contrast to the convective flow of hot air over the radiators. The inlet and radiators zones were the places where the velocities of air and the temperature gradients reached the highest values. Therefore the analysis of aerosol transport and deposition was also supported by calculation of both thermal and velocity fields.

The void fraction fields served for quantifying the contamination of the air. The most contaminated environment was in the case when the room was heated by the radiator, the outdoor temperature being 3.6°C and the ventilation air heated to 10°C while the least contaminated environment was simulated for the same thermal conditions but the ceiling heating was present. Qualitative pictures of particles transport and distribution showed the same. Figure 2 shows the cases of the same thermal conditions but different heating systems. The spatial view of the model and the dispersed aerosol in it brings about the most objective image of the indoor air contamination.

301 local regions were defined on all surfaces of the room to compute the deposition on each of them. The area of each region was approximately 0.8 x 0.8 m^2. These areas were joined together into sub-regions where the deposition was evaluated for individual aerosol sizes and outdoor and indoor temperatures. Sub-regions were based on the same area as boundary conditions were defined. The deposition analysis on each of them shows both the relative percentage of total amount of particles released into the domain and the surface flux in μg/m^2.

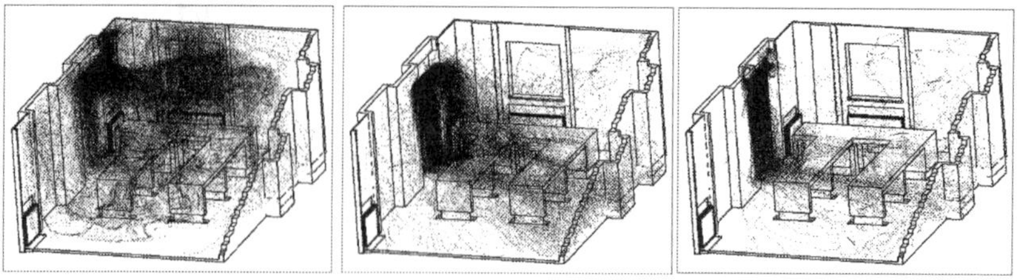

Figure 2: Spatial view of dispersed particles in the room for $t_{out} = 3.6°C$, $t_{vent} = 10°C$ in the following order: radiator, floor and ceiling heating system.

A large number of particles have settled down directly on the window under the inlet of air. This was caused by the fall of cold air down along the window and the wall; right after this air entered the room. This number of settled particles was between 63-83% in the case of ceiling heating, 29-71% for floor heating and 18-55% for radiator heating depended on the assigned thermal conditions. The amount of settled particles in that area was getting lower with the inlet of warmer ventilation air and was almost uniform for each size of particles with the exception of radiator heating. The deposition of 1 μm and 2.5 μm particles was lower (from 13 up to 26,7%) than the deposition of 10 μm particles on the window under the inlet. The effects of gravitational settling on 10 μm particles could be the cause why the deposition of them is higher while for the smaller particles it is easier to be yielded to convective fluxes. The same result about the deposition and gravitational settling was confirmed by Zhao [8]. Holmberg and Li [3] showed the difference between the amount of particles settled on the floor and on the walls for particles of size 2.5 and 4.5 μm too. The comparison can only be considered as an example.

Surface mass flux was always the highest in the outlet and on the window under the inlet. The highest values depend on the smallest area of the outlet and the high concentration of particles near the inlet area. The values of the surface mass flux were for the outlet in the order of 0.1 $\mu g/m^2s$ and for the window under the inlet in the range between 0.001 and 0.01 $\mu g/m^2s$. The flux on other surfaces varied from zero to order of 0.001 $\mu g/m^2s$. According to the highest mass flux which is independent of the surface, we could find the places for the air cleaner with highest efficiency of cleaning.

Due to minimal dispersion of the particles for the case of ceiling heating, the main area of the deposition was affected by the range of ventilation air. Therefore, next to the deposition under the inlet, particles were settled on the floor; nearly all the remaining particles settled there in the amount of 8-30%. The difference between the particles settled on the floor of larger size 10 μm and smaller particles of 1 μm and 2.5 μm was 10% in the case of the ventilation with the air of temperature 20°C. The warmer was the ventilation air entering the room, the higher was the distance reached and more particles of the size 1 μm and 2.5 μm were noticed in the area of the outlet situated opposite to the inlet in concentrations up to 8%. It corresponds to the lower deposition on the floor. Other surfaces did not gain more than 2% of particles of each particle size.

Floor heating was similar in deposition to ceiling heating. The deposition on the window under the inlet reached lower values. The values for the inlet area have already been mentioned. As for the floor values, these values were between 13-29%. 10 μm particles reached the highest values; these values were nearly constant for all thermal conditions. The velocities of the air seem be higher in that case than in the case of ceiling heating. Due to radiative heat transfer, more surfaces provide a better view on the heated surface, (e.g. the bottom of furniture and radiators), earn more heat and therefore are warmer. The same behaviour of particles was observed with ceiling heating. Also the rule that the warmer was the air entering the room, the higher was dispersion of particles in the air, took place. It could be shown on the balance of the deposition on the side of the room where the inlet is situated. The deposition on the wall increased while the deposition on the window under the inlet area decreased. It proceeded nearly proportionally considering the higher deposition on the furniture, outlet and slightly also on the other surfaces. The deposition on the furniture varied from 4 to 13% and at the outlet from 2% to 5% for 10 μm particles and up to 11% for the smaller particles. The deposition reached the values only between 2-4% on the side wall where the outlet was situated. The deposition on the other surfaces was more significant than in the case of ceiling heating where the deposition was mainly realised on the three of ten analysed surfaces. Therefore the values of these appropriate surfaces are lower in the case of floor heating.

The deposition on the ceiling was minimal in both of used heating systems taking into account the large area of ceiling. Mainly the particles of the size 1 μm were depositing. The highest values they reached were only 2.5% in the case of floor heating and 1% in the case of ceiling heating when the temperature of the ventilation air was 20°C. The highest deposition on the ceiling was in the case of radiator heating. Hot air rising over the radiators raises the particles to the upper part of the room as we can also see from the transport of the particles. The velocities of the air in the area over radiators increased 3 times by turning the radiator heating on. The deposition of the particles on the ceiling was connected with the intensity of heating. The deposition of fine particles was 28% while 10 μm particles reached 20% for the case of the highest heat flux. Values of deposition varied down to 14% for fine particles and to 7% for 10 μm particles for the case of the lowest heat flux. The difference between these values was nearly uniform.

The deposition on the floor was always a bit higher for 10 μm particles than for the others. The highest value 14% was in the case of incoming ventilation air preheated to 20°C otherwise the values reached around 5%. In [7] it is shown the deposition loss rate coefficient as the function of particle size and the velocity of the air. In our case it is not yet possible to discuss the deposition with air speed, as the simulation was run in steady state. We can only see the influence of airspeed on the aerosol dispersion. The movement of particles in ventilated areas is strongly influenced by airflow pattern [8]. Velocities of the air are mainly a response to the movement of ventilation air or movement caused by buoyancy over heated surfaces. We can also see the influence of the interaction between the convective flows and behaviour of ventilation air on aerosol. That is why the deposition of 1 μm and 2.5 μm particles in the outlet reached 25% in the case when the temperature of ventilation air was not the highest but only 10°C. The conclusion related to the incoming air, which governed the case of floor and ceiling heating did not seem to work for the case where more sources of momentum are presented. The deposition in the outlet decreased to only 6% when the temperature of incoming air raised to 20°C. The deposition on the wall and the area where the inlet was situated belonged to the most affected area as usual. The deposition of 10 μm particles is higher in the area of inlet opposite to a slightly higher deposition of fine particles on the wall around the inlet. The deposition on the radiators themselves was around 7% in the case when the heat flux was the lowest. The deposition on the ceiling was nearly always higher than the deposition on the floor except for the following case. It was proved that the effect of buoyancy on 10 μm particles was not so significant when the radiator heating is at the lowest simulated value. There were two times more of these particles settled on the floor (14%) than on the ceiling (7%).

The deposition data were also divided and compared in order to take into account the area of the room where the aerosol source was situated and the area where not. The result was that the deposition took place mainly in the area where the inlet was situated. This is shown on the ceiling heating when the deposition on the area of inlet window and on the floor underneath makes more than 90% of all settled particles. The most of values in the area without source did not reach even 1% of deposited particles. In one case the deposition reached 9% for 1 μm and 2.5 μm particles in the case of radiator heating and adjusted temperature of ventilation air at 3.6°C. In this case the deposition on the ceiling was uniform for the whole area concerned. Figure 3 shows the deposition on several surfaces for the conditions where the ventilation air is preheated nearly to the indoor temperature while outside temperature was reaching the wintertime average.

4 Conclusions

The highest contamination of the air by the dispersed aerosol was shown for the case of radiator heating. Convective flows caused by high temperature of radiator panels make the airflow in the room the most intensive. Thus the particles are dispersed most into several areas of the room. That caused the significant deposition on the ceiling for the case of radiator heating.

A different dispersion and deposition can be reached by suitable combination of ventilation air and the convective flows produced by radiators. It is shown that the least contamination of the air is only in the case of ceiling heating. Slightly different thermal conditions were reached by the floor heating, but aerosol dispersed more into the air than in the case of ceiling heating.

While the source of aerosol is the ventilation air, the most affected are the surfaces in the main field of its flow if there were no other sources of fluxes causing the vorticity of the air and the connected dispersion of aerosol. The warmer the air is the more straight direction it has and does not fall to the floor

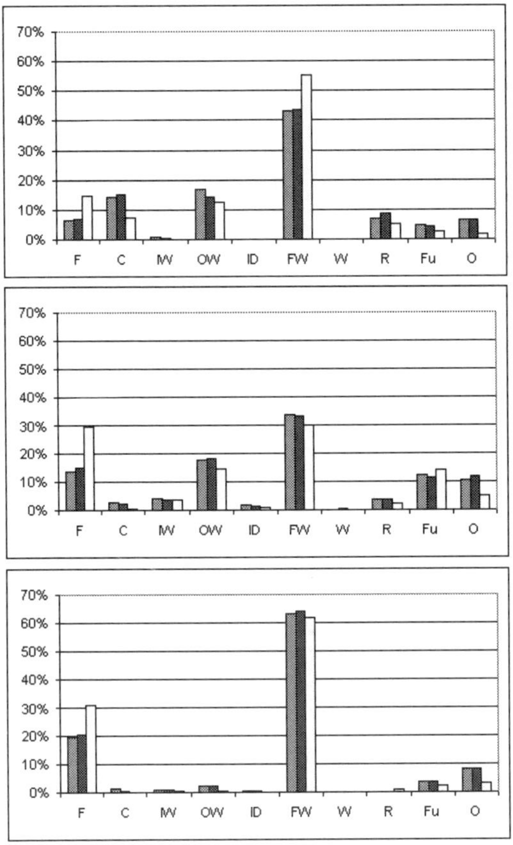

Figure 3: Relative deposition on several surfaces for the case of $t_{out} = 3.6°C$, $t_{vent} = 20°C$, in order: radiator (up), floor, ceiling (down) heating system. Values of three particle sizes 1 μm (blue/light grey), 2.5 μm (red/dark grey) and 10 μm (yellow/white) and surfaces in order from left to right: floor, ceiling, wall to indoor, wall to outdoor, indoor door, French window, window, radiators, furniture and outlets. F-floor, C-ceiling, IW-inner wall, OW-outer wall, ID-inner door, FW-French windows, W-window, R-radiator, Fu-furniture, O-outlet.

right after the inlet and therefore the dispersion of aerosol is higher too. Ventilating with the colder air can be the way of how to lower the dispersion of the aerosol coming from the outdoor air into the room. In general the effect of flow and buoyancy we can see on 1 μm and 2.5 μm particles. The deposition on the floor was typical of 10 μm particles resulting mainly from the gravitational settling. The sense of deposition on the floor is considerable while the dust on floor surfaces is available for re-suspension of particles, dermal contact and ingestion through hand-to-mouth transfer [6].

Mostly the smaller particles of 1 μm and 2.5 μm reached the outlet in larger amount in case the incoming ventilation air was heated to a higher value. The particles which deposited there left the room. So if there is no negative influence on another part of the house, we can state that the depositing in the outlet area belongs to the positive aspects of lowering the concentrations of aerosol in the indoor air. The outlets had the highest values of mass flux per surface area. Finding a similar place with high mass surface flux located in high deposition zone could be a good solution for where to locate the air cleaner. On the other hand, we could find the best place where to position a bed or an artwork to protect it from damage. It is necessary to choose the right place for mounting the ventilation holes not to negatively affect the inhabited area. The energy needs to heat the room is constantly increasing. So it is necessary to find a compromise between dustiness, needs of energy and final comfort when assuring the desired indoor climate.

Acknowledgement

Financial support from the grant COST P20, number OC166 of the Czech Ministry of education, youth and sports is gratefully acknowledged.

References

[1] Abadie M., Limam K., and Allard F. (2001). Indoor particle pollution: Effect of wall textures on particle deposition. Building and Environment, 36:821–827.
[2] Abt E., Suh H. H., Allen G., and Koutrakis P. (2000). Characterization of indoor particle sources: A study conducted in the metropolitan Boston area. Environmental Health Perspectives, 108(1):35–44.
[3] Holmberg S. and Li Y. (1998). Modelling of the indoor environment - particle dispersion and deposition. Indoor Air, 8:113–122.
[4] Jones A. P. (1999). Indoor air quality and health. Atmospheric Environment, 33:4535– 4564.
[5] Morawska L., He C., Hitchins J., Gilbert D., and Parappukkaran S. (2001). The relationship between indoor and outdoor airborne particles in the residential environment. Atmospheric Environment, 35:3463–3473.
[6] Thatcher T. L., Layton D. W., (1995). Deposition, resuspension and penetration of particles within a residence. Atmospheric Environment, 29:1487-1497.
[7] Thatcher T. L., Lai A. C. K., Moreno-Jackson R., Sextro R.G., Nazaroff W. W. (2002). Effects of room furnishing and air speed on particle deposition rates indoors. Atmospheric environment, 36:1811-1819.
[8] Zhao B., Zhang Y., Li X., Yang X., and Huang, D. (2004). Comparison of indoor aerosol particle concentration and deposition in different ventilated rooms by numerical method. Building and Environment, 39:1–8.

Characterization of organic functional groups, water-soluble ionic species and carbonaceous compounds in PM10 from various emission sources in Songkhla Province, Thailand

K. Thumanu[1], S. Pongpiachan[2,3], K. F. Ho[4], S. C. Lee[4] & P. Sompongchaiyakul[5]
[1]*Synchrotron Light Research Institute (Public Organization), Thailand*
[2]*Faculty of Environmental Management, Prince of Songkhla University, Thailand*
[3]*National Center of Excellence for Environmental and Hazardous Waste Management-PSU Satellite Center, Prince of Songkhla University, Thailand*
[4]*Research Center for Urban Environmental Technology & Management, Department of Civil & Structural Engineering, The Hong Kong Polytechnic University, Hong Kong, China*
[5]*Department of Marine Science, Faculty of Science, Chulalongkorn University, Bangkok, Thailand*

Abstract

PM_{10} samples were collected at nine sampling stations using a high volume (hi-vol) air sampler during the period of June–November 2007. Using the ATR-FTIR technique, the chemical compositions of organic and water-soluble ionic species (WSIS) PM_{10} aerosols from each emission source were identified. WSIS such as SO_4^{2-}, NO_3^-, CO_3^{2-} , NH_4^+ were mainly found in PM_{10} aerosols. The highest concentrations of NO_3^-, NH_4^+ and CO_3^{2-} were detected in aerosols collected from sampling sites adjacent to traffic roads. This can be explained by the heterogeneous reaction of SO_2/NO_2 mixtures with carbon soot that lead to the highest contribution of WSIS in combustion particles from vehicle exhausts. In addition, the ratios of organic carbon/elemental carbon (OC/EC) collected at heavy traffic road, bus terminal and traffic demonstrated the lowest values of

WIT Transactions on Ecology and the Environment, Vol 123, © 2009 WIT Press
www.witpress.com, ISSN 1743-3541 (on-line)
doi:10.2495/AIR090271

1.677 ± 0.198, 2.329 ± 0.570 and 2.770 ± 1.234 respectively. This indicates that the fine aerosols originating from vehicular emission are fresh particles. The relative contribution of organic functional groups like organic nitrate was highly detected in aerosols collected from industrial sampling sites. This could be ascribed to the intensive use of heavy oil and wood materials during the manufacturing process of animal feed and rubber sheet drying, respectively. More important it should be noted that from biomass burning sampling sites the relatively high intensity of carbonyl bands and aliphatic hydrocarbon IR absorption band illustrated the highest values in those PM_{10} associated with high OC/EC ratios. In this study, the oxidation state of sulfate aerosols were detected by using X-ray absorption near edge spectroscopy (XANES). The results show only S_6^+ peak of S-K edge at 2481 eV observed in all samples.

Keywords: PM_{10}, ATR-FTIR, organic carbon, elemental carbon, water-soluble ionic species, organic functional group, secondary organic carbon, XANES.

1 Introduction

It is well known that aerosols affect the Earth's radiation budget [1, 2]. Recent studies have also elucidated the adverse health impact of fine particulate matter associated with respiratory and cardiovascular diseases, affecting the morbidity and mortality in urban areas [3, 4]. The impact of aerosols on both the climate system and public health greatly depends on the chemical characteristics of organic carbon (OC), elemental carbon (EC), water-soluble ionic species (WSIS) and organic functional groups (OFG) such as polycyclic aromatic hydrocarbons (PAHs). Although there has been considerable confusion and debate over the role of carbonaceous aerosol on climate change over the past decades, its negative influences on visibility degradation and adverse ecological impacts are widely recognized [5, 6]. Since EC occurs mainly from imperfect combustion source of carbon based materials and fuels and is solely primary in nature, it appears reasonable to employ EC as an indicator of primary anthropogenic air pollutants. In contrast, the primary OC can react with trace gases and can also be generated as secondary organic carbon (SOC). Furthermore, more recent studies indicated the significant contribution of organic functional groups as precursors of secondary organic aerosols (SOA) [7].

In spite of its great impacts on both the atmospheric system and human health, published papers related to OC, EC, WSIS and OFG in the fine tropical aerosols are extremely limited. At present, it is difficult to accurately predict patterns of climate change and to precisely conduct a risk assessment without knowing the relative contribution of these chemical compounds in fine aerosols, particularly in the tropical atmosphere. Several studies reported significant contributions of WSIS (e.g. SO_4^{2-}, NO_3^{2-}, NH_4^+), and organic species (e.g. nitro aromatic compounds) in PM_{10} using Fourier Transform Infrared Spectroscopy (FTIR). However, these findings showed only the chemical characteristics of fine particles collected as a mixture of aerosols emitted from various sources [8, 9]. In this study, the determination of OC, EC, WSIS and OFG in PM_{10}

aerosols was conducted in order to provide insights into the origin of air pollution problems at the study area.

2 Methodologies

2.1 Sampling sites and descriptions

Songkhla Province is located 950 km south of Bangkok with a population over 1.32 million people. This province is situated on the eastern side of the Malayan Peninsula facing to the Gulf of Thailand. Hat-Yai, an economic center of Songkhla, has a complex urban environment with a mixture of commercial, residential and industrial establishments. All samples were collected during the period of June-November 2007. The sites descriptions and abbreviations were listed in Table 1.

Table 1: Sampling positions and periods.

Site	Sampling Period	Latitude				Longitude			
Group1: Background									
SL1	27/07/07-29/07/07	7°	10'	02.92"	N	100°	35'	11.36"	E
SL2	20/10/07-22/10/07	7°	10'	02.92"	N	100°	35'	11.36"	E
KHH	03/11/07-05/11/07	7°	00'	57.92"	N	100°	31'	12.76"	E
Group 2: Traffic									
BT	05/08/07-07/08/07	6°	59'	42.78"	N	100°	28'	58.02"	E
PR	27/08/07-29/08/07	7°	00'	52.99"	N	100°	28'	20.50"	E
Tesco	05/07/07-07/07/07	7°	00'	30.81"	N	100°	29'	39.21"	E
Group 3: Industry									
CPF	24/07/07-26/07/07	6°	54'	16.38"	N	100°	28'	05.15"	E
RMF1	30/07/07-01/08/07	7°	03'	19.97"	N	100°	37'	58.90"	E
RMF2	02/08/07-04/08/07	7°	03'	06.28"	N	100°	24'	07.77"	E
Group 4: Biomass									
RSB	16/11/07	7°	27'	00.52"	N	100°	25'	19.02"	E
PTB	18/11/07	6°	57'	40.45"	N	100°	33'	06.68"	E

SL: Songkhla Lake, KHH: Kor Hong Hill, BT: Bus Terminal, PR: Petkrasam Road, Tesco: Traffic Intersection in front of Tesco-Lotus, CPF: Charoen Pokphand Factory (Fish Can Factory), RMF: Rubber Manufacturing Factory, RSB: Rice Straw Burning, PTB: Para Tree Burning.

Four groups were categorized based on the different nature of emission sources. The site descriptions and abbreviations are clearly displayed in Table 1 and given as follows:

Group 1: This group was carefully selected as a representative of background sampling sites namely:

Songkhla Lake sampling stations (SL): It was situated about 13 km far away from the northern side of Prince of Songkla University (PSU) at the south of Songkhla Lake and approximately 14 km away from the western side of the Gulf of Thailand. This sampling station is far away from many industrial and traffic emission sources, including metallurgy factories and power plants, and the residential areas. Therefore it seems reasonable to consider SL as a representative of rural background sampling station. SL1 and SL2 represent the sampling period of July (27th to 29th July, 2007) and October (20th to 22nd October, 2007) in Songkhla Lake sampling station respectively.

Kor Hong Hill sampling stations (KHH): It was located on the top of Kor-Hong hill with the elevation of 356 m. This site represents as a mixture of all emission sources in urban area. Therefore, the samples collected at KHH can be considered as a representative of urban background air mass. The sampling was conducted during the period of 3rd-5th November, 2007.

Group 2: This group is categorized as a mixture of diesel and gasoline engine exhausts. Three sampling stations were classified into this category namely:

Bus terminal (BT): This site was located at the southwestern side of PSU and approximately 1.4 km far away from the campus. Since the majority of vehicles are diesel engine buses, it appears reasonable to consider BT as a source of diesel emissions. The air samples were collected during the period of 5th-7th August 2007.

Petkrasam road (PR): This site was situated at the heart of Hat-Yai city. This site suffers from air pollutions caused by heavy traffic congestions with a mixture of diesel and gasoline exhaust emissions. The sampling was conducted on 27th – 29th August, 2007.

Tesco: This station was located at the ion in front of the main gate of PSU adjacent to Tesco-Lotus supermarket. Tesco locates on the eastern side and approximately 2.5 km far away from the Hat-Yai city center encompassed with urban residential zones. It seems plausible to regard Tesco as a traffic emission source influenced by complex emissions from trucks, buses, cars and motorcycles. The air samples were collected from 5th to 7th July, 2007.

Group 3: This group was selected as a representative of industrial emission sources. Three sampling stations were classified into this group namely:

Charoen Phokphand Factory (CPF): This site was located at the animal feed factory of CP. This factory is the largest business conglomerate in Thailand. CPF can be considered as an emission source of diesel oil burning. The sampling was conducted from 24th to 26th July 2007.

Rubber Sheet Manufacturing Factory (RMF): RMF was situated at Tumbol Tungwan, Hat-Yai district. Wood materials were used as fuel for the rubber sheet drying process. The rubber sheet was treated with steam of high temperature and high pressures and then purified with sulfuric acid solution. RMF1 and RMF2 represent the sampling period of 30th July-1st August 2007 and 2nd-4th August 2007 respectively.

Group 4: This group can be classified as a biomass burning site and further categorized into two groups namely:

Para Rubber Tree Burning (PTB): This station was located at Namom district, Songkhla Province and can be recognized as an emission source of Para rubber tree burning. The air samples were collected on 18th November, 2007.

Rice Straw Burning (RSB): This station was situated at the rice filed in Satingpra district, Songkhla Province. Agricultural burning is the practice of using fire to reduce or dispose of vegetative debris from an agricultural activity. Although this practice is considered as the most effective measure to prevent plant disease, it produces a large amount of smoke that causes air pollution problems in Thailand every year. This station was regarded as a source of biomass burning. The sampling was conducted on 16th November, 2007.

2.2 Sample collection and analysis

A total of 33 samples were collected by Graseby-Anderson high volume air sampler with PM_{10} sampling inlet (TE-6001). The high volume air sampler was operated at the ground level at the flow rate of 1.4 m^3 min^{-1}. PM_{10} samples were continuously collected from group 1, group 2 and group 3. Samples from each location were collected for 24 h for three consecutive days. In addition, the group 4-PM_{10} was monitored for 3 h and collected for three times per day in order to avoid any overloading of air particulate matters due to heavy smoke from biomass burning. PM_{10} aerosols were collected on 47 mm Whatman quartz microfibre filters (QM/A). All the quartz fiber filters were preheated at 800°C for 12 h prior to sampling.

2.2.1 (ATR)-FTIR Spectroscopy analysis

The experiments were performed at 4 cm^{-1} resolution and 32 scan from 450 to 4000 cm^{-1}. The Infrared spectra were collected using the Attenuated Total Reflectance (ATR)-FTIR Spectroscopy with single reflection ATR sampling module containing a deuterated triglycine sulfate (DTGS) detector. Each spectrum was subtracted from the blank filter spectrum to remove the background infrared spectrum off the target sample. Spectra were collected by averaging 160 co-added of each samples. The normalized spectra of PM_{10} collected from each emission source will subsequently be averaged and employed as the representative normalized spectra of aerosol samples. To resolve overlapping bands in the mid IR regions, band fitting analysis was performed using OPUS 5.5 software (Bruker optic, German).

2.2.2 OC/EC analysis

The samples were analyzed for OC and EC using DRI Model 2001 (Thermal/Optical Carbon Analyzer) with the IMPROVE A thermal/optical reflectance (TOR) protocol. The protocol heats a 0.526 cm^2 punch aliquot of a sample quartz filter stepwise at temperatures of 140°C (OC1), 280°C (OC2), 480°C (OC3), and 580°C (OC4) in a non-oxidizing helium atmosphere, and 580°C (EC1), 740°C (EC2), and 840°C (EC3) in an oxidizing atmosphere of 2% oxygen in a balance of helium.

2.2.3 XANES analysis

XANES measurements were carried out using synchrotron radiation source at the Siam Photon Laboratory. The experiment was performed with transmission mode of S-K edge measured between 2460 and 2580 eV. Each sample was mounted on the holder and employed to measure the current signals, I_0 and I, before and after passing through the samples, respectively. The results were expressed as the sample absorption which is the ratio of $\ln(I_0/I)$.

3 Results and discussions

All the assigned bands are indicated in Table 2. Samples collected from transportation and industry sampling sites mainly consist of WSIS and OFG such as SO_4^{2-}, NO_3^-, CO_3^{2-}, NH_4^+, organonitrates and aromatic nitro compounds. According to the measurement results from transportation sampling sites, the integral areas of IR spectra of NO_3^-, CO_3^{2-}, NH_4^+ were approximately 7.0%, 11.8%, 6.4% respectively. Interestingly, those observed at the background sites were similar order at approximately 4.7%, 8.7 %, 6.3 % in that order (Fig. 1). There is a general tendency that the sum of relative integral areas of these three WSIS in traffic aerosols (26%) was higher than those of background sampling sites (19%). (Table3) The heterogeneous reaction of NO_2 on soot particles occurring inside the internal combustion engine may be responsible for the relatively high contribution of NO_3^- and NH_4^+ detected in the emitted particles from vehicle exhaust. Since the traffic sites are spaciously enclosed with the

Table 2: Infrared absorption bands and vibrational modes of observed species on PM_{10} aerosols collected from each emission sources.

Absorbance bands (cm^{-1})	Vibration mode	Species
1660-1620	NO_2 asymmetric stretching (R-ONO_2)	Organonitrates
1550-1500	NO_2 asymmetric stretching (Arom-NO_2)	Aromatic nitro compound
1485-1390	NH_4^+	Ammonium ions
1345-1315	νasym (NO_3^-)	Nitrate ions
1390	νasym (CO_3^{2-})	Carbonate
879	νasym ($CaCO_3^{2-}$)	Calcium carbonate
1180-900	S=0 stretching - SO_4^- - HSO_4^-	Sulfate species - Sulfate ions - Bisulfate ions
3000-2800	R-H	Aliphatic carbons
3750-3500	R-OH	Alcohols
1750-1700	C=O	Carbonyl species - Hemicellulose - Pectin - Lectin

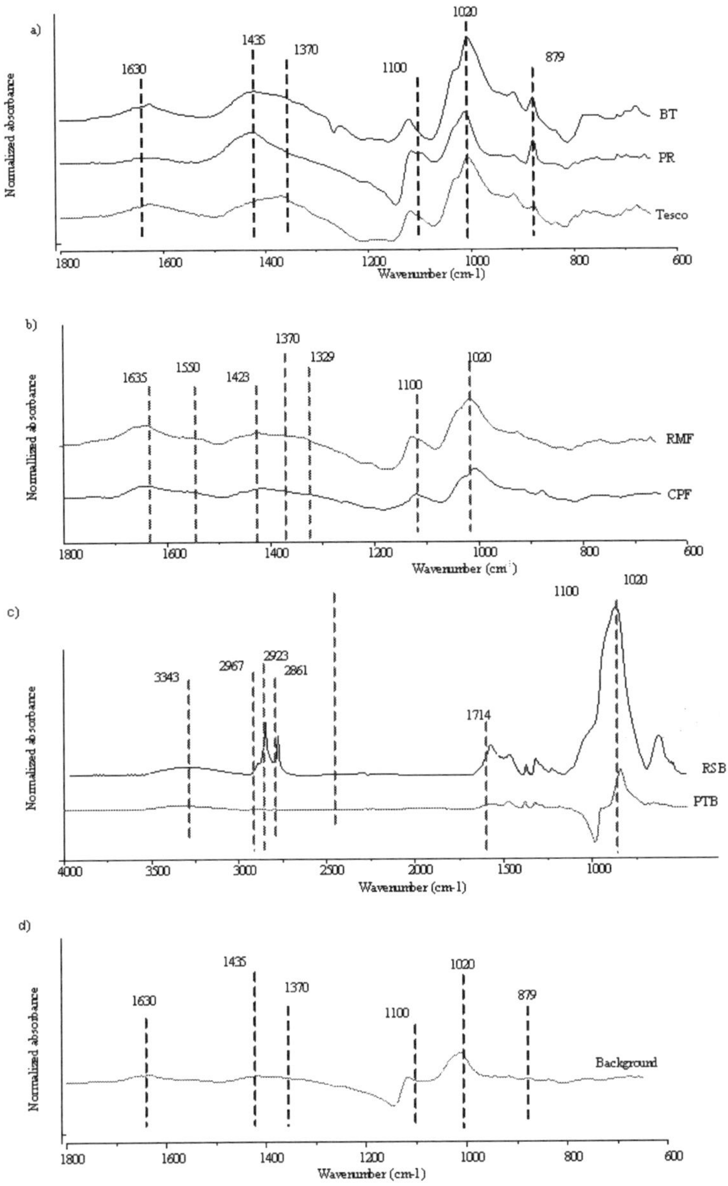

Figure 1: a) FTIR spectra of PM_{10} samples from each transportation sampling site, b) FTIR spectra of PM_{10} samples from each industrial sampling site, c) FTIR spectra of PM_{10} samples from each biomass burning sampling site d) average FTIR spectra of PM_{10} samples from each background sampling site. Subtract spectra were manipulated by spectra of PM_{10} aerosols minus spectra of quartz fiber filter (blank filter).

under-construction buildings, it is plausible to ascribe the relatively high level of CO_3^{-2} in traffic aerosols to the building materials (i.e. $CaCO_3$ in cement).

Also, it is important to note that the industrial aerosols clearly demonstrate the highest level of IR absorption band of aromatic nitro compound and organo nitrate in comparison with those of other emission sources (Fig. 1).

Table 3: Relative integral area (%) of Infrared absorption bands of averaged representative spectra of PM_{10} samples collected from each emission source.

Species	Relative Integral area (%)			
	Transportation	Industry	Biomass burning	Background
$CaCO_3$	9.5	2.4	-	6.3
SO_4^{2-}	57.0	54.6	62.1	63.6
Nitrate ions	7.0	4.5	-	4.8
Carbonate	11.9	8.1	3.3	8.7
Ammonium ions	6.7	4.3	5.8	6.4
Aromatic nitro compound	3.4	5.7	8.3	4.5
Organonitrates	4.4	18.2	11.8	5.7
Carbonyl	-	2.2	3.9	-

This can be explained by the imperfect combustion of diesel oils and wood materials employed during the heating and drying processes in the CPF and RMF respectively. In addition, it is well known that the polycyclic aromatic hydrocarbons could occur from industrial processes such as petroleum refining, coal coking and thermal power generation. On the other hand, the IR spectrum of biomass burning aerosols showed the highest intensity in the wavenumber ranged from 2,800 to 3,000 cm^{-1} indicating the aliphatic C-H bond absorption bands. There was also some evidence for the presence of CH_2 aliphatic carbon stretching (2,924 and 2,958 cm^{-1}), CH_3 aliphatic carbon stretching (2,850 cm^{-1}), alcohol (3,750-3,500 cm^{-1}), hemicellulose, pectin and lignin (1750-1700 cm^{-1}) measured in the biomass burning samples. It is also interesting to note that the band fitting results of background aerosols showed the highest integral area of sulfate species with 64% of relative contribution, plausibly influenced by dimethyl sulfide in maritime aerosols.

In order to categorize the PM_{10} particles emitted from different sources, the cluster analysis was applied to discriminate the samples into class on the basis of a distance mean based on peak shape, intensity, and peak position by using IR spectrum. As expected, Fig 2 shows the results of hierarchical tree (dendogram) categorizing PM_{10} aerosol into four clusters namely traffic (BT and Tesco),

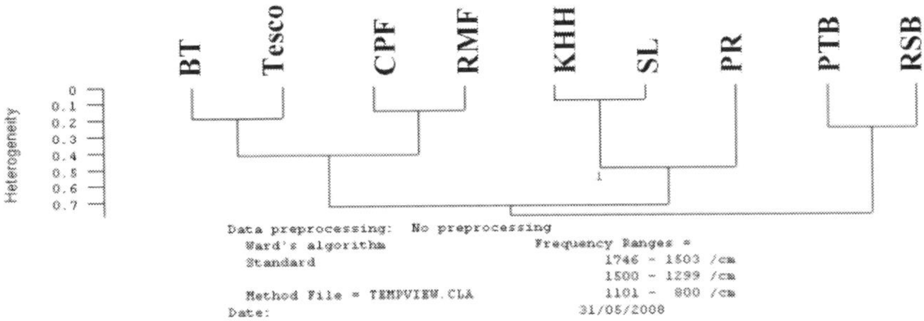

Figure 2: Hierarchical Cluster Analysis (HCA) dendrogram calculated from normalized mean spectra in the range of 1747-1298 cm^{-1} and 1101-800 cm^{-1} of PM_{10} aerosols from each emission sources Cluster Analysis was performed using the Ward's Algorithm.

Table 4: Concentration of TC, OC, EC, OC/EC ratio at different sampling sites.

Sampling sites		PM_{10} ($\mu g\ m^{-3}$)	TC* ($\mu g\ m^{-3}$)	OC ($\mu g\ m^{-3}$)	EC ($\mu g\ m^{-3}$)	OC/EC ($\mu g\ m^{-3}$)
Background	1. SL1	13.8 ± 2.88	4.30 ± 2.00	3.06 ± 1.43	1.23 ± 0.115	2.48 ± 1.15
	2. SL2	11.6 ± 3.68	0.977 ± 1.85	0.757 ± 0.397	0.221 ± 0.401	3.43 ± 6.48
	3.KHH	9.57 ± 3.97	1.85 ± 1.22	1.35 ± 0.894	0.499 ± 0.256	2.71 ± 1.79
Transportation	4. Tesco	46.9 ± 30.6	14.8 ± 37.4	8.57 ± 10.6	6.25 ± 13.8	1.37 ± 3.45
	5.BT	42.8 ± 24.9	14.1 ± 29.4	8.06 ± 8.15	6.05 ± 11.0	1.33 ± 2.77
	6. PR	25.1 ± 9.19	9.65 ± 22.9	5.43 ± 8.69	4.21 ± 7.38	1.29 ± 3.06
Industry	7. CP	24.5 ± 5.37	7.38 ± 15.4	5.16 ± 4.44	2.21 ± 4.22	2.33 ± 4.86
	8. RMF1	34.4 ± 8.58	15.8 ± 31.7	10.9 ± 17.2	4.97 ± 6.09	2.18 ± 4.36
	9. RMF2	36.7 ± 15.7	11.0 ± 24.8	6.92 ± 10.4	4.10 ± 6.83	1.68 ± 3.79
Biomass burning	10. RSB	218 ± 96.1	80.4 ± 91.9	65.0 ± 52.0	15.4 ± 12.6	4.22 ± 4.83
	11. PTB	83.7 ± 23.2	48.4 ± 66.5	38.6 ± 35.2	9.81 ± 10.1	3.93 ± 5.40

*Total carbonaceous (TC) aerosol was calculated by the sum of organic matter and elemental carbon.

industry (CPF and RMF), background (SL and KHH) and biomass burning (RSB and PTB). It is worth noting that PR shows distinctively a separation group of the cluster. This result can be explained by the relative highest integral area of

$CaCO_3^-$ band (shown at 879 cm^{-1}). These findings encourage policy makers to consider IR spectrum as an alternative parameter to conduct the source ratios of aerosols are displayed in Table 4. In general, the OC/EC ratios were observed in the range of 1-4, which is in good agreement with those of other cities around the world. In this study, the highest OC/EC ratios were found with the value of 3.9-4.2 in biomass burning aerosols. It is also interesting to note that the OC/EC ratios of PM_{10} collected at Tesco, PR and BT were lower than two, indicating the overwhelming influence of traffic exhaust over the three sampling stations. On the contrary, the relatively high OC/EC ratios (3.9-4.2) were analyzed from biomass burning samples. This can be due to the relatively high loadings of aliphatic hydrocarbon, ester carbonyl components such as hemicelluloses, pectin, and lignin as previously mentioned.

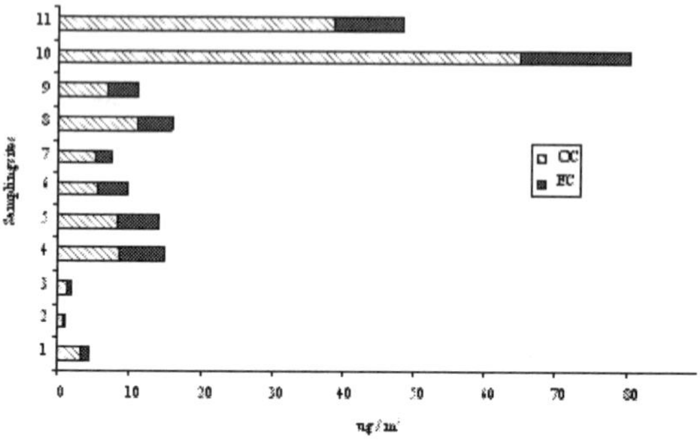

a)

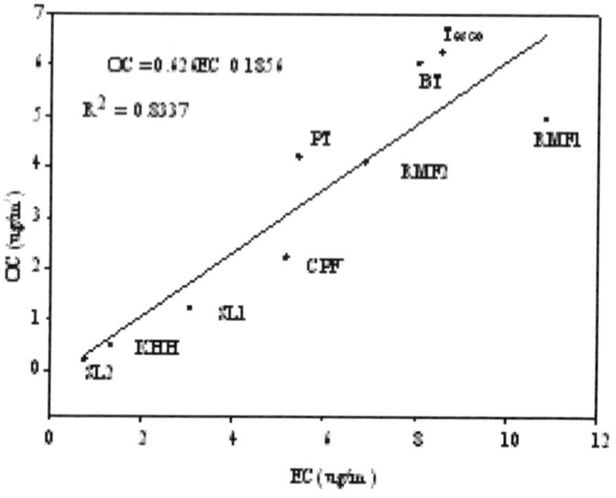

b)

Figure 3: a) Concentrations of OC and EC at different emission sources. b) Relationship between OC and EC in PM_{10} aerosols collected at Hat-Yai, Songkhla Province.

In this work, XANES technique was used to study the oxidation state of sulfur on the surface of PM_{10}. Since the sulfuric aerosol can be formed through a series of complex multi-phase interactions, involving the balance between acidic and basic ionic components in cloud and rain water, it is therefore important to recognize the oxidation state of particulate sulfuric compounds. Assuming that the aerosols collected from various emission sources are different in nature, it seems plausible to presume the distinctive differences of sulfuric oxidation state among air samples. Surprisingly, the S K-edge with the range of 2460-2520 eV was clearly identified in all samples. More importantly, the aerosols collected at the transportation and industrial sampling sites demonstrate predominantly only S_6^+ peak at 2481 eV in comparison with other oxidation states of sulfuric compounds (Fig. 4). Thus, in the future, we plan to use the XANES technique for determining the specification of sulfate species by comparing the results of standard reference material such as $(NH_4)_2SO_4$, $CaSO_4$ and $MnSO_4$ to ensure the correct data interpretation of sulfuric oxidation state associated with aerosol samples.

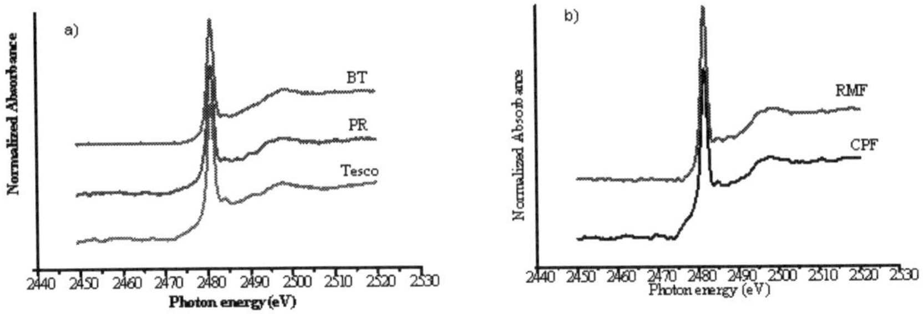

Figure 4: Sulfur K- edge XANES in aerosols collected from a) transportation and b) industry sampling sites.

Acknowledgements

The authors would like to thank the Synchrotron Light Research Institute (Public Organization), Ministry of Science of Technology, Royal Thai Government and also National Center of Excellence for Environmental and Hazardous Waste Management-PSU Satellite Center, Prince of Songkhla University for their continual financial support of this work. The authors appreciate Dr. Kemthong Sinwongsuwat for her contribution on English grammar correction.

References

[1] Badarinath, K.V.S., & Latha. K.M. Direct radiative forcing from black carbon aerosols over urban environment. Advances in Space Research, 37, pp. 2183-2188, 2005.

[2] Kiehl, J.T., & Briegleb, B.P. The relative roles of sulfate aerosols and greenhouse gases in climate forcing. Science, 260, pp. 311-314, 1993.

[3] Kampa, M & Castanas, E. Human health effects of air pollution, Environmental Pollution, 151 (2), pp. 362-367, 2008.

[4] Taus, N., Tarulescu, S., Idomir, M & Taus, R. Respiratory exposure to air pollutants, Journal of Environmental Protection and Ecology, 9 (1), pp. 15-25, 2008.

[5] Penner, J.E., Chuang, C.C & Liousse, C. The contribution of carbonaceous aerosols to climate change, Nucleation and Atmospheric Aerosols, pp. 759-769,1996.

[6] Penner, J.E., Chuang, C.C & Liousse, C. The contribution of carbonaceous aerosols to climate change, *Nucleation and Atmospheric Aerosols*, pp. 759-769, 1996.

[7] Kroll, J.H & Seinfeld, J.H. Chemistry of secondary organic aerosol: Formation and evolution of low-volatility organics in the atmosphere, Atmospheric Environment, 42(16), pp. 3593-3624, 2008.

[8] Adam, R., Barbara, J. T., John, H. O., Clifford, P. W., Jim, Z., Maria, M., Thomas, S., Steven, C., & Arthur, W. A functional group characterization of organic PM2.5 exposure: Results from the RIOPA study, Atmospheric Environment. 41, pp. 4585-4598, 2007.

[9] Antoine, G., Pierre-Alexandre, D., Véronique, J., & Patrick, B. Use of FTIR spectroscopy coupled with ATR for the determination of atmospheric compounds. Talanta, 68, pp. 1294-1302, 2

Section 6
Air pollution effects and environmental health

The relationship between air pollution caused by fungal spores in Mexicali, Baja California, Mexico, and the incidence of childhood asthma

R. A. de la Fuente-Ruiz[1], M. Quintero-Núñez[2], S. E. Ahumada[2] & R. O. García[2]
[1]*UABC, Mexicali, BC, Mexico*
[2]*Department of Environment of the Institute of Engineering, UABC, BC, Mexico*

Abstract

Air pollution has affected human life progressively since its sources of origination. It is present in all cities around the world. It has caused harm to humans and other living organisms. This situation has led studies to determine the impact of concentrations of air pollutants in the respiratory ailments. Mexico has developed similar investigations, but unlike other countries' advanced studies on pollutants aerobiology, Mexico's studies have not been conducted in depth, and there is virtually no research work that identifies the air pollutants of fungal spores and their association with respiratory disorders in children. The purpose of this article is to analyze the relationship between air pollution by fungal spores, and the incidence of childhood asthma in Mexicali, Baja California, Mexico. Diagnostics of asthma on children living in the city of Mexicali that received medical care were studied. The records of weather and climatic variables were provided by the Department of Meteorology at the Engineering Institute of UABC. The data of air pollution caused by fungal spores were obtained in three monitoring stations with collectors' rotation and impact of sampling. Mainly four groups of different spores (*Alternaria sp., Cladosporiun sp., Bipolaris and Stemphylium*) were registered in the atmosphere of Mexicali, with an average of 515 spores/m^3 and a standard deviation of 248 spores/m^3. The average annual relative humidity was 42.3%. The average temperature showed two different facets; a hot season with temperatures between 22°C and 44°C, and a cold season with temperatures between 8.9°C and 22°C. The incidence of childhood asthma presented a rate twice the national average (505 cases/one hundred thousand children). The multiple linear regression model proposed showed a significant relationship with a R2=0.86 (p=0.5), which established a direct association between air pollution by fungal spores, and the incidence of childhood asthma.

Keywords: childhood asthma, aerobiological contamination, fungal spores, asthma and weather conditions, Mexicali.

WIT Transactions on Ecology and the Environment, Vol 123, © 2009 WIT Press
www.witpress.com, ISSN 1743-3541 (on-line)
doi:10.2495/AIR090281

1 Introduction

Air pollution is a problem impacting all cities around the world, especially those places with a high production of pollutants due to technological development. It causes various diseases in the respiratory system. The concern about the harmful effects caused by air pollution is higher every day; an affection that has led to investigate the relation between the concentrations of air pollutants and respiratory tract problems in different age groups, as well as to find meaningful relationships in epidemiological terms [1, 2]. In such studies are taken in consideration different concentrations of air pollutants, weather variations, time of exposure and its health impacts on different age groups [3, 4].

Within this group, it is of great interest to address the effects of aerobiological pollutants in children which asthma is presented as a major disease caused by these types of pollutants that show the immune response characteristic to the allergens that float in the air [5]. In atopic children is usually presented as a chronic disease, which can last up to adolescence, and in some cases to adulthood which represents a problem of public health [6]. Pulimood (2008) conducted in 2002 an investigation in the United Kingdom on children during an increased in medical consultations due to asthma attacks which exceeded the usual medical consultation for this type of attacks, and found a direct correlation between levels of atmospheric pollution by *Alternaria* and *Cladosporium* spores, and the increase in the incidence and severity of the symptoms [7]. In Mexico similar correlations have been obtained [8–10]. Unfortunately, studies covering the weather conditions and aerobiological pollutants focus on respiratory infections have not been performed in Mexicali, Baja California, Mexico as of today [11] and it may be said that there are no conclusive studies on how to analyze the relationship between the atmospheric concentration of particles of biological origin such as pollen or fungal spores and its harmful effect on the airways in children.

Mexicali is the capital of state of Baja California, Mexico, with a population of approximately eight hundred and fifty-five thousand people [12], located on the northern border of Mexico adjacent to Imperial Valley, California, United States of America, sharing the same atmospheric basin [13]. On the other hand its location north of the Sonora's desert determines an extreme climate with long summers, with maximum temperatures reaching 42ºC to record temperatures of up to 53°C, contrasting with short winters with low temperatures that can reach below 7ºC, with a desert ecosystem that benefits from the tributaries of the Colorado River which contributes to the development of a strong agricultural activity, whose main crops are; wheat, barley, cotton, alfalfa, oats, sesame, safflower, sorghum forage, rye grass, vegetables and crops, and in a smaller scale watermelons, cantaloupes, corn, grapes, beans and prickly pear [14]. The objective of this study was to analyze the relationship between air pollution caused by fungal spores and the rising incidence of childhood asthma in the city of Mexicali.

2 Material and methods

The period of investigation covered one year (1/09/2004 to 31/08/2005). The studied population was composed by three different children's ages: One year and under, from 1 to 4 years, and 5 to 14 years old, all residents in different areas of the city of Mexicali and who received medical attention in governmental health facilities. The data was obtained from medical reports of all medical units of the Mexican Institute of Social Security (IMSS), Institute of Social Security of the State Workers (ISSSTE), Institute of Social Security for the State Workers of Baja California (ISSSTECALI), and Secretary of Health and Assistance (SSA). It was considered the total number of medical consultations from the first time by bronchial asthma among children between the ages selected. The cases were identified by age group and sex, according to medical diagnostics focused on the weekly report of new cases of disease (SUIVE) provided by the management of Epidemiology of the Health Secretariat of the State of BC, following the International Classification of Diseases, in its tenth revision (ICD-10). To facilitate the handling of official reports, were consolidated into a single record cases of asthma and status asthmaticus (CIE10: J45 and J46).

2.1 Meteorological data and air pollution

The meteorological variables were selected with an epidemiological approach, taking in consideration the records of the Relative Humidity Average (RHA) and Monthly Mean Temperature (MMT) data that was provided daily by the Department of Meteorology of the Institute of Engineering at the Universidad Autonoma de Baja California (UABC).

The atmospheric concentration of fungal spores data was obtained from the study conducted by Quintero (2006) [15], based on a sampling system using three monitoring stations, and strategically distributed in the city of Mexicali. The sampling of air pollutants was performed using a collector of rotation and impact type, Rotorod 40. It was estimated the rate of lung exposure (RLE) for fungal spores, considering the average exposure for 24 hours, as the product of the average concentration of spores/m^3 (ACS) in relation to the average daily volume of breathing in (AVB): RLE = [(ACS) (AVB)].

2.2 Statistical analysis

A descriptive analysis was performed from the obtained data: atmospheric concentration of fungal spores, the monthly mean temperature (MMT), relative humidity average (RHA) and the incidence of childhood asthma by age and sex, using graphic time series and frequency distribution for quality control. It was performed an independent analysis of the relationship between each of the variables proposals and cases of childhood asthma, taking into account each of the three age groups for the study. Finally, it was made an analysis focusing all cases of childhood asthma in a single group having its variations by sex. To interpret the association between the variables studied, it was designed a multiple linear regression model which allowed a reduction to a minimum statistical error and limitation of partiality. The handling of the data was performed with SPSS 12.

3 Results

During the years of the study there were a total of 1,212 new cases of childhood asthma registered, with a rate of 505 cases/one hundred thousand children, of whom 57% were males. The cases of asthma were more frequent in the 5 to 14 years age group (63%), with 766 cases, and in second place, the 1 to 4 group aged (31%,) and 373 cases, and with less involvement the children age of one year under, with 73 cases (six percent). The overall incidence of asthma showed a higher frequency in the months of March with 125 cases, 102 cases in June, 130 cases in October, 143 cases in November and December with 156 cases, as shown in table 1.

Table 1: Incidence of childhood asthma by age group and sex.

Months	>1	1 to 4	5 to 14	Total	Sex	
					F	M
S2004	2	31	63	96	50	57
O2004	10	34	86	130	57	72
N2004	8	49	86	143	51	77
D2004	6	66	84	156	47	68
E2005	6	37	55	98	42	48
F2005	6	3	67	76	40	46
M2005	7	50	68	125	54	76
A2005	5	27	58	90	38	49
M2005	9	23	67	99	39	52
J2005	8	29	65	102	46	62
J2005	3	15	33	51	25	39
A2005	3	9	34	46	34	43
Total	73	373	766	1212	523	689

Source: Health Secretary of Mexico 2006.

The incidence of asthma in general showed a cyclical behavior of bimodal type, with an initial plateau in December (156 cases), and a second plateau in March with 121 cases.

Regarding the atmospheric concentration of fungal spores, mainly 4 different groups of spores were identified, they were selected by its atmospheric concentration and its epidemiological significance; *Alternaria sp., Cladosporiun sp., Bipolaris y Stemphylium,* table 2. The general behavior of the atmospheric concentration of fungal spores with a cyclical behavior, showed an increased in concentration spores/m^3 in three stages: the first in March (503 spores/m^3), the

Table 2: Spores found in the sampling of a year, in Mexicali`s atmosphere.

Months	*Alternaria*	*Cladosporium*	*Stemphylium*	*Bipolaris*	Total
S2004	107	0	10	63	180
O2004	62	85	12	129	288
N2004	127	94	56	78	355
D2004	29	110	168	35	342
E2005	300	287	201	82	870
F2005	171	151	251	31	604
M2005	278	86	96	43	503
A2005	146	133	40	23	342
M2005	186	156	67	56	465
J2005	167	172	48	84	471
J2005	443	126	65	147	781
A2005	1203	303	183	297	1986
Total	3219	1703	1197	1068	7187

Source: Atmospheric sampling 2006.

second was in August (986 spores/m^3), and the third in January (870 spores/m^3) to February (604 spores/m^3). The descriptive statistics showed that the atmosphere of Mexicali, presents an annual average of 515 spores/m^3 with a standard deviation (SD) of 248 spores/m^3, and an average daily. As a result of calculating the rate of lung exposure to fungal spores (RLE), which may happen in a typical day, there is a possibility that on average 164 spores/day make contact with the mucus of the airways. Concentrations which become very important considering the potential of pathogenic fungi spores, and the atmospheric concentration limit to produce a case clinical of asthma.

$$RLE = [(ACS)\ (AVB)] \quad (1)$$

As a result

$$RLE = [(19\ \text{spores/m}^3)\ (8.64\ \text{m}^3\text{/day})] = 164\ \text{spores/day} \quad (2)$$

The annual relative humidity average (RHA) was 42.3%, with monthly range from 25% to 63% and an upward trend from august to February with a record of increase in humidity as shown in table 3.

The monthly mean temperature (MMT) records showed two different facets, one in the cold season that starts in November and continues until February, reached its minimum 8°C average, and a warm season, which starts virtually in may (28°C), increasing gradually up to august (35°C), with a record peak in the month of July that reached 44°C, as shown in table 4.

This means that in a typical day, a child resident in the city of Mexicali has an 86% probability to present the clinical symptoms of bronchial asthma.

The application of the multiple linear regression model was developed to analyze the behavior of the incidence of childhood asthma based on the epidemiological variables studied (monthly average spores/m^3, MMT and RHA) reflects a percentage of reliability with an $R^2 = 0.86$ ($P < .05$).

Table 3: Relative humidity averages in Mexicali.

Months	R.H. Maximum	R.H. Minimum.	R.H. Average
S2004	57.9	15.3	32.9
O2004	66	23.5	42.9
N2004	72.8	26.4	49
D2004	74	30.8	52.5
J2005	83.2	37.9	62
F2005	93.2	33.5	63.6
M2005	61.2	37.8	44.5
A2005	51.8	13.7	28.9
M2005	53.7	14.3	30.9
J2005	48.1	12.5	25.8
J2005	62.1	14.4	34.1
A2005	69.9	21.7	41.2

Source: Department of Meteorology of the Institute of Engineering, UABC, 2006.

Table 4: Monthly mean temperature in Mexicali.

	T°. Maximum.	T°. Minimum.	T°. Average
S2004	37.9	22.6	30.25
O2004	32.9	17.6	25.25
N2004	24.1	10.7	17.4
D2004	21.7	8.9	15.3
J2005	22.3	10.6	16.45
F2006	22.3	11.4	16.85
M2006	27.7	14.0	20.85
A2005	32.5	15.0	23.75
M2005	36.0	20.9	28.45
J2005	39.7	23.4	31.55
J2005	44.0	28.6	36.3
A2005	42.8	28.7	35.75

Source: Department of Meteorology of the Institute of Engineering of the UABC, 2006.

4 Discussions

There are several epidemiological studies realized at a number of cities in Mexico's northern border, where the relationship between changes in human health and the concentration of air pollutants was analyzed, which only few considered; aerobiological pollutants and the increased incidence of childhood asthma in the specified form were found [16, 17]. It may be said that there are virtually no studies that specifically analyzed the air pollutants such as fungal spores and the incidence of childhood asthma, despite the high concentrations of air pollutants of this type, reported in taxonomic studies carried out in different parts of Mexico, [18–20]. Papers published in other regions such as Europe and USA, analyzed this type of pollutant and its potential to cause health damage [21–24].

Mexicali is a border town in northern Mexico, with an economy based on the in bond industry and agriculture, where plants are grown mainly for family of *Gramineas* such as wheat or corn, whose harvests leaves are easily invaded by fungi. On the other hand the waste of most agricultural products are a fertile ground for the incubation of fungi, as *Alternaria* and *Cladosporium* [25], coupled with weather and specific topographical conditions (on average 11 meters below sea level), contribute to an higher concentration of fungal spores in the atmosphere that can be associated with the high index of lung exposure of 164 spores/day, combined with the type of spores observed and its association with the incidence of asthma. Special Comment demand registration of 1986 spores/m^3 observed in August, which takes place in the fallowing of farmland, which involves the removal of land by releasing the *Ascospores* and *Conidias,* specialized structures that were adapted by fungi order to remain in latent form in different time periods in dry soil, waiting for moisture and a biological substrate to reproduce and germinate.

Halomen (1997), established a direct relationship between the spores of *Alternaria* and childhood asthma, in similar studies which showed a positive relationship between the atmospheric concentration of spores, and the presence of specific alterations of human health [26–28], which helps to explain the presence of a rate of childhood asthma 505/one hundred thousand children, that represents approximately the national average two folded [29]. The largest percentages of these were observed in males at the age before puberty, as it was reported by Sears (2008) in its study on the epidemiology of asthma [30].

By observing the incidence of childhood asthma it has been appreciated a bimodal type behavior with a well-defined subsequent relationship to the maximum levels of air pollution by fungus spores. The subsequent relationship is explainable as a delayed response caused by the amount of dispersion time and deposition in the lung epithelium, [31], joined to a physio-pathological response with a latency period of 5 days on average, and a belated reaction family, which could last up to 7 days after contact, before the severity of the clinical data required a visit to a physician [32]. In relation to the application by sex shows a higher percentage of males, similar to the reported in national records.

Finally the implementation of the proposed regression model showed a significant relationship with an R2 = 86 (p = 0.05) among the epidemiological variables (average relative humidity, temperature and atmospheric concentration of fungal spores/m^3) and the increase in the incidence of childhood asthma in the city of Mexicali

5 Conclusions

This study provides an analysis of the impact of air pollution caused by fungal spores of *Alternaria, Cladosporiun, Bipolaris* and *Stemphylium* on children's respiratory system in Mexicali, as a response to a combination of the unique characteristics of Mexicali and its Valley such as extreme temperatures, moisture levels, diverse crops plantation, type of plants and bushes and the topographical features of the region. The results of the linear regression model proposed, allowed to establish a 86% probability, a positive relationship between levels of atmospheric pollution by fungal spores, the increase in the temperature rates, air humidity and behavior of the incidence of childhood asthma.

It is important to note that the findings of air pollutants by fungal spores encountered in this work derived from the only study carried out in the region.

References

[1] Rosales, C.J.A., Torres, M.V.M., Olaiz, F.G. & Borja, A.V.H., Los efectos agudos de la contaminación del aire en la salud de la población: evidencia de estudios epidemiológicos. *Salud Pública Méx.,* **43**, pp.544-555, 2001.

[2] Romero, P.M., Más, B.P., Lacasaña, N.M., Téllez Rojo, S.M.M., Aguilar, V.J. & Romieu, I., Contaminación atmosférica, asma bronquial e infecciones respiratorias agudas en menores de edad de La Habana. *Salud Pública Mex*, **46,** pp. 222-233, 2004.

[3] Dales, R.E., Cakmak, S., Judek, S., Dan, T., Coates, F., Brook, J.R. & et al., Influence of outdoor aeroallergens on Hospitalization for asthma in Canada. *Asthma and immunology*, **10 (1016**), pp. 303-306, 2004.

[4] Chardon, B., Lefranc, A., Granados, D. & Gremy, I., Air pollution at doctors' house call for respiratory diseases in the Greater Paris area. *Occupational& Environmental Medicine*, **64 (5),** pp. 320-324, 2007.

[5] Van, H.C.L., Maes, T. Joos, G.F. & Tournoy, K.G., Chronic inflammation in asthma: a contest of persistence vs resolution. *Allergy*, **63**, pp. 1095-1109, 2008.

[6] Spahn, J.D. & Covar, R., Clinical assessment of asthma progression in children and adults. *J. Allergy clin immunol,* **121 (39),** pp. 548-557, 2008.

[7] Pulimood, B.T., Corden, M.J., Bryden, C., Sharples, L. & Nasser, M.S., Epidemic asthma and the role of the fungal mild Alternaria alternate. *J Allergy Clin Immunol*, **120 (3)**, pp. 610-617, 2008.

[8] Hernández, C.L., Téllez, R.M.M., Sanín, A.L.H., Lacasaña, N.M., Campos, A. & Romieu, I., Relación entre consultas de urgencias por

enfermedad respiratoria y contaminación atmosférica en Ciudad Juárez, Chihuahua. *Salud Pública Méx.*, **42**, p.p. 288-297, 2000.

[9] Romieu, I., Meneses, F. & Ruiz, R., Effects of air pollution on the respiratory health asthmatic children living in México City. *Am. J Respir Crist Med.*, **154**, pp. 300-.307, 2007.

[10] Villa, I.M., Vargas, Z.J., Gortáre, M.P., Flores. S.A., Badii, Z.M., González, H.R. & et all., Efectos de la contaminación atmosférica sobre la población infantil en Cd.. Obregón Sonora. *ITSON-DIEP.*, **3 (10)**, pp. 19-28, 2001.

[11] Rosas, I., Mc Cartney, H.A., Calderón, C., Lacey, J., Chapela, R. & Ruiz, V.S., Analysis of the relationships between environmental factors (aeroallergens, air pollution, and weather) and asthma emergency admissions to a hospital in Mexico City. *Allergy,* **53 (4),** pp. 394-401, 1998.

[12] Sistema Nacional de Información Estadística y Geografía, México. www.inegi.gob.mx/lib/buscador/ busqueda.aspx.

[13] Sweedler, A., Fertig M., Collins, K. & Quintero, N.M., *Air Quality in California-Baja California. Border Region,* SCERP: California US, 34-57, 2003.

[14] Vengas, F.R. *Manual para el reconocimiento de los árboles y arbustos más comunes en la ciudad de Mexicali*, UABC: Baja California México, pp. 65-75, 1991.

[15] Quintero, N.M., Ahumada, V.S.E. & Núñez, P.P.G. Mapeo Polínico. Cuarto reporte del proyecto. Online. http://scerp.org/apps/cont.mgt/doc files/ Mapa PolinicoMexicali. %202004-2%20 (espa_ol).pdf. Mexicali.

[16] Reyna, C.M.A., Quintero, N.M. & Collins, K. Correlation Study of the Association of PM_{10} with the Main Respiratory Diseases in the Populations of Mexicali, Baja California and Imperial County, California. *Revista Mexicana de Ingeniería Biomédica*, **26 (1)**, pp. 22-36, 2005.

[17] Hernández, C.L., Barraza, V.A., Ramírez, A.M., Moreno, M.H. & Miller, P., Carbajal-Arroyo, L.A. Romieu, I., Morbilidad infantil por causas respiratorias y su relación con la contaminación atmosférica en Ciudad Juárez, Chihuahua, México. *Salud Pública Mex.*, **49**, pp. 27-36, 2007.

[18] Arrequín, S.M.L., Fernández, N.R., Palacios, CH.R. & Quiroz, G.L. Morfología de las esporas de Pteridofitas. Isosporas del estado de Querétaro, México. *Poli botánica,* **2**, pp. 10-60, 1996.

[19] Gamboa, A.M.M., Escalante, E.F. Alejos, G.F., Garcia, S.K., Delgado, L.G. & Peña, R.L.M. Natural Zinniol derivatives from Alternaria Tagetica. Isolation, synthesis and structure-activity correlation. *Journal of Agricultural and Food Chemistry,* **50**, pp. 1053-1058, 2000.

[20] Gamboa, A.M.M., García, S.K., Alejos, G.F., Escalante, E.F., Delgado, L.G. & Peña, R.L.M. Tagetolone and tagetenolone, two phytotoxic polyketides from Alternaria T., *Journal of Agricultural and Food Chemistry*, **49**, pp. 1228-1232, 2001.

[21] Gioulekasn, D., Damialis, A., Papakosta, D., Spieksma, S., Giouleka, P. & Patakas, D. Allergenic fungi spore records (15 years) and sensitization in

patients with respiratory allergy in Thessaloniki-Greece. *J Invest Allergol Clin Immunol*, **14 (3)**, pp. 225-231, 2004.

[22] Bartra, T.J., Mapa Fúngico y estudio multicéntrico de sensibilización a Hongos en Cataluña. *Alergol Inmunol Clin*, **18 (3)**, pp. 106-121, 2003.

[23] Sanz, C., Isidro, G.M., Dávila, I., Moreno, E., Laffond, E. & Lorente F., Analysis of Polymorphisms in Patients with Asthma. *J. Investig Allergol Clin Immunol,* **16 (6)**, pp. 331-337, 2006.

[24] Ghosh, N., Camacho, R., Saadeh, C., Gaylor, M. & Smith, D. W., Study on the Fungal Aeroallergen Concentration in the Texas Panhandle Using a Buckard Volumetric Spore Trap. *J Clin Alergia Immunol*, **113**, S91, 2004.

[25] Sabariego, R.S., Díaz de la Guardia G.C., Alba S.F. (2004) Estudio aerobiológico de los conidios de Alternaria y Cladosporium en la atmósfera de la ciudad de Almería (SE de España). *Rev Iberoam Micol*; **21**: 121-127

[26] Halonen, M., Stern, D.A., Wright, A.L., Taussig, L.M. & Martinez, F.D. Alternaria as a major allergen for asthma in children raised in a desert environment Am. J. Respir. *Crit. Care Med,* **155 (4)**, pp.1356-1361, 1997.

[27] Del Palacio, A. P.S.M., Arbi, A., Valle, A., Perea, S. & Rodríguez, N.A., Bipolares en un enfermo español con sinusitis alérgica crónica. España. *Rev Iberoam Micol*, **14**, pp.191-193, 1997.

[28] Gómez, S.A., Torres, R.J.M., Alvarado, R.E., Mojal, G.S. & Belmonte, S.J., Seasonal Distribution of Alternaria Aspergillus, Cladosporium and Penicillium Species Isolated in Homes of Fungal Allergic Patients. *J Investig Clin Inmunol*, **16 (6)**, pp. 357-363, 2006.

[29] Secretaria de Salud (S.S.). Office of Sistema Nacional de Vigilancia Epidemiológica, Mèxico DF. www.dgepi.salud.gob.mx/anuario/html/anuarios.

[30] Sears R.M., Epidemiology of asthma exacerbations. *J Allergy Clin Immunol*, **122 (4)**, pp. 662-668, 2008.

[31] Sáenz, C.L. & Gutiérrez, B.M., *Esporas atmosféricas en la comunidad de Madrid.* Industrias Gràficas MAE: Madrid, pp. 9-12, 2003.

[32] Hernández, C.L., Téllez, R.M.M., Sanín, A.L.H., Lacasaña, N.M. Campos, A. & Romieu, I., Relación entre consultas de urgencias por enfermedad respiratoria y contaminación atmosférica en Ciudad Juárez, Chihuahua. *Salud Pública Mex*, **42,** pp. 288-297, 2000.

Dioxin and furan blood lipid concentrations in populations living near four wood treatment facilities in the United States

C. Wu, L. Tam, J. Clark & P. Rosenfeld
Soil Water Air Protection Enterprise, USA

Abstract

To evaluate historical exposure from wood treatment facilities, blood samples were collected from 65 current and past residents of four communities surrounding wood treatment facilities throughout the United States. The pattern of dioxin/furan congeners detected in blood samples was found to be consistent with exposure to contaminants generated during the wood treatment process. The levels of 2,3,7,8-tetrachloro-p-dibenzodioxin toxic equivalents (2,3,7,8-TCDD TEQs) for all 17 carcinogenic dioxin/furan congeners, octa-chlorinated dibenzo-p-dioxins (OCDD) adjusted to its TEQ value and 1,2,3,4,6,7,8-hepta-chlorinated dibenzo-p-dioxin (1,2,3,4,6,7,8-HpCDD) adjusted to its TEQ value in the U.S. population were compared to the TEQ levels of the combined dataset for all 4 communities and of the datasets for each individual community. TEQ concentrations in these communities were found to be significantly greater than the TEQ data for the general U.S. population. These findings reveal that a very significant potential for contaminant-related health risks exists in communities surrounding wood treatment facilities.
Keywords: wood treatment, blood analysis, dioxins, furans, creosote, pentachlorophenol.

1 Introduction

In this study, total 2,3,7,8-tetrachloro-p-dibenzodioxin toxic equivalents (2,3,7,8-TCDD TEQs) (all 17 carcinogenic dioxin/furan congeners), octa-chlorinated dibenzo-p-dioxin (OCDD) (adjusted to its TEQ value), and 1,2,3,4,6,7,8-hepta-chlorinated dibenzo-p-dioxin (1,2,3,4,6,7,8-HpCDD) (adjusted to its TEQ value) in blood for residents living in four communities including Alexandria,

WIT Transactions on Ecology and the Environment, Vol 123, © 2009 WIT Press
www.witpress.com, ISSN 1743-3541 (on-line)
doi:10.2495/AIR090291

Louisiana; Pineville, Louisiana; Grenada, Mississippi; and Florala, Alabama surrounding wood treatment facilities in the United States were evaluated. The facilities used pentachlorophenol (PCP) and creosote as insecticides to treat wood and released dioxins and other hazardous substances into the surrounding communities. The Alexandria, LA facility has been in operation since 1926, the Pineville, LA facility has been in operation since 1948, the Florala, AL facility has been in operation since the early 1900s, and the Grenada, MS facility has been in operation since 1904.

Dioxins and furans are by-products and impurities generated and released during human activities such as industrial, municipal, and domestic incineration/combustion processes and during the manufacture of chlorinated phenols and other chlorinated chemicals like PCP [1–4]. Polychlorinated dibenzo-p-dioxin (PCDD) and polychlorinated dibenzofuran (PCDF) are toxic chlorinated compounds that are usually released as mixtures into the environment. The use and incineration of PCP and creosote-treated wood products creates highly chlorinated dioxin and furan congeners, such as the signature congeners OCDD and 1,2,3,4,6,7,8-HpCDD [5–9].

There are 210 different dioxin and furan congeners. Seventy-five are possible dioxin congeners and 135 are possible furan congeners. The dioxin and furan congeners thought to be most toxic to humans are the seven dioxins and ten furans with a particular pattern of chlorines known as the 2,3,7,8-congeners. These 17 congeners are reported to cause cancers, and have endocrine and reproductive effects. The different PCDD/F congeners are structurally similar and have a similar mechanism of action. These chemicals are typically evaluated as 2,3,7,8-TCDD TEQs [10].

Human blood sampling is used for evaluating and assessing historical exposure to contaminants [6, 7, 11]. Dioxins and furans have relatively long half-lives in human blood, therefore, sampling human blood can also be used to assess historical exposure. Pirkle *et al.* [12] estimated the serum half-life of 2,3,7,8-TCDD in humans to be 7.1 years (range of 2.9-26.9 years) in a group of 36 Vietnam veterans. In a subsequent study, Michalek [13] estimated the serum half-life of 2,3,7,8-TCDD in humans to be 8.7 years (95% CI of 8.0–9.5 years) in a group of 343 Vietnam veterans. A half-life range from 3.5 to 15.7 years was estimated for dioxin congeners other than 2,3,7,8-TCDD in a study performed by Flesch-Janys *et al.* [14].

To evaluate whether exposure contamination in the communities surrounding wood treatment facilities were higher than the general US population, blood samples were compared to the Center for Disease Control (CDC) National Health and Nutrition Examination and Survey (NHANES) dataset for 2003 to 2004. The Centers for Disease Control and Prevention (CDC) conducts a survey, the National Health and Nutrition Examination Survey (NHANES), every two years through the National Center for Health Statistics (NCHS). The NHANES data set contains health and nutritional information on the U.S. population. The 1999-2000 survey collected data on 116 chemicals and was the first to include PCDD/F serum analyses. The 2001-2002 NHANES survey collected data on 135 chemicals. The data was yet again updated in 2003-2004 [15, 16].

The NHANES dataset was used as descriptive reference statistics for 2,3,7,8-TCDD TEQ in blood-lipid in the U.S. population. The results of blood sampling in the communities adjacent to the wood treatment facilities were compared to the NHANES dataset and found to be significantly greater than the 2,3,7,8-TCDD TEQ concentrations in the NHANES dataset.

2 Materials and methods

2.1 Blood sampling and analysis

Blood samples were collected from 65 current and past residents of the communities immediately surrounding wood treatment facilities and analyzed to evaluate levels of dioxins and furans. Samples were then shipped to Severn Trent Laboratories (STL) in Sacramento, CA for analysis.

In accordance with USEPA Method 8290, STL used high-resolution gas chromatography (HRGC)/mass spectrometry (HRMS) to analyze the blood samples. Each serum sample was spiked with $^{13}C_{12}$-labeled internal standards prior to extraction. A DB-5 capillary column was used to separate the target analytes. Each sample batch included a Method Blank (MB), a Laboratory Control Sample (LCS), and the unknown serum samples. Blood lipid content was determined gravimetrically. Data was reviewed using comprehensive multi-tiered quality assurance and quality control procedures.

To efficiently evaluate mixtures of PCDD/Fs, Toxic Equivalency Factors (TEFs) were developed [17]. TEFs establish the toxicity of the different congeners in relation to 2,3,7,8-TCDD for use in evaluating human health concerns. The EPA has determined that TEFs are currently the best method for evaluating complex mixtures of PCDD/Fs. The concentration of each PCDD/F is multiplied by its respective TEF to obtain a 2,3,7,8-TCDD toxic equivalents (TEQ) value [18]. These individual TEQs are then summed to provide a total dioxin TEQ value [19]. TEQ values for individual congeners can also be calculated by multiplying the concentration of the congener by the TEF value. The two individual congeners evaluated in this study are OCDD and 1,2,3,4,6,7,8-HpCDD with TEF values of 0.0003 and 0.01 respectively.

2.2 NHANES data set analysis

The 2003 to 2004 NHANES data for dioxins and furans was downloaded from the CDC website [15]. The data was downloaded in SAS format, but converted to Microsoft Excel format using SYSTAT 11.0 statistical software package. The detection limit over the square root of two was substituted for concentrations below the detection limit in all calculations. The CDC uses this method to analyze the NHANES dataset [20]. 2,3,7,8-TCDD TEQs were calculated for the data using the WHO 2005 TEF values and total dioxin TEQs were summed for each individual. Congener specific analysis was also conducted by calculating

TEQs for OCDD and 1,2,3,4,6,7,8-HpCDD. The data was then filtered for males and females 25 to 88 years of age to correspond with the complete dataset of all four communities. The data was also filtered for individuals 44 to 88 years of age for the Grenada dataset, 34 to 72 years of age for the Pineville dataset, 37 to 79 years of age for the Alexandria dataset, and 25 to 76 years of age for the Florala dataset to correspond with the individual community datasets. The use and incineration of PCP and creosote-treated wood products creates highly chlorinated dioxin and furan congeners, such as the signature congeners OCDD and 1,2,3,4,6,7,8-HpCDD [5–9]. Therefore, these congeners were isolated from the NHANES dataset for statistical analysis. TEQs were then used to calculate the sum, mean, maximum, minimum and other summary statistics for all congeners and specific congeners in lipid-adjusted serum concentration data.

2.3 Statistical analysis

Statistical analysis of the datasets was used to determine if the cohorts' total 2,3,7,8-TCDD-TEQ, OCDD (adjusted to its TEQ value) and 1,2,3,4,6,7,8-HpCDD (adjusted to its TEQ value) blood lipid concentrations are statistically different and greater than the general US population of the same age range. The Statistics Online Computational Resource (SOCR) software [21] developed by the University of California, Los Angeles was used for the analysis of the data. The combined four communities, individual community and NHANES datasets are not normally distributed. Therefore, the Wilcoxon rank-sum test, a non-parametric test for assessing whether two samples of observations come from the same distribution, was used to evaluate the datasets.

The Wilcoxon rank-sum tests the null hypothesis that the two sample sets are drawn from a single population, and therefore their probability distributions are equal. The samples must be independent, and the observations must be continuous measurements. The Wilcoxon rank-sum test generates a z-score and *p*-value for the datasets. A large positive or negative z-score indicates that group A (or B) values exceed the group B (or A) values respectively. The *p*-value is the probability that two groups of datasets come from the same population [22].

Statistical analysis was performed for total 2,3,7,8-TCDD TEQ (all 17 carcinogenic dioxin/furan congeners), OCDD (adjusted to its TEQ value), and 1,2,3,4,6,7,8-HpCDD (adjusted to its TEQ value) blood lipid concentrations for the combined four communities and the individual communities against the NHANES dataset. The age-range of the NHANES dataset was adjusted to match each comparison group.

3 Results and discussion

The results of the statistical analysis of the total 2,3,7,8-TCDD TEQ (all 17 carcinogenic dioxin/furan congeners), OCDD (adjusted to its TEQ value), and 1,2,3,4,6,7,8-HpCDD (adjusted to its TEQ value) blood lipid concentrations demonstrate that the populations surrounding the wood treatment facilities combined and individually have statistically higher TEQs in blood lipid than the general population of the U.S. of the same age range ($p<0.05$). Table 1 presents the data statistical summary for the communities and NHANES TEQ blood lipid data. Table 2 presents the Wilcoxon rank-sum test data outputs comparing the NHANES dataset to the combined four communities and the four individual community TEQ blood lipid datasets.

Table 1: Data statistical summary.

All 4 Sites Compared To NHANES 2003-4 Data (24-88)	*Group Total TCDD TEQs*	*NHANES Total TCDD TEQs*	*Group OCDD TEQs*	*NHANES OCDD TEQs*	*Group 1,2,3,4,6,7,8-HpCDD TEQs*	*NHANES 1,2,3,4,6,7,8-HpCDD TEQs*
Count	65	1159	65	1140	65	1155
Maximum	381.95	103.70	3.68	0.98	9.09	4.56
Mean	102.67	16.45	0.44	0.11	1.66	0.42
50th Percentile	80.66	13.38	0.27	0.08	0.97	0.32
75th Percentile	131.96	21.48	0.47	0.13	2.20	0.55
90th Percentile	214.14	30.47	0.82	0.22	3.42	0.83
95th Percentile	237.39	38.26	0.95	0.29	4.50	1.06
Standard Devation	80.82	11.89	0.52	0.09	1.68	0.38
Grenada, MS Compared To NHANES 2003-4 Data (44-88)	*Grenada, MS Total TCDD TEQs*	*NHANES Total TCDD TEQs*	*Grenada, MS OCDD TEQs*	*NHANES OCDD TEQs*	*Grenada, MS 1,2,3,4,6,7,8-HpCDD TEQs*	*NHANES 1,2,3,4,6,7,8-HpCDD TEQs*
Count	22	737	22	724	22	734
Maximum	332.51	103.70	1.69	0.98	4.67	4.56
Mean	53.41	20.75	0.39	0.13	1.26	0.51
50th Percentile	31.42	17.92	0.28	0.10	0.85	0.41
75th Percentile	50.88	25.44	0.39	0.17	1.19	0.66
90th Percentile	84.99	34.70	0.69	0.26	2.19	0.95
95th Percentile	123.86	44.62	0.88	0.32	3.74	1.27
Standard Devation	67.66	12.69	0.35	0.11	1.07	0.43
Pineville, LA Compared To NHANES 2003-4 Data (34-72)	*Pineville, LA Total TCDD TEQs*	*NHANES Total TCDD TEQs*	*Pineville, LA OCDD TEQs*	*NHANES OCDD TEQs*	*Pineville, LA 1,2,3,4,6,7,8-HpCDD TEQs*	*NHANES 1,2,3,4,6,7,8-HpCDD TEQs*
Count	11	751	11	740	11	749
Maximum	381.95	76.54	3.68	0.98	9.09	4.56
Mean	129.29	15.32	0.62	0.10	1.69	0.42
50th Percentile	105.49	13.70	0.26	0.08	0.82	0.33
75th Percentile	155.30	19.43	0.35	0.12	1.30	0.54
90th Percentile	224.05	25.53	0.96	0.20	2.94	0.80
95th Percentile	303.00	30.73	2.32	0.26	6.02	1.04
Standard Devation	102.03	9.17	1.04	0.09	2.57	0.37
Alexandria, LA Compared To NHANES 2003-4 Data (37-79)	*Alexandria, LA Total TCDD TEQs*	*NHANES Total TCDD TEQs*	*Alexandria, LA OCDD TEQs*	*NHANES OCDD TEQs*	*Alexandria, LA 1,2,3,4,6,7,8-HpCDD TEQs*	*NHANES 1,2,3,4,6,7,8-HpCDD TEQs*
Count	11	773	11	761	11	770
Maximum	127.80	103.70	0.59	0.98	3.54	4.56
Mean	61.23	17.21	0.26	0.11	1.05	0.45
50th Percentile	58.35	15.31	0.21	0.08	0.70	0.36
75th Percentile	70.87	22.07	0.29	0.14	0.95	0.60
90th Percentile	82.37	29.16	0.47	0.22	2.40	0.89
95th Percentile	105.08	34.92	0.53	0.28	2.97	1.11
Standard Devation	27.52	10.59	0.15	0.09	1.01	0.38
Florala, AL Compared To NHANES 2003-4 Data (25-76)	*Florala, AL Total TCDD TEQs*	*NHANES Total TCDD TEQs*	*Florala, AL OCDD TEQs*	*NHANES OCDD TEQs*	*Florala, AL 1,2,3,4,6,7,8-HpCDD TEQs*	*NHANES 1,2,3,4,6,7,8-HpCDD TEQs*
Count	21	1003	21	987	21	1000
Maximum	267.23	76.54	1.72	0.98	7.60	4.56
Mean	162.04	14.32	0.50	0.09	2.37	0.39
50th Percentile	159.69	12.23	0.42	0.07	1.87	0.30
75th Percentile	202.59	18.62	0.71	0.11	2.70	0.50
90th Percentile	234.35	25.43	0.83	0.19	3.50	0.76
95th Percentile	238.15	30.75	0.91	0.24	6.62	0.96
Standard Devation	52.95	9.23	0.38	0.08	1.77	0.35

Table 2: Wilcoxon rank-sum test summary.

All 4 Sites Compared To NHANES 2003-4 Data (24-88)	*Group Total TCDD TEQs*	*NHANES Total TCDD TEQs*	*Group OCDD TEQs*	*NHANES OCDD TEQs*	*Group 1,2,3,4,6,7,8-HpCDD TEQs*	*NHANES 1,2,3,4,6,7,8-HpCDD TEQs*
Sample Size	65	1159	65	1140	65	1155
Mean	102.67	16.45	0.44	0.11	1.66	0.42
Rank Sum	73614	676087	69254	657361	68345	676465
Test Statistics	3867	71469	6991	67109	8875	66200
Z-score		-12.19		-11.02		-10.37
P(T<=t) one-tail		0.000		0.000		0.000
P(T<=t) two-tail		0.000		0.000		0.000
Grenada, MS Compared To NHANES 2003-4 Data (44-88)	*Grenada, MS Total TCDD TEQs*	*NHANES Total TCDD TEQs*	*Grenada, MS OCDD TEQs*	*NHANES OCDD TEQs*	*Grenada, MS 1,2,3,4,6,7,8-HpCDD TEQs*	*NHANES 1,2,3,4,6,7,8-HpCDD TEQs*
Sample Size	22	737	22	724	22	734
Mean	53.41	20.75	0.39	0.13	1.26	0.51
Rank Sum	13260	275161	14248	264384	13807	272340
Test Statistics	3208	13007	1934	13995	2595	13554
Z-score		-4.84		-6.06		-5.43
P(T<=t) one-tail		0.000		0.000		0.000
P(T<=t) two-tail		0.000		0.000		0.000
Pineville, LA Compared To NHANES 2003-4 Data (34-72)	*Pineville, LA Total TCDD TEQs*	*NHANES Total TCDD TEQs*	*Pineville, LA OCDD TEQs*	*NHANES OCDD TEQs*	*Pineville, LA 1,2,3,4,6,7,8-HpCDD TEQs*	*NHANES 1,2,3,4,6,7,8-HpCDD TEQs*
Sample Size	11	751	11	740	11	749
Mean	129.29	15.32	0.62	0.10	1.69	0.42
Rank Sum	8275	282428	7621	274755	6739	282441
Test Statistics	52	8209	585	7555	1566	6673
Z-score		-5.63		-4.88		-3.53
P(T<=t) one-tail		0.000		0.000		0.000
P(T<=t) two-tail		0.000		0.000		0.000
Alexandria, LA Compared To NHANES 2003-4 Data (37-79)	*Alexandria, LA Total TCDD TEQs*	*NHANES Total TCDD TEQs*	*Alexandria, LA OCDD TEQs*	*NHANES OCDD TEQs*	*Alexandria, LA 1,2,3,4,6,7,8-HpCDD TEQs*	*NHANES 1,2,3,4,6,7,8-HpCDD TEQs*
Sample Size	11	773	11	761	11	770
Mean	61.23	17.21	0.26	0.11	1.05	0.45
Rank Sum	8380	299340	7386	290993	6540	298832
Test Statistics	189	8314	1052	7320	1997	6474
Z-score		-5.45		-4.27		-3.01
P(T<=t) one-tail		0.000		0.000		0.001
P(T<=t) two-tail		0.000		0.000		0.003
Florala, AL Compared To NHANES 2003-4 Data (25-76)	*Florala, AL Total TCDD TEQs*	*NHANES Total TCDD TEQs*	*Florala, AL OCDD TEQs*	*NHANES OCDD TEQs*	*Florala, AL 1,2,3,4,6,7,8-HpCDD TEQs*	*NHANES 1,2,3,4,6,7,8-HpCDD TEQs*
Sample Size	21	1003	21	987	21	1000
Mean	162.04	14.32	0.50	0.09	2.38	0.39
Rank Sum	21294.00	503506.00	19161.50	489374.50	20.79	500937.00
Test Statistics	0	21063	1797	18931	437	20563
Z-score		-7.85		-6.49		-7.53
P(T<=t) one-tail		0.000		0.000		0.000
P(T<=t) two-tail		0.000		0.000		0.000

Studies correlate exposure to chemicals such as dioxins, furans, and PAHs, with increased risk of developing a variety of diseases. Exposure to dioxins and furans can lead to endocrine disruption, reproductive and developmental defects, immunotoxicity, hepatotoxicity, neurotoxicity, and a variety of cancers [2, 3, 23, 24]. Exposure to PAHs can increase the risk of developing breast, lung, and skin cancer, leukemia, respiratory toxicity, and reproductive toxicity [25, 26]. Exposure to a mixture of dioxins, furans, and PAHs can significantly increase the risk of developing adverse health effects since these chemicals may have additive and synergistic properties [27].

Similar to other studies that have investigated exposure from residing near wood treatment facilities, the levels of dioxins, furans, and PAHs found in human blood in this study further demonstrate that the residential areas have been and are being exposed to potentially unsafe levels of these contaminants due to past management practices of these wood treatment facilities. The pattern of dioxins and furans found in the blood samples is consistent with dust generated during the incineration of PCP and creosote-treated wood. The residents near the wood treatment facilities also have statistically higher concentrations of 2,3,7,8-TCDD TEQs in blood for all carcinogenic dioxin/furan congeners and for the specific congeners associated with PCP than the general population of the U.S. of the same age range. Furthermore, considering dioxin's long half-life in blood, these concentrations are even more significant. Comparing the exposure of residents around these similar sites gives insight into the pattern of exposure that communities adjacent to other wood treatment facilities might experience.

References

[1] Dougherty, R.C. Human exposure to pentachlorophenol. *Pentachlorophenol chemistry, pharmacology, and environmental toxicology, Environmental Science Research 12*, ed. K. R. Rao, Plenum Press: New York, p. 355, 1978.

[2] Agency for Toxic Substances and Disease Registry (ATSDR). Toxicological Profile for Chlorodibenzofurans. ATSDR, Atlanta, GA. 1994.

[3] Agency for Toxic Substances and Disease Registry (ATSDR). Toxicological Profile for Chlorinated Dibenz-p-Dioxins (CDDs). ATSDR, Atlanta, GA. 1998.

[4] Webster, T.F. & Commoner, B. Overview: The Dioxin Debate. *Dioxin and Health, Second Edition*, eds. A. Schecter and T.A. Gasiewicz, John Wiley & Sons, Inc.: Hoboken, NJ, pp. 1-53, 2003.

[5] Paepke, O., Ball, M. & Lis, A. Various PCDD/PCDF patterns in human blood resulting from different occupational exposures. *Chemosphere*, **25(7-10)**, pp. 1101-1108, 1992.

[6] Dahlgren, J., Warshaw, R., Horsak, R.D., Parker, F.M. & Takhar, H. Exposure assessment of residents living near a wood treatment plant. *Environmental Research,* **92**, pp. 99-109, 2003.

[7] Dahlgren, J., Takhar, H., Schecter, A., Schmidt, R., Horsak, R., Paepke, O., Warshaw, R., Lee, A. & Anderson-Mahoney, P. Residential and biological exposure assessment of chemicals from a wood treatment plant. *Chemosphere,* **67**, pp. S279-S285, 2007.

[8] Agency for Toxic Substances and Disease Registry (ATSDR). Toxicological Profile for Pentachlorophenol. ATSDR, Atlanta, GA. 2001.

[9] Harnly M., Petreas M., Flattery J. & Goldman L.. Polychlorinated dibenzo-p-dioxin and polychlorinated dibenzofuran contamination in soil and home-produced chicken eggs near pentachlorophenol sources. *Environmental Science and Technology,* **34**, pp. 1143-1149, 2000.

[10] United States Environmental Protection Agency (USEPA). Interim Procedures for Estimating Risks Associated with Exposures to Mixtures of Chlorinated Dibenzo-p-dioxins and dibenzofurans (CDDs and CDFs) and 1989 update. U.S. Environmental Protection Agency, Risk Assessment Forum, Washington, DC, EPA/625/3-89/016. 1989.

[11] Hensley, A.R., Scott, A., Rosenfeld, P.R. & Clark, J.J., Attic Dust and Human Blood Samples Collected Near a Former Wood Treatment Facility. *Environmental Research*, **105(2)**, pp. 194-199, 2007.

[12] Pirkle, J., Wolfe, W., Patterson, D., Needham, L., Michalek, J., Miner, J., Peterson, M. & Phillips, D., Estimates of the half-life of 2,3,7,8-tetrachlorodibenzop- dioxin in Vietnam veterans of Operation Ranch Hand. *J. Toxicol. Environ. Health,* **27(2)**, pp. 165-171, 1989.

[13] Michalek, J.E., Pharmacokinetics of TCDD in veterans of Operation Ranch Hand: 10-year follow-up. *J. Toxicol. Environ. Health*, **47**, pp. 209-220, 1996.

[14] Flesch-Janys, D., Becher, H., Jung, D., Konietzko, J., Manz, A. & Papke, O., Elimination of polychlorinated dibenzo-p-dioxins and dibenzofurans in occupationally exposed persons. *J. Toxicol. Environ. Health*, **47**, pp. 363-378, 1996.

[15] National Center for Health Statistics (NCHS). National Health and Nutrition Examination Survey Data. Hyattsville, MD: U.S. Department of Health and Human Services, Centers for Disease Control and Prevention. 2003.

[16] National Center for Health Statistics (NCHS). National Health and Nutrition Examination Survey Data. Hyattsville, MD: U.S. Department of Health and Human Services, Centers for Disease Control and Prevention. 2005.

[17] United States Environmental Protection Agency (USEPA). Health effects assessment for creosote. United States Environmental Protection Agency Vol: EPA/600/8-88/025 p. 27. 1987.

[18] United States Environmental Protection Agency (USEPA). Exposure and Human Health Reassessment of 2,3,7,8-Tetrachlorodibenzo-p-Dioxin (TCDD) and Related Compounds. December 2003.

[19] Chen J., Wang S., Yu H., Liao P. & Lee C., Body burden of dioxins and dioxin-like polychlorinated biphenyls in pregnant women residing in a contaminated area. *Chemosphere,* **65**, pp. 1667-1677, 2006.

[20] Centers for Disease Control and Prevention (CDC). Third National Report on Human Exposure to Environmental Chemicals. Department of Health and Human Services. National Health and Nutrition Examination Survey (NHANES). 2005.

[21] Statistical Online Computational Resource (SOCR). UCLA Department of Statistics. Los Angeles, California. 2007.

[22] Statistica. StatSoft, Tulsa, Oklahoma. 2007.
[23] Mandal, P.K., Dioxin: a review of its environmental effects and its aryl hydrocarbon receptor biology. *J. Comp. Physiol. Biol.*, **175**, pp. 221-230, 2005.
[24] Schecter, A. & Gasiewicz, T.A., (eds). *Dioxin and Health, Second Edition*, John Wiley & Sons, Inc.: Hoboken, NJ.
[25] Agency for Toxic Substances and Disease Registry (ATSDR). Toxicological profile for polycyclic aromatic hydrocarbons (PAHs). ATSDR, Atlanta, GA. 1995.
[26] Bostrom, C.E., Gerde, P., Hanberg, A., Jernstrom, B., Johansson, C., Kyrklund, T., Rannug, A., Tornqvist, M., Victorin, K. & Westerholm, R., Cancer risk assessment, indicators, and guidelines for polycyclic aromatic hydrocarbons in the ambient air. *Environ. Health Perspect.*, **110(suppl 3)**, pp. 451-489, 2002.
[27] Carpenter D.O., Arcaro K. & Spink D.C., Understanding the Human Health Effects of Chemical Mixtures. *Environ. Health Perspect.* **110 (suppl 1)**, pp. 25-42, 2002.

GHG intensities from the life cycle of conventional fuel and biofuels

H. H. Khoo[1], R. B. H. Tan[2] & Z. Tan[2]
[1]*Institute of Chemical and Engineering Sciences, Singapore*
[2]*Department of Chemical and Biomolecular Engineering, National University of Singapore, Singapore*

Abstract

Among all the various air pollution issues, greenhouse gases are the key environmental and global concern that the world is facing today. The European Union's project on Carbon Measurement Toolkit was developed to determine the greenhouse gas (GHG) intensities of products. This kind of evaluation is important for fast growing industrial nations, especially in assessing sources of alternative energy. Life cycle assessment or LCA is used to model the following fuels delivered to Singapore: foreign conventional fuel production; biofuels from palm oil grown in neighbouring countries (with 'worst' and 'best' cases of direct land use change); and biodiesel produced from used cooking oil in Thailand. The life cycle approach used in this article is similar to the method developed by the European Union's Carbon Measurement Toolkit. The case studies involve raw material production/plantation, processing and final delivery by long-distance transportation. The investigation highlights that despite being labelled as a "green" or "carbon neutral" source of renewable energy, the actual ability of biofuels (especially those made from crops) to reduce GHGs hangs delicately on several crucial factors, namely, direct land use change.
Keywords: greenhouse gases, life cycle assessment (LCA), biofuels, cradle-to-pump, Carbon Measurement Toolkit, direct land use change (LUC).

1 Introduction

Due to the depletion of fossil fuels, the use of biofuels as a renewable source of energy has gained worldwide attention. The focus on these 'green' energy carriers has accelerated along with the need to mitigate global warming.

WIT Transactions on Ecology and the Environment, Vol 123, © 2009 WIT Press
www.witpress.com, ISSN 1743-3541 (on-line)
doi:10.2495/AIR090301

However, long-term plans or strategies to replace biofuels with conventional ones should be implemented with caution. The authors assert that biomass-to-bioenergy production systems are not without social and ecological risks since all fuel production industries require resource extraction and energy inputs [1].

Biofuels cannot be considered a sustainable solution to energy security if their production results in environmental destruction, deforestation, or food shortage [2, 3]. The "Carbon Footprint" of a product is generally the sum of carbon dioxide (CO_2) and other greenhouse gases – carbon monoxide (CO), methane (CH_4), and nitrous oxide (N_2O) – generated throughout the life cycle of a product. The European Union's Carbon Measurement Toolkit [4] was developed to calculate the total GHGs of products, taking into account all the phases of the life cycle (fig. 1).

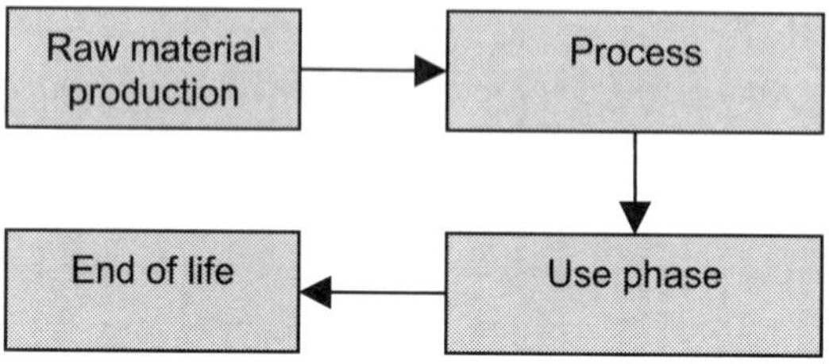

$GHG = EA1$*a1 + $EA2$*a2 + ... EAn*an

Where:

- **GHG** is the total emission of greenhouse gases expressed in mass of CO_2 equivalent or CO_2-eq
- ***EAi*** are the values of the environmental aspects provided by the applicant
- ***ai*** are the conversion factors that will be provided in the tool after this project

Figure 1: Life cycle diagram of toolkit [4].

2 Production of conventional fuel and biofuels

Based on the Carbon Measurement Toolkit concept, the following life cycle stages of fuels delivered to Singapore are modeled:

1) foreign conventional fuel production, which is locally refined into petro-diesel
2) biodiesel from palm oil with 'best' reported case of land use
3) biodiesel from palm oil with 'worst' reported case of land use
4) biodiesel produced from used cooking oil in Thailand.

The life cycle system boundary is from 'cradle-to-pump'. The cradle starts with crude oil mining for case 1, biomass plantation for cases 2 and 3, and used cooking oil collection for case 4. The life cycle ends at the pump, which is selected as 1 MJ diesel/biodiesel ready for use.

2.1 Conventional fuel production (case 1)

Conventional fuel production is a matured industry and several reports describing its operations are widely available. The first life cycle stages of conventional fuel is foreign crude oil extraction and separation, next, storage and handling before it is shipped by ocean tanker to a refinery in Singapore (fig. 2). Fossil fuel inputs of oil and natural gas is required in the process [5]. Electrical energy is used for pumping and transferring crude oil out of wells, and injecting excess water back in. Compared to shipment distance from Middle East to Singapore (8,000 km), the emissions from local transportations for oil production, separation or storage/handling operations are considered minimal. At the refinery, the hydrocarbon chains are separated by fractional distillation at about 200°C and 350°C, followed by hydrocracking to produce petro-diesel.

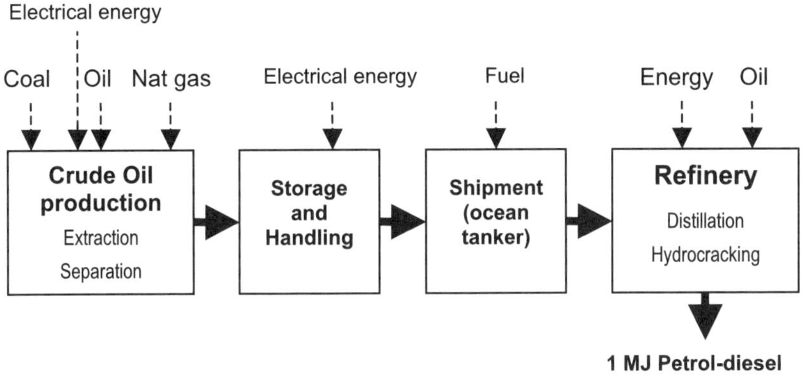

Figure 2: Life cycle of foreign fuel production.

The diesel yield from crude oil is taken to be 20%. It is also assumed that no losses occur during crude oil handling, storage and shipment. The energy value of the final diesel product is 42.9 MJ/kg [5].

2.2 Biofuel from palm oil (cases 2 and 3)

Biodiesel is a form of biofuel which can be used to replace petro-diesel for a wide range of applications, including transportation and high-temperature industrial processes. For the case study, biodiesel produced from palm oil grown from a neighbouring country is considered. Several reports have described the processes involved in the production of palm oil [6, 7]. The life cycle stages start with plantation, harvesting of fresh fruit bunches (FFB), milling of FFB to produce crude palm oil (CPO), which is sent to the biorefinery (fig. 3).

Inputs of fertilizers, fuel, and electrical energy are required in palm oil cultivation and harvesting. The FFB yield is reported to be 18,870 kg per hectare (kg/ha) based on a reference year (2007); and 0.2 kg CPO can be derived from 1 kg FFB [8]. At the biorefinery, 95% of CPO is processed into refined, bleached and deodorized (RDB) oil and finally, 100% RDF is converted to biodiesel. The final biodiesel has a value of 37.9 MJ/kg [9]. Sequestration of CO_2 occurs via photosynthesis during the growth of palm oil trees.

This article attempts to highlight the important fact that despite being labeled as 'green' or 'carbon neutral' fuels, the actual ability of biofuels (made from crops) to reduce greenhouse gases hangs delicately on several crucial factors, namely direct land use change [1–3]. Environmental scientists have been critical over biofuels made from palm oil primarily due the clearing of forestland and loss of biodiversity [3, 10, 11].

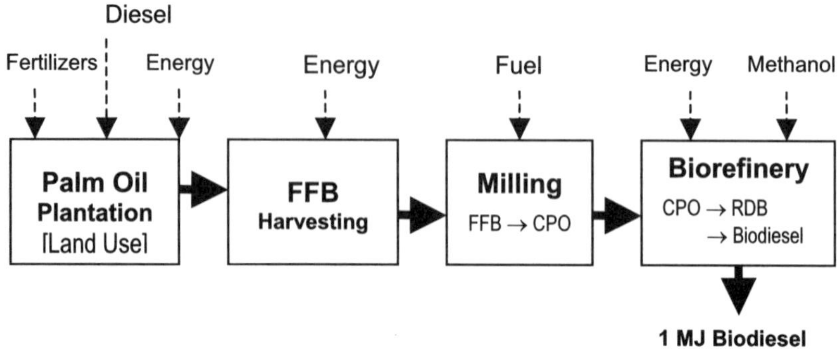

Figure 3: Life cycle of palm oil to biodiesel.

2.3 Biodiesel production from used cooking oil in Thailand (case 4)

The fourth case study is from Thailand, where used cooking oil (UCO) is collected and converted into biodiesel via a process known as transesterification (fig. 4). The process is semi-continuous, with a maximum operating capacity of 3 batches per day [12]. The production output is reported to be 1000 litres per batch.

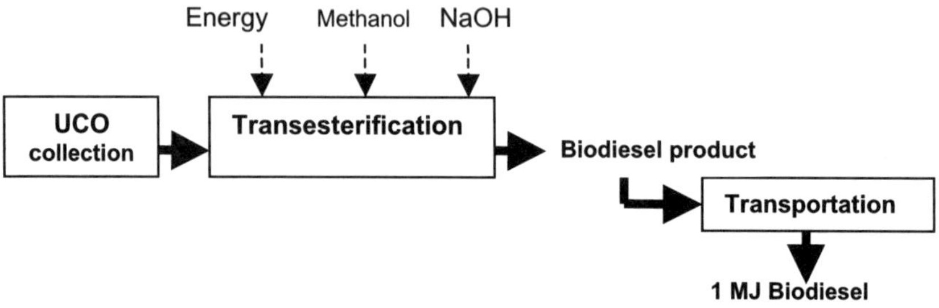

Figure 4: Life cycle of UCO to biodiesel.

The reactants for the process are methanol (mainly) and sodium hydroxide or NaOH as catalysts. Energy requirements are supplied as electrical power from the Thai national electricity grid [12]. The energy content of the biodiesel is expected to be 42.65 MJ/kg [13].

3 GHG emission data and calculations

3.1 Greenhouse gas emission data

The data for the case studies are obtained from personal communication with companies and institutes involved in palm oil biodiesel research, and supplemented by several reports [5, 7–9, 12]. They are compiled in Tables 1, 2 and 3. In case studies 2 and 3, fertilizer inputs are required during the plantation stage.

Table 1: Inputs and associated GHG for conventional petro-diesel production.

CASE 1: Conventional production of 1 MJ petro-diesel				
Life cycle stages	**Resource input (R)**		**Energy input (E)**	
Oil Extraction (output: crude oil)	Input (kg)	g CO_2-eq per kg input	Type of energy	g CO_2-eq per kWh/MJ
	Oil (0.0811)	256.83	Electrical energy (0.012 kWh)	970 [a]
	Natural gas (0.00051)	512.20		
Storage and Handling	-	-	Electrical energy (2.33 x10^{-6} kWh)	970 [a]
Refinery (output: petro-diesel)	Heavy oil (0.000256)	947.76	Electrical energy (0.0473 kWh)	460 [b]
Transportation	Total 2.2 g CO_2-eq for 8,000 km by ocean tanker			

[a] Coal-fired electricity. [b] Singapore electrical grid mix.

Table 2: Inputs and associated GHG for palm oil to biodiesel production.

CASE 2/3: The production of 1 MJ biodiesel from palm oil				
Life cycle stages	**Resource input (R)**		**Energy input (E)**	
Palm oil plantation (output: FFB)	Input (kg)	g CO_2-eq per kg input	Type of energy	g CO_2-eq per kWh/MJ
	Fertilizer (0.000224)	205.5	Diesel (0.0153 MJ)	0.0096
			Electricity (0.00273 kWh)	540 [c]
Milling (output: CPO)	-	-	Diesel (0.205 MJ)	0.0096
Biorefinery (output: biodiesel)	Methanol (0.00260 kg)	786	Electricity (0.00091 kWh)	540 [c]

[c] Malaysian electricity grid mix.

Table 3: Inputs and associated GHG for UCO to biodiesel production.

CASE 4: The production of 1 MJ biodiesel from used cooking oil (UCO)				
Life cycle stages	**Resource input (R)**		**Energy input (E)**	
UCO collection	Input (kg)	g CO_2-eq per kg input	Type of energy	g CO_2-eq per kWh
	-	-	-	-
Transesterification	Methanol (0.005)	260	Electrical energy (0.000703 kWh)	740 [d]
	NaOH (0.00023)	1180		
Transportation	Total 0.023 g CO_2-eq for 1,275 km by coastal tanker			

[d] Thai electricity grid mix.

3.2 Total GHG calculations

For each GHG calculation, the CO_2-equivalent conversion factors for global warming are made in accordance with the IPCC Third Assessment Report, *Climate Change 2001*. For all cases, the calculations for GHG intensities for the processing stages, from cradle-to-gate, are as follows:

$$\sum GHG = \sum_{p}\sum_{i}(R_{p,i} * CO_2eq1_i) + \sum_{p}\sum_{j}(E_{p,j} * CO_2eq2_j) + T_{ship} + [NetGHG_{land}] \ldots(1)$$

where,

Indices

i : resource/material input
j : energy/fuel input
p : a process in the supply chain

Variables

$CO_2.eq1_i$: amount of GHG (in grams) per kg of R_i
$CO_2.eq2_j$: amount of GHG (in grams) per kWh of E_j
$R_{p,i}$: amount of resource i (in kg) in a given process p
$E_{p,j}$: amount of energy input (kWh) in a given process p
T_{ship} : GHG due to shipment of crude oil/biodiesel from Middle East/Thailand to Singapore

T_{ship} is only considered for cases 1 and 4, where ocean and coastal tankers are used respectively. Any greenhouse gases from land transportation within the region are considered negligible. Direct land use change impacts [Net GHG_{land}] is discussed in the next section.

3.3 Net GHG from direct land use change (LUC)

Direct land use change or LUC is associated with the conversion of land area for the purpose of the supply chain production of a biofuel product [13]. Estimations of GHG emissions from direct LUC have been discussed in many reports [3, 9, 11]. But to date, there is no internationally recognized "standard model" or data relating to land use practices. Therefore, the authors emphasize that in this article, the 'best' and 'worst' case for LUC does no represent a globally accepted benchmark for GHG emissions for any agriculture-related land use applications.

The estimation of direct LUC applied for case studies 2 and 3 is the projected GHG emissions due to the clearing of a forestland area (deforestation) for growing palm oil biomass required per 1 MJ biodiesel. For each case, the net GHG from palm oil LUC is calculated as:

$$NetGHG_{land} = LUC_{CO2\text{-}eq} - Palm_{CO2} \quad (2)$$

where

$LUC_{CO2\text{-}eq}$: The total GHG emissions (per year) due to the clearance of tropical forest area to produce crude palm oil

$Palm_{CO2\text{-}eq}$: The amount of CO_2 absorbed by the palm tree (also per year).

Chen [7] estimated that 0.87 kg CO_2 can be absorbed by palm oil plantations via photosynthesis per kg CPO produced. Since 0.026 kg biodiesel is equivalent to 1 MJ biodiesel, the amount of CPO required in its production is 0.0273 kg. This can be translated to 0.87 x 0.0273 = 0.002375 kg CO_2-eq sequestered per MJ biodiesel. For 'best' reported case, palm oil is grown on existing cropland and hence: $LUC_{CO2\text{-}eq} = 0$.

The focus for 'worst' case is on GHG emissions from deforestation. It has been projected that for every kg of CPO produced, about 6.65 kg of CO_2-eq is released to the atmosphere due to carbon losses from direct LUC (deforestation) [9]. This corresponds to CO_2 emissions of:

$\Rightarrow$ 6.65 x 0.0273 = 0.1815 kg CO_2-eq released for every MJ biodiesel.

$NetGHG_{land}$ is considered zero for cases 1 and 4.

4 Results and discussions

The results are displayed in figures 5 to 8. It can be observed from fig. 5 that most of the GHG are first of all from crude oil extraction (68% of the total) and next, the refinery (about 32%). Very minimal emissions are observed from the delivery of crude oil to Singapore and even less from crude oil storage and handling.

Fig. 6 illustrates the environmental benefit of palm oil grown on dedicated cropland to produce biofuels (where $LUC_{CO2\text{-}eq} = 0$). Any contributions to global warming from plantation, milling and the biorefinery can be considered insignificant compared to large benefits of sequestering CO_2 by palm trees via photosynthesis. The total net GHG generated is *negative* for the entire biodiesel life cycle. For case 2 biodiesel can be regarded as an environmentally friendly and sustainable option for replacing conventional fuels.

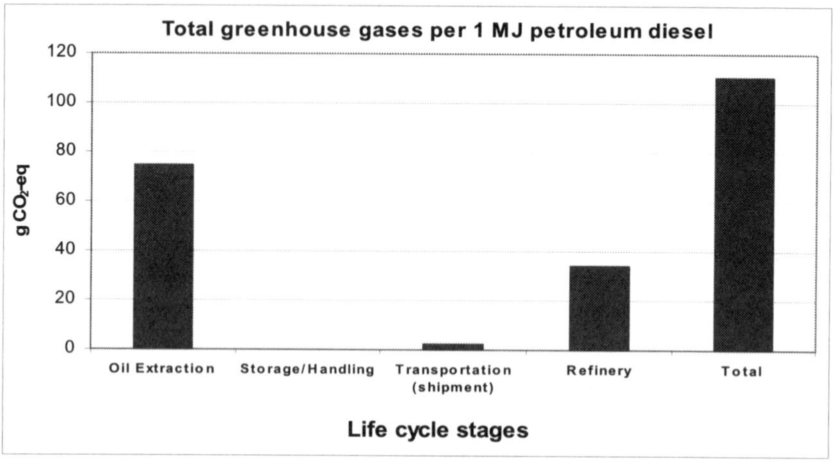

Figure 5: GHG results for the conventional production of 1 MJ petro-diesel.

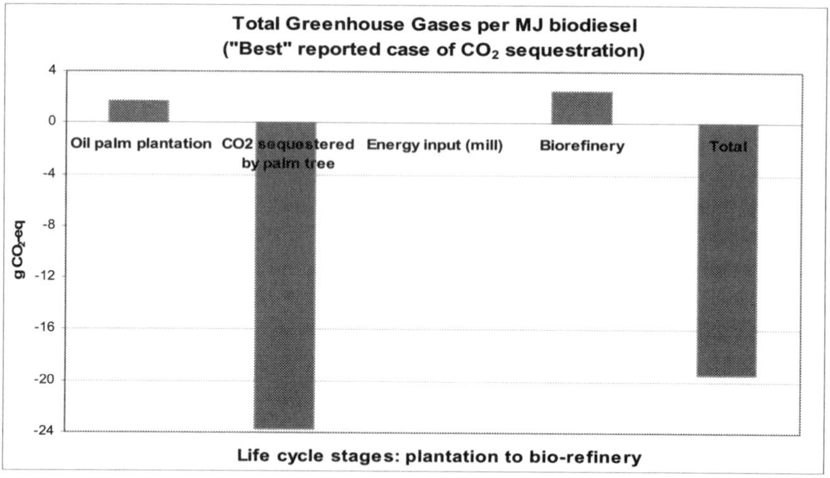

Figure 6: GHG results for 1 MJ biodiesel from palm oil ('best' case of land use).

Fig. 7 displays one of the 'worst' case of land use change for growing palm oil to produce biodiesel [9]. The attempt to sequester CO_2 from palm trees pales in comparison to the huge amounts of GHG emissions caused by deforestation. All other stage (plantation, milling, biorefinery) for palm oil to biodiesel is taken

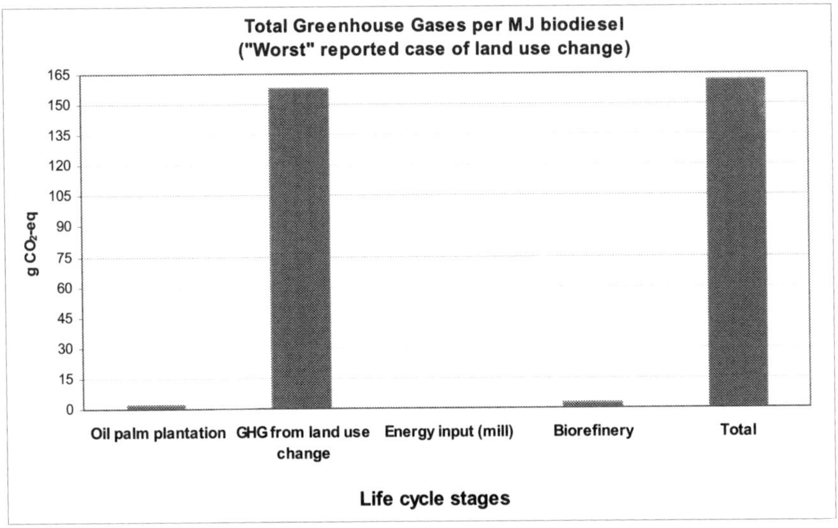

Figure 7: GHG results of 1 MJ biodiesel from palm oil ('worst' case of LUC).

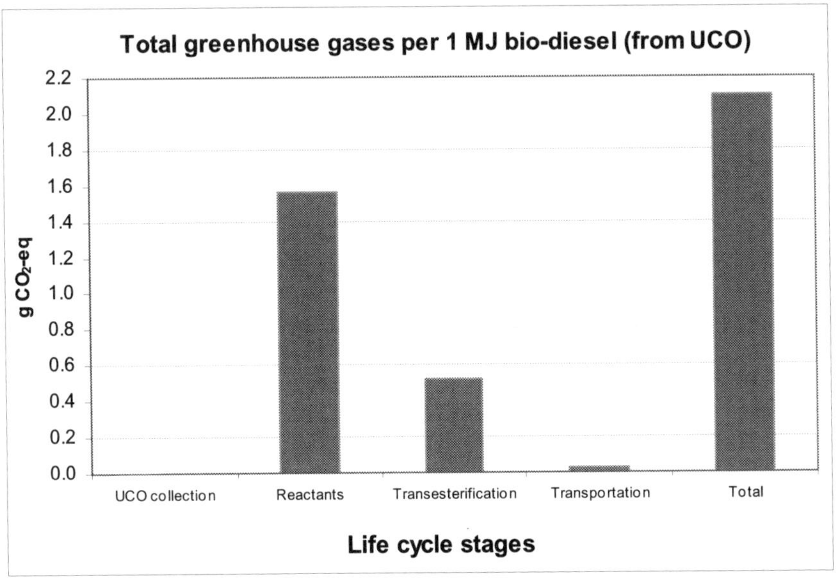

Figure 8: GHG results of 1 MJ biodiesel from UCO.

to display the same trend as fig. 6 since the same practise for palm oil processing is considered.

Finally, fig. 8 displays the GHG for the life cycle stages of UCO to biodiesel produced in Thailand [12]. The greenhouse emissions are mostly from the use of reactants (75% of the total) and the rest are from electrical energy inputs for the process. Very minimal emissions are shown to be from transportation.

A final comparison of all four cases is displayed in fig. 9. As expected, the least preferred option is case 3. The generation of GHGs from the conventional production of petro-diesel (case 1) is less than those from case 3. These two examples highlight that if not produced sustainably, biodiesel are worse than conventional fuels in their contribution to global warming.

The huge contrast between the environmental performances of cases 2 and 3 show that biofuels can potentially be a carbon negative source of energy, but their actual ability to offset GHGs depend on how the biomass feedstock is produced.

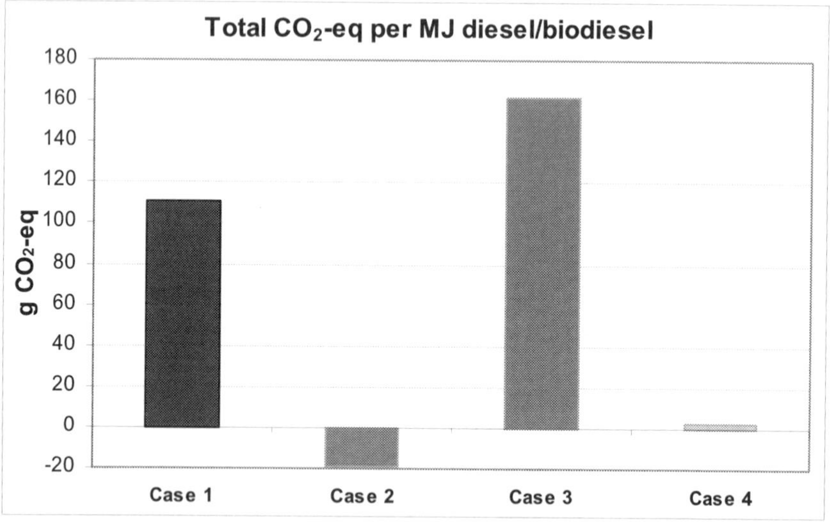

Figure 9: Comparison of total GHG for all 4 cases.

The fourth option, which is the conversion of UCO into biodiesel, exhibits great prospective for the recycling of used or waste oils, as well as, for reducing global warming impacts. Such sustainable biodiesel production strategies, from non-edible sources, are expected to result in improved air quality and greenhouse gas mitigation [13].

4.1 Further discussions on direct LUC

The sustainability of biofuel production systems is under intense scrutiny as their contribution towards, or ability to reduce, greenhouse gases remain unclear and controversial [10, 11]. In this article, the authors attempt to demonstrate the two extreme ends of the global warming impacts of biodiesel based on 'best' and 'worst' reported cases of palm oil biomass resources, and compare their environmental performances with conventional fossil fuel and another source of biodiesel.

Case studies 2 and 3 were based on the amounts of CO_2 sequestration by palm trees [7], and the net GHGs arising from direct LUC [9]. It should be highlighted again that presently there are no global standards or models of assessing greenhouse gases from direct LUC [14]. The method of calculation presented in this article does not necessarily represent a global or "standard" measure of GHG emissions for any agriculture-related land practises. For example, Reijnders and Huijbregts [15] reported that GHGs from palm oil plantations which replace tropical forests can range between 2.8 to 19.7 kg CO_2-eq per kg palm oil. By applying these values in Eqn (2), the total global warming results for case 3 will decrease slightly (with 2.8 kg CO_2-eq per kg palm oil) or increase tremendously (19.7 kg CO_2-eq per kg palm oil).

However, it is worth noting that generally many scientists unanimously agree that biofuels produced from converted land, particularly tropical rainforests, have greater GHG impacts than the fossil fuels they replace [9, 11, 15].

5 Conclusions

Worldwide, the primary focus of air pollution mitigation is on greenhouse gas reduction or minimization. Biofuels are expected to help moderate global warming impacts as well as reduce the dependence on fossil fuels. However, caution should be employed in the move towards the large scale replacement of conventional energy resources with renewable ones. Although claimed to be "carbon neutral", the actual potential of any renewable fuel to absorb CO_2 should be properly measured through its life cycle, or, from cradle-to-pump.

Tools to model or measure the carbon footprint of products are necessary to ensure that any strategies established for promoting renewable energy sources will properly mitigate – rather than contribute to – greenhouse gases in the atmosphere [16]. This kind of work demonstrates the importance of carbon footprint measurement tools, based on LCA, to provide relevant information necessary in establishing the right policies and strategies aimed at minimizing greenhouse gas pollution.

References

[1] Janulis, P., Reduction of energy consumption in biodiesel fuel life cycle, *Renewable Energy,* **29**, pp. 861-871, 2004.

[2] Haverkort, A., Bindraban, P. & Bos, H. (editors), *Food, fuel or forest: Opportunities, threats and knowledge gaps of feedstock production for bio-energy*, Proceedings: Wageningen, Netherlands, March 2, 2007.

[3] Koh, L.P. & Wilcove, D.S., Oil palm: disinformation enables deforestation, *Trends in Ecology & Evolution*, **24**, pp. 67-68, 2009.

[4] Carbon Measurement Toolkit, Swedish Environmental Management Council, Stockholm, Sweden, 2009, Online. http://www.msr.se/en/About-us/Projects/carbon-footprint/

[5] Sheehan, J., Camobreco, V., Duffield, J., Graboski, M. & Shapouri, H., *Life cycle inventory of biodiesel and petroleum diesel for use in urban*

bus, Final Report, National Renewable Energy Laboratory (NREL), U.S. Department of Energy, 1998.

[6] Kim, S.S. & Dale, S.B., Life cycle assessment of various cropping system utilized for producing biofuels: bioethanol and biodiesel, *Biomass and Bioenergy*, **29(6)**, pp. 426-439, 2005.

[7] Chen, S.S., LCA approach to illustrate palm oil's sustainability advantage, *International Palm Oil Sustainability Conference*, Sabah, Malaysia, April 13-15, 2008.

[8] Schmidt, J.H., *Life cycle assessment of rapeseed oil and palm oil*, PhD thesis, Department of Development and Planning, Aalborg University, Aalborg, Denmark, 2007.

[9] Wicke, B., Dornburg, V., Junginger, M. & Faaij, A., Different palm oil production systems for energy purposes and their greenhouse gas implications, *Biomass and Bioenergy*, 32, pp. 1322-1337, 2008.

[10] Fitzherbert, E.B., Struebig, M.J., Morel, A., Danielsen, F., Bruhl, C.A., Donald, P.F. & Phalan, B., How will oil palm expansion affect biodiversity?, *Trends in Ecology & Evolution*, 23 (10), pp. 538-545, Oct 2008.

[11] Fargione, J., Hill, J., Tilman, D., Polasky, S. & Hawthorne, P., Land clearing and the biofuel carbon debt, *Science*, **319(5867)**, pp. 1235-1238, 2008.

[12] Pleanjai, A., Gheewala, S.H. & Garivait, S. Greenhouse gas emissions from production and use of used cooking oil methyl ester as transport fuel in Thailand, *Journal of Cleaner Production*, 2009. In print

[13] Demirbas, A. Biodiesel from waste cooking oil via base-catalytic and supercritical methanol transesterification, *Energy Conversion and Management*, **50(4)**, pp. 923-927, 2009.

[14] Kim, H., Kim, S. & Dale, B.E., Biofuels, Land Use Change, and Greenhouse Gas Emissions: Some Unexplored Variables, *Environmental Science and Technology*, **43(3)**, pp. 961-967, 2009.

[15] Reijnders, L. & Huijbregts, M.A.J., Palm oil and the emissions of carbon-based greenhouse gases, *Journal of Cleaner Production*, **16(4)**, pp. 477-482, 2008.

[16] von Blottnitz, H. & Curran, M.A., A review of assessments conducted on bio-ethanol as a transportation fuel from a net energy, greenhouse gas, and environmental life cycle perspective, *Journal of Cleaner Production*, **15(7)**, pp. 607-619, 2007.

Some aspects on air pollution in historical, philosophical and evolutionary context

A. A. Berezin[1] & V. V. Gridin[2]
[1]*Department of Engineering Physics, McMaster University, Canada*
[2]*Department of Chemistry, Technion-Israel Inst. of Technology, Israel*

Abstract

This paper discusses import of physical ideas on management and human response to Air Pollution (AP). It looks into some historical, evolutionary, ecological and philosophical aspects of AP. Paper emphasizes a need for more extended philosophical reflections on AP which should go beyond mere technical issues of monitoring and control.
Keywords: air pollution, biosphere, evolution, adaptation, hormesis, quantum physics, chaos, ecology, singularity theory, management of air pollution, fuzzy logic, random models.

1 Introduction

A standard orthodox view of Air Pollution (AP) is that it is, so to say, "by definition" an unquestionably negative happening, which – ideally – should be eliminated or, at the very least, minimized and controlled. Such a thesis, in-spite that it is almost universally accepted as an Axiom of Truth, obfuscates the fact that AP (as well as other forms of pollution) is an inevitable and, in a sense an integral, component of modern technological civilization and, perhaps, can even exhibit some redeeming aspects. In other words, upon a more critical look it should be recognized that modern AP is a direct result of technological and social development of humankind.

In this paper we look at AP within a more encompassing philosophical vista. Instead of looking at AP as a singularly negative phenomenon, we outline a more pragmatic and adaptive approach to the whole gamut of AP aspects as well as air quality in general. As an objective phenomenon, AP is a (practically inevitable) consequence of such key aspects as industrialization, transportation and

WIT Transactions on Ecology and the Environment, Vol 123, © 2009 WIT Press
www.witpress.com, ISSN 1743-3541 (on-line)
doi:10.2495/AIR090311

urbanization. These are coupled with an (often excessive, wasteful and poorly regulated) use of natural resources (oil, gas, water, etc).

The authors, who are physicists, are looking at the above issues through prism of possible physical connections and analogies [1, 2]. This is a typical way of how methods of physics (especially, theoretical physics) make their "intrusion" into other areas of technology, science and social knowledge. Numerous examples of that can be found is such diverse areas as global economy and dynamics of stock market, psychology and psychiatry, conflict resolution studies, population dynamics, ecology, and even in inter-human relationships and sexology.

This paper has a two-fold goal. One aim, in extension of our previous work [1, 2], is to discuss possible relevance of a variety of physical effects to AP. The other aim is to discuss import of physical mind-frame and some ideas of quantum and non-linear physics for various aspects of human response to AP. The latter include issues of regulatory and managerial nature, such as methods and strategies of decisions making regarding AP.

2 Brief history of air pollution

Paraphrasing title of the bestselling book by the physicist Stephen Hawking "Brief History of Time" here is, a somewhat grotesque, "brief history of air pollution" in a concise hindsight.

2.1 Air pollution before the Industrial Age

Apparently, every generation tends to over-rate problems it faces. The Past almost invariably bears a flavor of the "golden Age". So is the mythology that in the past the "water was clean and the air was fresh". The question here, of course, what precisely "clean" and "fresh" mean? As argued below, these definitions are not always a clear-cut matter.

In a pre-industrial age, almost all energy was produced by burning of wood, primarily for heating and cooking purposes. Early metallurgy also used coal, but due to a general scarcity of the population, these activities did not generate a significant contribution to global (and in most cases – even local) AP levels.

2.1.1 Natural versus human-induced air pollution

The very meaning of the word "pollution" implies that there is some departure from the norm. Hence, in order to define what is AP, one must first define the base-line, that is what is the "air without AP". Apparently, the best we can take for that is a standard chemical composition of the air. In terms of volume percentage, Earth's atmosphere is 78% nitrogen, 21% oxygen, 0.9% argon, and 0.04% carbon dioxide with very small percentages of other gases. Our atmosphere also contains water vapor, normally about 1%. In addition, (natural) Earth's atmosphere contains traces of dust particles, pollen, plant grains and other solid particles. These latter additives can, in a certain sense, be referred as "natural pollution", although this term is not commonly used.

2.1.2 Climate oscillations

It is an established fact that the Earth's climate constantly experiences changes at a variety of time scales. Some of these data are based on the measuring the relative concentrations of stable isotopes in ancient ice. Isotopic methodology is also a powerful investigative tool in several major areas of science and technology [3] and its potential significance for AP-related issues is still underappreciated. One such method can be the use of sensitive Doppler spectroscopic monitoring of the polluted atmosphere because thermal velocities of various isotopic species (such as, e.g., isotopically different CO_2 molecules) are slightly different.

2.1.3 Volcanic and seismic activity

Volcanic eruptions produce local variations in air composition due to ejection of sulfur dioxide, other gases, and solid particulate. It is logically consistent to view such variations as naturally induced AP. When these eruptions are of major magnitude, changes in the atmosphere can become temporarily global. Possible examples are historical Thera eruption in Greece ca. 1650 BCE, which is believed to destroy the Minoan civilization, or Krakatau eruption of 1883. In spite that the atmosphere as a whole "cleans itself" after such eruptions in a few short years, the changes in opacity affect vegetation, which, in turn, may produce shifts in CO_2 concentration. This fact even further complicates the present-day Global Warming controversy [4]. Another factor is the effect of ocean plankton which plays a key role in a CO_2 balance in the atmosphere. This dynamics can be affected by major seismic activity at the ocean floor, as well as climatic changes leading to re-arrangements of ocean currents.

2.1.4 Cosmic factors

There are multiple factors, external to the planet, that have direct and indirect effect on the types and levels of AP on a global scale. Variations in Solar activity (Sun spots, etc), fluxes of high energy cosmic particles, density of inter-planetary dust and occasional asteroid hits (e.g., Tunguska impact of 1909) – all affect AP to a variable degree. At the same time, such aspects as possible gravitational effects on AP coming from the Moon or other planets have not been studied systematically. Since effects from the Moon on Earth are quite significant (e.g., daily tides or female monthly cycle which is believed to be period-locked with Moon's orbital motion), similar direct gravitational effects on AP are in all likelihood can also be noticeable (e.g., tide-like waves in AP patterns).

2.2 Industrial era

Ina context of AP the beginning of industrial era can be traced to advent of a massive use of coal and (later) an introduction of the first steam engines which later led to the steam-powered ships and railways.

While with some stretch of imagination it is possible to count the Modern Era from about 1860-1870 (e.g., Civil War in USA and Franco-Prussian War in Europe), the most common "starting date" conveniently coincides with the beginning of the 20-th century. The first recorded air flight on a heavier-than-air machine (Brothers Wright, 1903) is what we usually take as the beginning of modern ("our") civilization in a proper sense.

2.2.1 Massive industrialization

Along with the continuing use of coal, beginning of the 20-th century saw a rapid advent of oil, which within a few decades has replaced coal as the Number One commodity in the world economy. Presently, AP due to the exhausts of billions of internal combustion engines in cars and several hundred thousands air flights every day, is the prime contributing factor to the world-wide AP. In parallel with this, there are at least three other factors contributing to AP in a significant way, namely, chemical industry, agriculture and dumping sites.

2.2.2 Chemical industry

While coal keeps contributing to AP (now mostly through CO_2 and NO_x from steel production and coal electro-power plants), chemical industry has become a major contributor too. The diversity of products of chemical industry is enormous and beyond the scope of this article. It is sufficient to mention plastics, paints, fragrances, pharmaceuticals, industrial chemicals, agricultural fertilizers and all kinds of modern food processing industries. Another (growing) outlet is the semiconductor and computer industry which produces a variety of potentially toxic waste (e.g., computer batteries, metal parts, liquid crystal panels, etc).

2.2.3 Agriculture industry

A separate category is the contribution of the intestine gases (mostly, methane) from zillions of agricultural animals all across the globe. While by obvious reasons this peculiar contribution may be somewhat downplayed in public perception, it, on a balance, contributes quite significantly to AP. In the same category, one can mention such an unpalatable (but inevitable) issue as numerous sewage treatment facilities, many of which are far from perfect quality in terms of purification of escaping gases and poor water treatment.

2.2.4 Dumping sites

Problems related to handling of ever-growing heaps of trash is taking greater and greater place among public concerns. It is becoming a hotly debated social and political issue. Municipal and regional elections are often run primarily on these concerns. Nobody wants trash and garbage around, yet we all have little choice but to produce more and more of it. Underground trash burials is only a palliative solution, firstly, because greater and greater scarcity of "free land", and, secondly, because decomposition gases tends to leak even from beneath many meters of the ground.

2.2.5 Waste incineration

Incineration of combustible waste goes on at various scales – from private local incinerators to major centralized incineration factories. In-spite that the exhausts of incinerators are usually equipped with a variety of filters and electrostatic precipitators, some fraction of gases (primarily, VOC – volatile organic compounds) inevitably escape to the atmosphere.

2.2.6 Other forms of pollution

While AP is one of the major points of public concern (especially in some megalopolises as, e.g., Los Angeles), other forms of pollution are entering into the focus of attention as well. Among them, pollution of waters (both inland waters and oceans) and soils are the most obvious forms, but some more subtle and less recognized forms of pollution are also taking a growing part, primarily, electromagnetic and informational pollution [1]. There are several mechanisms of how these forms of pollution may interact with AP.

3 Ecological and physical aspects of air pollution

3.1 Human adaptation to Air Pollution

One of the prime features of human race and the biology at large is the idea of adaptation to the changes in the environment. This adaptation is highly interactive and includes both positive and negative feedback. It also proceeds at a variety of characteristic time scales and, hence, using the terminology of theoretical physics, can be called time scale invariant. This is one of the reasons for massive use of ideas of chaos and nonlinear dynamics in the present-day ecological and evolutionary studies (e.g., [5]). In this context AP is not an exception, and human adaptation to it (as well as its practically possible control) is an on-going reality which should be analyzed and accepted.

3.1.1 Evolutionary (Darwinian) adjustments

According to the standard Darwinian concept, the two prime aspects of biological evolution are adjustment to the (changing) environments and survival of the fittest. At this moment, there is no singular explanation of what exactly drives the evolution. Apart from theological and mystical explanations, the most scientifically sound line of arguing goes along theory of spontaneous pattern formation and complexity emergence. These ideas are taken from physics, especially from Dynamics of Non-Linear Systems, which include such versatile analytical tools as the Theory of Chaos, Theory of Catastrophes and the notion of Strange Attractors. The latter illustrates the idea that the (evolving) system has some “goal” which it is striving to reach (“attracted to”).

3.1.2 Hormesis effect (chemical, radioactive, electromagnetic)

Hormesis effect is an umbrella term for a non-linear dose response of organisms to various (defined as harmful) environmental factors. It is a part of social trivia that harmfulness or usefulness of various factors often depends on the dose. Many heavy metals (manganese, vanadium, chromium, etc) are poisonous, yet some of them are included in poly-vitamin pills in micro doses.

In a proper sense, hormesis is a dose-related, allegedly beneficial, response to a number of (nominally harmful) environmental factors. Examples are chemical, radioactive, and electromagnetic hormesis. It is established that small doses of radioactivity can stimulate human immune system and lead to a better adaptability and strengthening. At the same time, it is known that natural radioactivity is one of the prime factors leading to genetic mutations (atomic scale changes in the genetic code) and this, in turn, fosters evolutionary changes. Likewise, because AP in general has many chemically diverse contributors, it is quite likely that some of them in low doses may be more beneficial than harmful. People living in large cities usually adapt to moderate levels of AP. At the same time, when people permanently living in remote areas with a pristine environment and clean air suddenly come to large cities with their smogs and car exhausts, they often experience adverse health reactions. While this is in no way an advocacy for more AP, in view of these authors the role of AP as adaptability factor should be recognized and studied in a more focused way.

3.1.3 Societal reflection and controversies

For all practical purposes no form of pollution can be completely eliminated. Therefore, excessive demands of "zero-tolerance" to AP are untenable. Likewise, the ideas of Deep Ecology and sustainability ("back to nature" movements) are, for the most part, naive and illusive. At the same time, because these views look straightforward and appeal to large segments of the population, they are becoming a part of a broad environmental discourse and generate within them some good ideas. Some spin-off include growing activity in a search for alternatives to fossil fuels (e.g., electric cars) and alternative forms of power generation (solar, wind, ocean tides, etc) or more environmentally friendly goods, like small wearable electronics, etc.

3.2 Critical effects in air pollution

In physical terms AP is a distributed phenomenon which most commonly treated as a (quasi-continuous) diffusion process. It is known, both phenomenologically and theoretically, that quasi-continuous dynamics often leads to singular manifestations with a highly non-uniform concentrations of energy and particles. An example is a soliton wave which propagates as a localized crest. Similar effects of self-induced coagulation of particulate in AP may happen either spontaneously or be triggered by outside factors, as, for example, coherent electromagnetic waves [1]. Some more subtle effects as gravitational reduction of the wave functions of micro-particulate and aerosols [2] may further contribute to coagulation dynamics.

3.2.1 Bifurcations and Chaos Theory

Chaos Theory is presently a significant segment of physics with applications to many diverse areas ranging from climatology to economics. One of its key ideas is utter sensitivity to the initial conditions which is a characteristic feature of many non-linear systems. This is the so called "Butterfly Effect" which a code name for a fast (exponential) divergence of the phase trajectories in a phase space of the system. This feature often manifest itself in AP, for example a slight change in a wind pattern can drastically affect the deposition of hard particulate and air quality at a particular locality.

The said Butterfly Effect has two consequences with opposite implications to the human response to AP and its management. On one side, it drastically lowers the level of possible predictability of AP dynamics, especially on longer time scales (say, many days). On the other side, high sensitivity to the initial conditions allows very small perturbations to affect the dynamics in a macroscopic way. This situation is commonly known in physics when very weak forces radically change the outcome of the process. An example of the latter is a beta-decay of some isotopes. In case of AP this, potentially, leads to a possibility to affect patterns of AP propagation by energetically small (but informationally structured) pulses [1]. One example can be application of Laser pulses similar to the methodology of Laser Doppler cooling. This could selectively affects various categories of pollutants due to differences in their spectroscopic properties and dispersion in their thermal velocities.

3.2.2 Theory of Catastrophes

Theory of Catastrophes (Rene Thom) is one of the precursors of the modern Chaos Theory. Both have ideas of bifurcation (forking paths) and both are applicable to systems with jump-wise transitions between dynamical regimes. In case of AP, the Theory of Catastrophes can describe sharp transitions in AP descriptors (pollution level and pollution content in time and spatial variations).

A historical example is the so called *Toba Catastrophe* – a eruptions of a super-volcano some 70,000 years ago. It lead to a catastrophic increase of AP and a drop of a global temperature by several centigrade. Analysis of human genetic code indicates that the proto-human population dropped to a few thousand breeding pairs worldwide. As a result, this was a likely bottleneck happening in the emergence of modern humans. Thus, one can, somewhat metaphorically, say that the origin of (modern) human species is a direct result of a sharp singular increase of AP level for a short time interval.

Another finding of Bifurcation Theory and Chaos is the existence of windows of Order within Chaos. In terms of AP patterns and dynamics this can manifest itself in the formation of AP clusters showing some spatial regularity across the Globe, in some analogy with the known electrostatic problem of the arrangement of N charges over the surface of a sphere.

3.2.3 Ergodicity Principle and dynamical singularities

Ergodicity is a term which originated in statistical physics and thermodynamics (e.g., Boltzmann's "H-theorem") which posits that a dynamical system with macroscopic number of degrees of freedom (e.g., a mole of a gas), given a sufficient time, will pass through all *microscopic* states in its phase space. As a complimentary term to the notion of entropy, Ergodicity Principle demonstrates the existence of fluctuations of any scale and shape. Examples of the latter are the formation of extremely dense pockets of AP in some localities (e.g., notorious smogs in Los Angeles area), which may be in excess of what actual emissions entail. In terms of the Chaos Theory, these dynamical regimes (quasi singularities) can be described as Strange Attractors (sets of points in the phase space to which system periodically returns).

3.3 Air pollution in evolving biosphere

Within a view of the human race as a part of the evolving biosphere, AP which is due to human activity, should be considered as an organic part of the biosphere in whole. As such, AP (both natural and human made) is neither “bad” nor “good”, but should be treated in pragmatic terms. The latter requires balancing of monitoring with technologically and economically available control and adaptability.

3.3.1 Self-regulating systems

The above described Ergodicity Principle stays in some dialectical opposition to the Second Law of Thermodynamics (law of entropy increase), because it points to an inherent ability of non-linear systems to self-organization and complexity emergence. This ability of self-regulation is thought to belong to the biosphere at large (“Gaia” hypothesis), as well as to separate and adjacent components of it, of which Earth’s atmosphere (and hence, AP) is a part.

3.3.2 Gaia Hypothesis

Within the context of Gaia Hypothesis (James Lovelock), atmosphere is not just an outside non-organic (non-living) entity in which we happen to be embedded, but an immanent part of the biosphere itself. This idea can be seen as an extension of Le Chatelier-Braun principle to the entire global system comprising of biosphere and atmosphere. For that matter, a relatively small (in terms of bulk concentration) additives to the air which are brought by AP may also play their role in Gaia, similarly to microelements in human nutrition.

3.3.3 Air pollution in toxicity context

At the present moment there is no sufficiently developed understanding of the effects of toxicity of chemical elements and compounds versus the adaptational effects of hormesis (Section 3.1.2). In a crude sense, the first are seen as “bad” while the latter (hormesis) as “good” (boosts the immune system). However, the juxtaposition of these two is more subtle, and proper understanding of mutual interactions in this tandem may call for the application of some ideas of theoretical physics.

One of the perils (not yet properly appreciated) of the modern technological age lies in the use of ultra-pure materials. Currently available technological methods allow production of materials purified to a degree several orders of magnitude higher than can be found in nature. For example, semiconductor industry uses such ultra-pure materials as GaAs or CdTe. All these elements are poisonous to human. And yet, trace amounts of them are all around and we all take them to our bodies through the nutrition and inhaling. So, where is the catch here?

In physics there are concepts of “dressed particles” and quantum states. This means that the particles (e.g., impurity atoms) surrounded by other particles are screened in their interactions with other parts of the system. Examples of these dressing (screening) effects are Cooper pairs in superconductivity, or polarons in crystals. Polaron is a "dressed" electron and its "dress" is a cloud of phonons (lattice vibrations).

Likewise, toxic elements in there natural (not pure) forms are surrounded by other compounds. This mitigates and passivates their toxicity action. In fact, in non-pure form their toxicity can even turn into a health benefit – many elements toxic in pure form (selenium, manganese, vanadium) are part of poly-vitamin supplements. Effects of super-pure materials (which do not have the benefit of this screening effect) are not well known, but a reasonable assumption is that their toxicity action can be enhanced by many orders of magnitude. It is well known that drinking exclusively distilled water is harmful. This is because distilled water is deprived of microelements and trace additives essential for our physiology. The latter are contained in natural water (even filtered). Similarly, a strive to an ultra pure air can, in fact, be more hazardous than beneficial. In other words, some level of AP (and chemical diversity brought by AP) should not only be tolerated, but should be an immanent part of our existence.

4 Future scenarios and human response

In this article we adopt the view that AP is an inherent part of the anthropogeneic ecology. This vista suggests a few possible future scenarios for the role of AP in a social fabric, as well as to the managerial response to it.

4.1 Self-Organized Criticality

Idea of SOC (Self-Organized Criticality) lies within modern Theory of Chaos [1] and points towards the formation of high level instabilities in non-linear systems. In a nutshell, SOC model accounts for a gradual build-up of instability in a non-equilibrium system and its subsequent discharge in a form of sudden avalanches. These avalanches can happen at a variety of spatial-temporal scales (scale invariance) and proceed in a stochastic non-coherent form (e.g., luminescence), or in a form of highly correlated singular discharge (e.g., coherent laser emission). A popular visual model for SOC is a collapse of a sand pile with a gradual adding of more sand. As such, AP is a complex heterogeneous system having gaseous components (nitrogen oxides, sulfur dioxide, VOC, etc), liquid aerosols and solid components (hard particulate). Therefore, dynamical modifications of AP patterns can happen at a broad range of scales including sudden singular-type phase transformations. This makes AP another test model for the application of SOC methodology.

Above mentioned (Section 3.2.1) triggering of macroscopic phase changes in AP by targeted laser pulses can be further extended to include upper layers of the atmosphere, in particular, ionosphere. At the moment, there is no comprehensive understanding of the links between processes in the ionosphere and ground-based AP. However, due to high responsivity of ionosphere to outside factors (example is Aurora Borealis) and the fact that is carries a distributed electrostatic charge, all effects of electrostatic phase transitions [2] and self-organization in charged plasma can be potentially induced by properly designed perturbations (e.g., laser beams). Redistribution of charge density in ionosphere can, in turn, reflect on AP at the ground level.

4.2 Singularity Theory

The ideas of an impending Singularity in technological, ecological and demographic fabric of the global society are getting a growing factual support [6]. From global economic competition and radical shifts in international markets to such technological trends as expected breakdown of the Moore's Law (doubling of computer chip power every 18 months) and emergence of Quantum Computing [7], one can foresee a radical (singular) shift. From local ecological patterns we come to a global ecological linkage of which AP is an inevitable component.

A corollary to the Singularity Theory are the package of ideas known as Trans-humanism – a transcendence of human realm into a post-human existence. The suggested scenarios here run from the apocalyptic pictures of global ecological collapse (to which AP will likely to contribute) to the visions of the ascendance of a New Golden Age when all acute problems of today are harmoniously solved. While either of these extremes is unlikely to come in a pure form, the suggested time-frames for them are surprisingly short (30 to 50 years hence).

4.3 Decision making in air pollution management

Managerial decisions concerning AP require balancing of numerous (often conflicting) objectives and vested interests (technological, social, economical, political, developmental, health-related, etc). Additionally, scientific (and even philosophical) aspects of AP issues are often complicated and controversial, with different (often utterly adversarial) viewpoints (e.g., Human-made Global Warming controversy). Below we suggest a few thoughts on these issues.

4.3.1 Fuzzy Logic

Fuzzy Logic (FL) nowadays is an accepted mainstream technique for the choosing a strategy in multiple-choice situations. In fact, numerous automata (like washing machines or photocopiers) have often build-in (embedded) FL microprocessors to select performance regimes.

4.3.2 Random strategies

Decision making by applying random strategies has a long history. Sometime, they are preferable to the "first choice" (most optimal) solutions. Dodges in Venice and judges in some jurisdiction were selected by a draw. Some earlier military manuals suggested random choice (e.g., by dice) of strategic moves (e.g., selecting mounting passes for the army). Such a strategy makes more difficult for the adversarial commander to guess the most likely route which would likely be taken on the basis of rational optimization. Randomness (dicing) removes such rational optimization and at best the adversary can do the same – to do his own dicing which renders his guessing purely probabilistic. Likewise, qualitative and quantitative decisions on AP management, may include random choice (perhaps, with some weighting factors) among several educated guesses.

4.3.3 Adoption of second and third-best options

Adaptational Modelling presumes an outlining of managerial decisions through highly interactive (non-linear) linkage to a dynamically changing situation. As such, AP exhibits a volatile, ever-changing, spatial-temporal pattern. This makes it prone to the application of the ideas of Strange Attractors of the Chaos Theory. In this context, an optimal (or quasi-optimal) managerial solution aimed on a particular (local) aspect of AP is envisioned as a point on a phase trajectory of a (multidimensional) space of possible solutions. This is a generalization of the decision making based on FL for the case when the number of parameters are much greater that the traditionally used in FL methods. In this vein, a correction (and possible improvement) of the standard FL method may be to choose the second (or third, etc) option as the actually accepted managerial decision.

Prime problem with most environmental problems, AP including, that there management requires a simultaneous taking into account of numerous, often contradictory, factors and interests. In the spirit of emergent thinking along the ideas coming from physics (such as superposition principle, mixed quantum states, quantum non-localities, classical and quantum chaos, least action principles, etc), the viable strategy may be sought along the holistic approach which takes a singular view of a situation as a whole, rather than weighting separate parts of the scenario. This is applicable to AP problems as well as a variety of environmental and ecological issues at large.

5 Conclusion

This article present AP as a multi-sided challenge which, far from being uniformly “negative”, may have some redeeming aspects (e.g., hormesis and ecological and genetic adaptation). Therefore, analysis of its structure and implications, as well as managerial response to it, require interdisciplinary approach, some aspects of which are outlined above. Ideas drawn from the work of such visionary scientists as Roy Kurzweil [6], Roger Penrose [8], James Lovelock [9], Rupert Sheldrake [10], or Seth Lloyd [11] (to name a few) can serve as a fertile ground for further contemplations of local and global solutions. In this regard it appears especially promising to explore the linkages of AP with other fundamental planetary forces and interactions (gravitational, electromagnetic, and quantum). This can further contribute to the understanding of key aspects of AP as well as foster the development of methods of its monitoring and control.

In particular, Lloyd [11] entertains the premise that the dynamics of large systems can, in fact, resemble the operation of quantum computers [7]. In-spite that such an idea may appear far fetched, it is gaining observational and theoretical support. Its application to the Earth’s atmosphere with its AP as a sort of embedded neural network (interactive particulates and aerosols), can turn out to be relevant for the deeper comprehension of processes of chemical self-regulation in the Earth's atmosphere.

References

[1] Berezin, A.A. & Gridin, V.V., Electromagnetic and informational pollution as a co-challenge to air pollution. *Air Pollution XVI*, ed. C.A. Brebbia & J.W.S. Longhurst, WIT Press: Southampton, Boston, pp. 533-542, 2008.

[2] Berezin, A.A., Quantum effects in electrostatic precipitation of aerosol and dust particles. *Air Pollution XIII*, ed. C.A. Brebbia, WIT Press: Southampton, Boston, pp. 509-518, 2005.

[3] Berezin, A.A., Isotopicity: Implications and Applications, *Interdisciplinary Science Reviews*, 17 (1), 74-80, 1992.

[4] Lomborg, B., *The Skeptical Environmentalist*. Cambridge University Press, 2001.

[5] Kauffman, S.A., *The Origins of Order – Self-Organization and Selection in Evolution*. Oxford University Press, New York/Oxford, 1993.

[6] Kurzweil, R., *The Singularity is Near (when humans transcend biology*. Penguin Books, New York, 2005.

[7] Berezin, A.A., Quantum computing and security of information systems, Safety *and Security Engineering II*, eds. M. Guarascio, C.A. Brebbia, F. Garcia, WIT Press: Southampton, Boston, pp. 149-159, 2007.

[8] Penrose, R., *The Road to Reality: A Complete Guide to the Laws of the Universe*. Jonathan Cape, London, 2004.

[9] Lovelock, J., *The Revenge of Gaia: Why the Earth is Fighting Back – and How We Can Still Save Humanity*. Allen Lane, Santa Barbara (California), 2006.

[10] Sheldrake, R., The *Presence of the Past: Morphic Resonance and the Habits of Nature*. Times Books (Random House), New York, 1988.

[11] Lloyd, S., *Programming the Universe: Is the Universe actually a Giant Quantum Computer?*. Vintage Books (Random House), New York, 2007.

Risk assessment of atmospheric toxic pollutants over Cairo, Egypt

M. A. Hassanien
Air Research and Pollution Control Department, National Research Center, Dokki, Cairo, Egypt

Abstract

The aims of this study were to investigate the distribution of toxic pollutants, primarily those that pose great risk for human health (Co, Cr, Cd, Pb, Mn, V, As, Sb, Ni, and Ti), in atmospheric air samples collected from various sites in Cairo, Egypt; to assess human health risk estimates derived from the metal inhalation of urban inhabitants; and to explore the relationship between potential exposure levels and risk estimates. Methods based on the integration of environmental modeling and Geographical Information System (GIS) were used in the present study. Samples of airborne particulate matter were collected during the summer season 2005 from seven sampling sites in Cairo, Egypt. The atmospheric mean concentrations ($\mu g/m^3$) of the measured metals in the atmosphere of Cairo were Co (0.0196), Cr (0.0113), Cd (0.0017), Pb (0.9485, Mn (0.0975), V (0.0310), As (0.0063), Sb (0.0165), Ni (0.0133), and Ti (0.3483). Cancer risks, as well as non-cancer effects, due to inhalation exposure were assessed for 10 toxic metals. Individually, in relation to carcinogenic risks, As, Cr, and Cd inhalation might potentially cause an increase of the cases of cancer more than 1E-6 for As and Cr in all investigated sites and Cairo as well. The current results suggest that, although in general terms the concentration of metals is not relatively high in summer for the area, attention should be paid to As, Cr, and Cd as carcinogenic materials.

Keywords: atmospheric air, Cairo, toxic metals, hazards, risk levels.

1 Introduction

In recent years, air pollution has become one of the most obvious and important environmental problems in Cairo City. Cairo, the capital of Egypt has

WIT Transactions on Ecology and the Environment, Vol 123, © 2009 WIT Press
www.witpress.com, ISSN 1743-3541 (on-line)
doi:10.2495/AIR090321

approximately 11 million inhabitants. Large numbers of motor vehicles congest the city and create noise and air pollution problems. Major sources of air pollution are the rapidly increasing number of motor vehicles traveling on the streets and roads, the many small factories, building construction, road modifications and transport system construction [1]. A series of environmental studies identified air pollution as the most critical environmental problem in Cairo with particulate matter and toxic metals as the two greatest health risks in Cairo's air [2–6]. It has been found that several toxic metals, including arsenic (As) cadmium (Cd), nickel (Ni), lead (Pb), vanadium (V), cobalt (Co), chromium (Cr), manganese (Mn), antimony (Sb), and titanium (Ti) and their compounds, are associated with the fine particulate matter in the ambient air. Toxic metal emissions from factories or car exhausts can result in serious environmental problems such as the restriction of atmospheric visibility, while their toxicity may present health problems to humans at certain concentrations [7–10].

The assessment of human exposure to airborne toxic metals is an important risk. Large segments of the Egyptian population have been, and are now being, exposed to toxic chemicals from all media and through all routes of exposure (inhalation, ingestion, and dermal absorption) [11]. Quantitative risk assessment, a relatively new endeavor in Egypt, requires good data on environmental concentrations, and the relation between these concentrations and human exposure [12]. It has become an important tool utilized in environmental programs that mandate the protection of humans from exposure to toxic chemicals [13]. Quantifying inhalation exposure necessitates identifying physical activities and their associated ventilation rate(s), and the time spent in each activity [14].

The objective of this paper is to develop and implement a methodology that can be used to assess the potential exposure of population targets to hazardous air pollutants (toxic metals). Methods based on risk assessment modeling and Geographical Information System (GIS) technology are used to determine: a) the geographical boundaries of areas potentially exposed to hazardous substances; and b) the nature and strength of the relationship between the spatial distribution of toxic metals and the estimated risk levels.

2 Methodology

2.1 Ambient sampling sites description

Seven sampling sites (Fig. 1) were selected to represent the city of Cairo, with different impact sources e.g., background, traffic, industrial, and residential impacts.

2.2 Sampling and analysis

The analysis of toxic metals in airborne particulate gives information on pollution levels, but the high cost limits these kinds of analysis to a few samples. Fifty-six air particulate samples were collected during summer 2005. Airborne

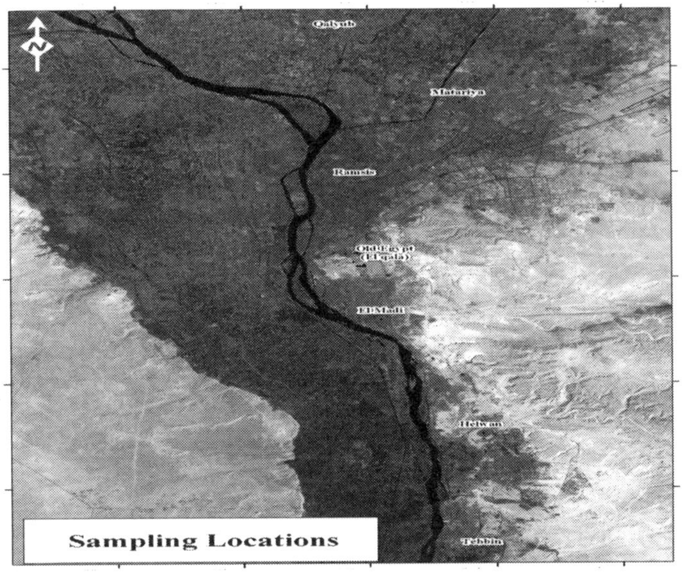

Figure 1: Map showing the sampling sites over Cairo.

particulate samples were collected once/week. Air sampler units were portable (mains electricity or battery operated), and water proofed. Particles in the fine fraction were collected by using critical orifice, keeping the volume of flow rate constant at 3L/min by using a flow meter (CT Playton LTD, UK). The duration of each sampling period was 24 hours. The samples were collected under varying weather and traffic conditions to represent the summer season. The filters were pre-dried at 105°C for 2h and pre-weighed. Loaded filters were weighed before and after sampling and total suspended particulate were evaluated. The filter was extracted into 0.5 M or 1.0 M HNO_3 contained in polyethylene vials and agitated within an ultrasonic bath [15–17, 12].

2.3 Instrumentation

The following AAS techniques were used: i) Shimadzu AA-670, flame AAS (FAAS); ii) Varian AA-1275 equipped with GTA-96 electro thermal atomizer, electro-thermal AAS (ETA-AAS); and iii) Varian VGA-76 hydride generator. These techniques were used to measure the concentration of the elements in the sample. Laboratory check samples were prepared from unexposed mineralized filters with the standard addition of elements. Furthermore, a standard reference material (SRM 1648) from NITS (USA) has been used to validate the analysis. A quality assurance was carried out according to recommended procedure.

2.4 Assessment of exposure

In this section, an attempt has been made to estimate the exposure of Egyptian citizens for 10 pollutants (toxic metals) in air. Distribution associated with inhalation rate, body weight, residency duration, time spent indoors and outdoors as well as the total hours at home and away from home has been characterized by numerous investigators [2, 12, 13, 18, 19]. The calculation of Hazard Quotient (H.Q.), and carcinogenic risk estimates were carried out by using the *Risk*Assistant* model – the description of this model has been written elsewhere [19–22].

2.5 Statistical analysis

The statistical analysis was performed by using Microsoft Excel 2000 and SPSS version 10 programs. This represents the descriptive statistics for comparability with air pollution standards and other exposure assessments. Statistical significant differences were computed *t*-test and used for the comparison of means. A probability of 0.05 or less was considered as significant. Correlation coefficients as well as the derived equations from the relationship between related parameters were calculated.

3 Results and discussion

The results of the current study showed that, in general terms, metal concentrations in air samples collected from different sites increased slightly according to the land use and anthropogenic activities. Table 1 summarizes the current metal concentrations in atmospheric air through the sampling period. A significant difference was found between collection sites with higher values $p < 0.05$ in the case of Cr, As, Sb, and Ni. The reported concentrations for the examined pollutants, in the current study, were found to be higher than those reported by Hassanien and Shakour [17] in the suburban resort area (Hurghada, Egypt). This variation referred to wind direction, which influences the elemental concentrations, especially when the wind flow comes across the industrial areas of the city [5]. The atmosphere of the investigated urban areas has been found to be heavily polluted with various metals, as attributed to heavy traffic, high population density and municipal buildings [5, 23–25]. Consequently, the elemental composition of particulate at all investigated sites is clearly affected by anthropogenic activities.

The present results showed that cadmium concentrations were in the range of 0.0004-0.0034, with an average 0.0017 $\mu g/m^3$ lower than the average annual concentration of 0.003 $\mu g/m^3$ measured in Minsk, Belarus and the WHO Air Quality Guideline for cadmium of 0.005 $\mu g/m^3$ [26, 27]. However, in the present study, the concentration of cadmium in air was higher than concentration range of 0.00005-0.0002 $\mu g/m^3$ measured in Northern Europe, 0.0002-0.0005 $\mu g/m^3$ in Central Europe, and 0.00006-0.0012 $\mu g/m^3$ in Southern Europe [28].

Lead levels in the ambient air in Cairo have decreased during recent decades between 1990 and 2003. The annual lead concentrations in particulate matter range from 0.5 $\mu g/m^3$ in the residential area northeast of the city to 3 $\mu g/m^3$ in the city center. Near lead smelting operations, the average lead concentration can reach 10 $\mu g/m^3$ [29]. In 2004, air quality in the "Shoubra El Kheima" industrial

Table 1: Atmospheric metals concentration (µg/m^3) in the investigated areas.

Site	Co	Cr	Cd	Pb	Mn	V	As	Sb	Ni	Ti
El Madi	0.0231	0.0112	0.0016	0.1577	0.0669	0.0233	0.0038	0.0084	0.0122	0.3471
Tebbin	0.0285	0.0194	0.0007	0.2246	0.1689	0.0380	0.0025	0.0109	0.0162	0.4730
Ramsis	0.0159	0.0091	0.0004	0.4278	0.0627	0.0217	0.0028	0.0117	0.0118	0.2691
Moustord-Matariya	0.0226	0.0119	0.0021	5.3535	0.1942	0.0655	0.0325	0.0686	0.0292	0.3813
Qalyub	0.0183	0.0068	0.0013	0.0385	0.0627	0.0206	0.0010	0.0035	0.0093	0.3586
Old Egypt (El qala)	0.0116	0.0064	0.0026	0.2511	0.0518	0.0217	0.0011	0.0112	0.0056	0.2854
Helwan	0.0175	0.0146	0.0034	0.1862	0.0750	0.0264	0.0002	0.0014	0.0086	0.3239
Average	0.0196	0.0113	0.0017	0.9485	0.0975	0.0310	0.0063	0.0165	0.0133	0.3483
Max	0.0285	0.0194	0.0034	5.3535	0.1942	0.0655	0.0325	0.0686	0.0292	0.4730
Min	0.0116	0.0064	0.0004	0.0385	0.0518	0.0206	0.0002	0.0014	0.0056	0.2691
Count	7	7	7	7	7	7	7	7	7	7
S.D.	0.006	0.005	0.001	1.946	0.058	0.016	0.012	0.023	0.008	0.068

area was improved, with lead levels dropping to 1.02 μg/m^3, which nearly meets the Law 4/1994 annual average of 1.0 μg/m^3 [3, 4]. The most interesting finding in this study was that the current results illustrated that the mean concentration of Pb was 0.9 μg/m^3 over Cairo, which is lower than the Egyptian standard limit values set by Law 4/1994 [30]. Comparing the present data with the international studies e.g., in Europe in 1990 the concentrations of lead in background air were mainly within the 0.01-0.03 μg/m^3 range. In 2003 the concentrations mainly ranged between 0.005 and 0.015 μg/m^3 [28]. In Minsk, Belarus, the average annual concentration of lead in 2004 was 0.083 μg/m^3. In cities and in the vicinity of industrial sources (i.e. 1–10 km distance) higher concentrations prevail, up to an order of magnitude [31-33]. Arsenic is a potential carcinogen at trace levels and induces cancer response depending on the mode of intake, i.e. inhalation or ingestion [34]. Mean arsenic levels in remote and rural areas range from 0.00002 to 0.004 μg/m^3, while in urban areas the level can range from 0.003 to 0.2 μg/m^3 and much higher in the vicinity of industrial sources [35] compared to 0.063 μg/m^3 measured in the current study.

It was found that the current measured data were slightly higher than the Hungarian data for V and Ni and lower in the case of Cr where Vaskövi et al. [36] measured the trace metals in airborne particles and the concentration of V, Ni and Cr were 0.02, 0.01, 0.02 μg/m^3, respectively in an urban area in Hungary. Moreover, comparing the current results with the others, e.g. Apostoli [37] in Italy, who reported that the results of Cr were 0.003-0.01; 0.01-0.9 μg/m^3 for rural and industrial areas, respectively. In the same way, the present work evaluated V and it ranged from 0.0206-0.0665 μg/m^3. Cohen [38] stated that the concentration of V ranged from 0.007-0.170 μg/m^3 for rural areas and in the industrial areas it was found to be in the range of 0.01-1 μg/m^3. He reported that air V content varies as a function of location, typical concentrations in rural areas ranged from 0.0025 to 0.0075 μg/m^3 and from 0.060 to 0.300 μ/m^3 in urban settings.

Correlation calculations, performed on the concentration of metals in the air, gave significant correlations between most studied elements, as illustrated in Table 2. This indicates that air pollution by metal originated from a common emission anthropogenic source – most probably automobiles influencing the whole studied areas, as consistent with other studies [39, 40].

Table 2: Correlation coefficient between metals concentration.

	Cr	**Cd**	**Pb**	**Mn**	**V**	**As**	**Sb**	**Ni**	**Ti**
Co	**0.79**	-0.38	0.22	**0.72**	**0.50**	0.28	0.24	**0.59**	**0.90**
Cr		-0.08	0.06	**0.64**	0.40	0.08	0.05	0.36	**0.74**
Cd			0.15	-0.13	0.08	0.10	0.07	-0.15	-0.25
Pb				**0.73**	**0.93**	**0.99**	**0.99**	**0.91**	0.19
Mn					**0.92**	**0.74**	**0.75**	**0.89**	**0.76**
V						**0.93**	**0.93**	**0.95**	**0.50**
As							**0.99**	**0.93**	0.23
Sb								**0.91**	0.21
Ni									**0.50**

Table 2 showed a distinct positive correlation more than 27% of correlation coefficient values (r) were higher than 0.9 and 55% ≥ 0.5. Strong correlation was found between Co, Pb, Mn, and V and most other concerned metals. However, in my study, cadmium with other metals was not correlated or showed poor correlations in agreement with other investigators [41, 39].

Initially, a screening of 10 toxic metals were identified that could present impact due to inhalation exposure. To screen for inhalation exposures, the *Risk*Assistant* model was used to estimate the non- and carcinogenic effects. Some toxic metals can cause cancer, e.g., arsenic, chromium, and cadmium in this study. The current study found pollutants in the air that have a concentration that poses significant risk for the health of the population in different sites under investigation, with a variation in the magnitude due to the concentration and population type. The results of total exposure of these metals for two target population groups were used to compute the associated cancer risk.

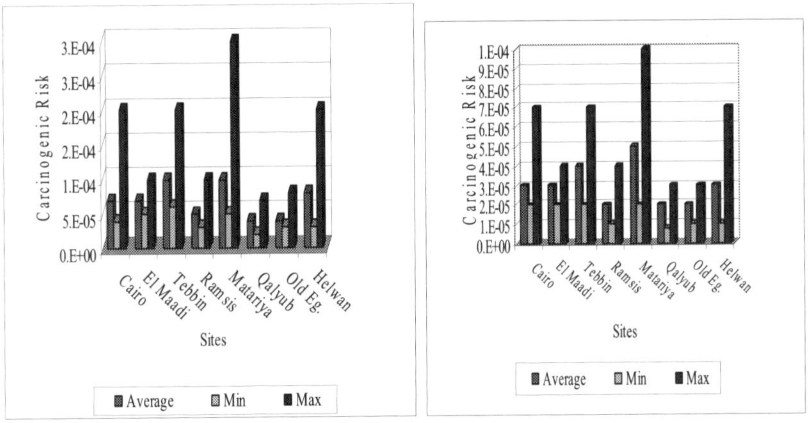

a (adult) b (children)

Figure 2: Carcinogenic estimates over the different investigated sites for adult and children populations.

Figures 2a, 2b and 3 clearly show the estimates developed by using the *Risk*Assistant* program for both children and adults in the different investigated sites and in Cairo for the average, minimum and maximum concentration values. For all sets of exposure estimates, the total risk and hazard quotients (H.Q.) were investigated for all of the investigated toxic metals.

4 Conclusion

The current work presented the results of this risk assessment to the decision makers in Cairo, Egypt. It was one of the few risk assessment studies performed in Egypt, and it is certainly the first one that deals with the assessment of multi-pollutant exposure based on routine environmental pollution monitoring data. In

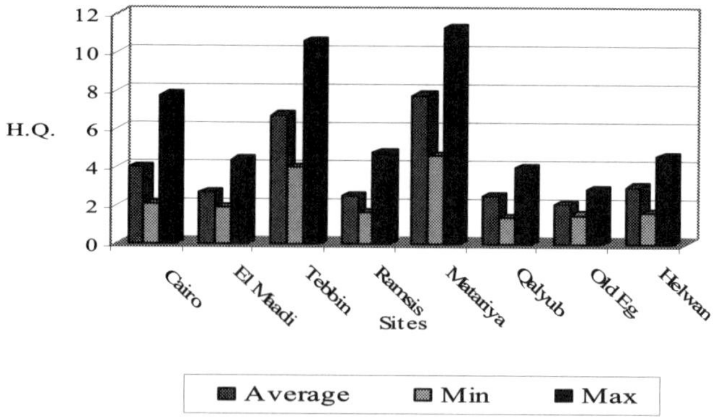

Figure 3: H.Q. values for the different investigated sites for the two population groups.

conclusion, the baseline human health risk assessment conducted for Cairo assessed the potential human health risks associated with carcinogenic and non-carcinogenic heavy metals of concern in the air. The non-carcinogenic H.Q. is more than 1, indicating that a potential health hazard may exist. The carcinogenic risk estimates in different sites ranged from 8E-6 to 1E-4 and 2E-5 to 3E-4 for children and adult populations, respectively, indicating that corrective action may be necessary. This cancer risk is due to inhalation exposure to arsenic, cadmium, and chromium in the air. This approach is intended to provide background information and guidance to governments in making risk management decisions, particularly in setting standards, regulatory programs and priorities.

Consequently, risk estimates in the present study can be used to serve as data synthesizers and "translators" to inform risk managers and, to identify critical knowledge gaps which research can address. For Cairo, in which there are emitting sources, the average toxic metals exposure is at or above the level at which the risk of adverse effects may begin to increase. Further research is needed on vehicle emission control technology, especially on diesel engines. In addition, a national discussion needs to be resurrected about the role of nuclear energy as a medium-term alternative to fossil fuels.

References

[1] LIFE-Lead (2005) Baseline Human Health Risk Assessment. Millennium Science and Engineering, Inc. USAID/ Egypt, Office of Environment.

[2] Hassanien, M. A.; Abdel-latif, N.M.; Shakour, A.A. and Saleh, I.A. (2006). Estimation of inhalation intake of metals in a traffic- affected residential area of east Cairo-Egypt. Cent. Eue. J. Occup. Environ. Med. 12(4):257–267.

[3] EEPP (Egyptian Environmental Policy Program). (2004a). Ambient Air Criteria Document, June 2004.

[4] EEPP (Egyptian Environmental Policy Program). (2004b). Health Assessment and Economic Evaluation, June 2004.
[5] Hassanien, M.A, Rieuwerts, J., Shakour, A.A., and Bitto, A. (2001). Seasonal and annual variations in air concentrations of Pb, Cd and PAHs in Cairo, Egypt. Int. J. Environ. Health Research. 11:1: 13–27.
[6] Schroeder, W. H.; Dobson, M.; Kare, D.M.; and Johnson, N.D. (1987). Toxic trace elements associated with airborne particulate matter. A review. JAPCA. 37(11):1267–1285.
[7] Gertler, A.W., Gillies, J.A., Pierson, W.R. (2000). An assessment of the mobile source contribution to PM10 and PM2.5 in the United States. Water, Air, and Soil Pollution 123, 203–214.
[8] Monaci F., Moni F., Lanciotti E., Grechi D., R. Bargagli. (2000). Biomonitoring of airborne metals in urban environments: new tracers of vehicle emission, in place of lead. Environmental Pollution 107: 321–327
[9] Pakkanen, T.A., Loukkola, K., Korhonen, C.H., Aurela, M., Makela, T., Hillamo, R.E., Aarnio, P., Koskentalo, T., Kousa, A., Maenhut, W. (2001). Sources and chemical composition of atmospheric fine and coarse particles in the Helsinki area. Atmospheric Environment 35, 5381–5391.
[10] Wrobel, A., Rokita, E., Maenhaut, W. (2000). Transport of traffic related aerosols in urban areas. Science of the Total Environment 257, 199–211.
[11] LIFE-Lead (2006) Evaluation of Remedial Alternatives. Final Draft. Millennium Science and Engineering, Inc. USAID/ Egypt, Office of Environment.
[12] Hassanien, M.A. (2001). Assessment of human exposure to atmospheric trace metals in a residential area of Cairo, Egypt. Cent. Euro. J Occup. Environ. Med. 7(3-4): 253–262.
[13] Funk, L.M, Sedman, R., Beals, J.A.J., and Fountain, R. (1998). Quantifying the distribution of inhalation exposure in human populations: 2. Distribution of time spent by adult, adolescents, and children at home, at work and at school. Risk Anal. 18: 1:47–56.
[14] EPA, (1989). "EPA exposure factors handbook" EPA/600/8-89 (Environmental Protection Agency, Washington, D.C.).
[15] Harrison M.R. and Williams, R.C. (1982). Airborne cadmium, lead and zinc at rural and urban sites in north-west England. Atmos Environ. 16: 2669–2681.
[16] Samara C., Vousta D., and Kouimtizis TH. (1990). Characterization of airborne particulate matter in Thessaloniki, Greece part 1: Sources related heavy metals concentrations within TSP. Toxicol and Environ Chem; 29:107–119.
[17] Hassanien, M.A., and Shakour, A.A. (1999). Assessment of trace elemental composition of air particulate matter at Hurgada, East Egypt. Cent Euro J Occup Enviro. Med. 5; (3-4):291–301.
[18] Silvers, A., Florence, B.T., Rourke, D.L., and Lorimor, R. J. (1994). How children spend their time: A sample survey for use in exposure and risk assessments. Risk Anal. 14; 6: 931–944.

[19] Hassanien, M.A., Dura, GY. and Karpati, Z. (1999). Potential of exposure to volatile organic compounds occurrence of natural origin in thermal water. Cent Eur J Occ and Environ Med. 5 (2): 160–172.
[20] Fugas, M., Sega, K., Sisovic, A. (1982). Study of personal exposure to airborne particles and carbon monoxide. Environ. Monit. Assess. 2:157–169.
[21] Sega, K., Fugas, M. (1992). Time budget as a factor relating indoor air pollution to total human exposure. In proceedings of the International Conference-Indoor Climate of Buildings, indoor Air Quality in Central and Eastern Europe, Strbske Pleso, September 29-October 2, 1992, Slovo Bratislava.
[22] Sexton, K., Spengler, J. D., and Trettman, R. D. (1992). Personal exposure to respirable particles: a case study in Waterbury, Vermont. Atmos. Environ. 18:1385–1398.
[23] AbdEL-Latif, N.M.M. (1993). An investigation on some combustion generated pollutants affected plant growth. MSc thesis. Zagazig University, Faculty of Science, Botany Department, Zagazig, Egypt.
[24] Rizk, H.F.S., Shakour, A.A., and Meleigy, M.I. (1995). Concentration of surface ozone, total oxidants and heavy metals in urban areas downwind of industrial area in Cairo, Egypt. In: International Conference of Heavy Metals in the Environment. Vol. 1. (Eds. R-D. Wilken, J. Forster & A. Knochel), Hamburg, pp. 164–167.
[25] Koliadima, A., Athanasopoulou, A., and Karaiskakis, G. (1998). Particulate matter in air of the cities of Athens and Patras (Greece): particle size distribution and elemental concentrations. Aerosol Sci. Technol. 28: 292–300.
[26] WHO/IPCS. Cadmium. Environmental Health Criteria 134. WHO, Geneva, 1992.
[27] WHO. Air Quality Guidelines for Europe. Second Edition. WHO regional Publications, European Series, No.91, 2000.
[28] Aas W., Breivik K. (2005). Heavy metals and POP measurements 2003, EMEP-CCC Report #9/2005, NILU, Kjeller, Norway, 103 pp.
[29] Bowen, J. L. (1995). Technical analysis monitoring and analysis component Cairo Air Improvement Project. Annex D submitted by Datex, Inc. Arlington, Virginia.
[30] EEAA (Egyptian Environmental Affair Agency) (1994). Environmental Protection Law No.4.
[31] Herpin U., Siewers U., Markert B., Rosolen V., Breulmann G., Bernoux M. (2004). Second German heavy metal survey by means of mosses, and comparison of the first and second approach in Germany and other European countries, Environ. Sci. Poll. Res. 11, 57–66.
[32] Wenzel K.D., Hubert A., Weissflog L., Kühne R., Popp P., Kindler A., Schüürmann G. (2006). Influence of different emission sources on atmospheric organochlorine patterns in Germany, Atmos. Environ. 40, 943–957.
[33] Landesumweltamt: Jahresdauswertung nach EU-Luftqualiätsrichtlinien, Monitoring data, Landesumweltamt Nordrhein-Westfalen, Essen, Internet

(2005): http://www.lua.nrw.de/index.htm?luft/immissionen/ aktluftqual/ eu_luft_akt.html

[34] Byrd, D. M., Roegner, M. L., Griffiths, J. C., Lamm, S. H., Grumski, K. S., Wilson, R., and Lai, S. (1996). Carcinogenic risks of inorganic arsenic in perspective. Int Arch Occup Environ Health. 68: 484-494.

[35] Hindmarsh, J. T. (2000). Arsenic, its clinical and environmental significance. J. Trace Elem. Exp. Med. 13: 165–172.

[36] Vaskövi, E., Muylle, E., Bacskai, J., Kertesz, M., Sipos, M., and Lászlo, A. (1995). National applying of a new sampling opportunity for the measurement of trace metals and sulfates in airborne particles. Proceedings of the 6th symposium on particle size analysis, environmental and powder technology. Gyor, Hungary, 7-8 June 1995.

[37] Apostoli, A. (1992). Criteria for the definition of reference values for toxic metals. Sci Total Environ. 120; 23–37.

[38] Cohen, M.T. (1996). Vanadium and its immunotoxicology. Toxicol Ecotoxcol. News 3; 5: 132–135.

[39] Chattopadhyay, G.; Lin, K. Chi-Pei, and Feitz, A. J. (2003). Household dust metals levels in the Sydney metropolitan area. Environ Res. 93:301–307.

[40] Karar K. and Gupta, A.K. (2006). Seasonal variations and chemical characterization of ambient PM10 at residential and industrial sites of an urban region of Kolkata (Calcutta), India. Atmospheric Research 81: 36–53.

[41] Mielke, H. W., Gonzales, C. R., Smith, M. K., Mielke, P.W. (1999). The urban environment and children's health: soils as an integrator of lead, zinc and cadmium in New Orleans, Louisiana, USA. Environ Res. 81:117–129.

PBDEs and PCBs in European occupational environments and their health effects

I. L. Liakos, D. Sarigiannis & A. Gotti
European Commission, Joint Research Centre,
Institute of Health and Consumer Protection,
Physical and Chemical Exposure Unit, Italy

Abstract

Flame retardants such as polybrominated diphenyl ethers (PBDEs) and polychlorinated biphenyls (PCBs) have been widely used in numerous applications for the retardation of fires. In this effort the indoor air concentrations of PBDEs and PCBs in European environments, obtained from various research studies, are gathered, analysed and evaluated. Specific micro-environments and materials used in indoor buildings appear to influence the concentration of flame retardants. Even though PBDEs and PCBs in Europe were found at low concentrations, there are some indoor environments presenting elevated levels of halogenated flame retardants (HFRs). Congener PBDE 209 is the most abundant in every studied environment. The extensive use of electrical devices increases the PBDEs concentration. High PBDE and PCB concentrations were found in the UK due to the strict fire regulations in this country. High PCB concentrations in indoor air were detected in buildings reinforced with concrete, as well as in schools, industrial and public buildings and in recycling plants. HFRs have shown that they are linked with various diseases including cancer, immune, neurological, endocrine, and reproductive effects and chlorance. Limitation and/or banning of HFRs is ongoing by many organisations and countries and the need for a universal approach is required.
Keywords: halogenated flame retardants, PBDEs, PCBs, emission sources, indoor air concentrations, health effects.

1 Introduction

Since the 1960s, flame retardants have been used as additives in products to reduce the danger of fire and consequently the risk of life when fires occur in

WIT Transactions on Ecology and the Environment, Vol 123, © 2009 WIT Press
www.witpress.com, ISSN 1743-3541 (on-line)
doi:10.2495/AIR090331

various indoor environments, where people spend most of their time. The annual world production of flame retardants is approximately 600,000 metric tons, of which about 60,000 are PCBs and 150,000 PBDEs (Darnerud *et al* [1]). HFRs are added to compartments in cars, trucks, trains and airplanes. Flame retardants are used in products such as insulating materials, electronic and electrical goods, upholstered furniture, textiles, sealants, plastics, building materials and carpets. However, there are potential risks from the toxicity and eco-toxicity of the usage of flame retardants. During environmental analyses, flame retardants were present in a wide range of environmental samples, including marine biota, mother's milk and sediments. (Kemmlein *et al* [2], McDonald [3])

As of 1st of July 2006 Directive 2002/95/EC (European Commission, 2003) (the "RoHS" Directive) restricts the use of flame retardant chemicals belonging to the group of PBDEs in electrical and electronic equipment. The maximum tolerated concentration for these chemicals is 0.1%, as set in Commission Decision 2005/618/EC. Certain applications, materials and components can be exempted from the restrictions if their elimination or substitution via design changes or materials and components is technically impracticable, or if the negative environmental, health and/or consumer safety impacts caused by substitution outweigh the environmental, health and/or consumer safety benefits thereof. Even though DecaBDE is no longer produced in the EU, 7,600 tonnes of it are imported each year in addition to 1,300 tonnes that are included in articles (Pakalin *et al* [4]).

PCBs have been domestically manufactured since 1929 but because of possible health implications and environmental impacts, their use and production were severely restricted in many countries. Sweden and UK restricted their use and production in 1972, the USA in 1977, Norway in 1980, Finland in 1985, and Denmark in 1986 ([5]). Due to their non-flammability, chemical stability, high boiling point, and electrical insulating properties, PCBs were used in numerous industrial and commercial applications. Japan banned the production and use of PCBs in 1972 (WHO [6]). The 24 countries of the Organization for Economic Co-operation and Development (OECD) adopted a Decision in 1973, limiting the use of PCBs to certain specific applications and asking for the control of the manufacture, import, and export of bulk PCBs, for adequate waste treatment and for a special labelling system for PCBs and PCB-containing products [7].

PCBs have been classified in the European Union as R33, N; R50-53 [8]. R33 has been assigned as danger of cumulative effects. When a preparation contains at least one substance assigned the phrase R33, the label of the preparation must carry the wording of this phrase as set out in Annex III to Directive 67/548/EEC, when the concentration of this substance present in the preparation is equal to or higher than 1%, unless different values are set in Annex I to Directive 67/548/EEC. In addition R50 has been assigned as very toxic to aquatic organisms and R53 that may cause long-term adverse effects in the aquatic environment [9].

Even though flame retardants are subjects to several restrictions in use they can still be found in a wide-range of indoor environments in the EU. In the

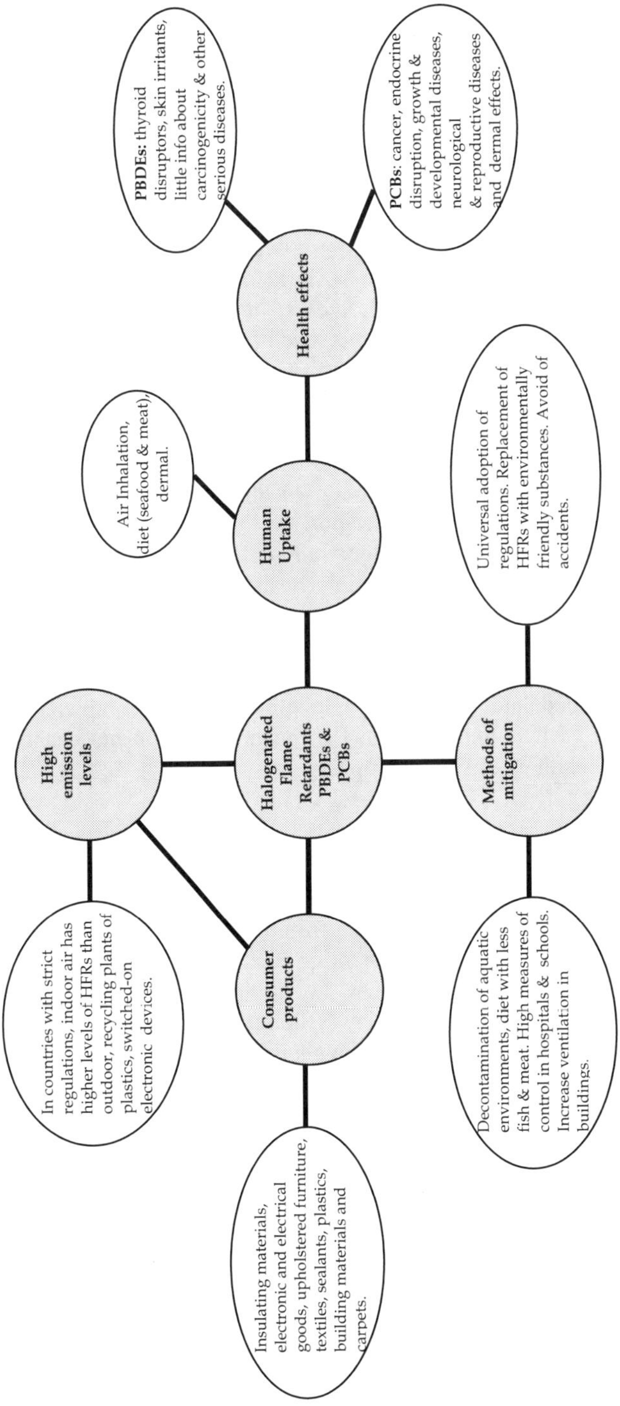

Figure 1: Graph summarising the current knowledge on the HFRs given in this paper.

present work, a study of the concentrations of HFRs in indoor air in Europe, their possible sources as well as their effects of those in human health is presented. A summary - histogram prior to analysis of the results and conclusions is illustrated in Figure 1.

2 Methodology

Firstly, a literature review was acquired. Studies that included indoor air measurements of PBDEs and PCBs in Europe were collected and their results were analysed. The data contained the specific congener, the sum of many congeners, the values of the measurements, the environment where the measurements took place, the literature reference of which the data were derived, the location and the number of the measurements. Then, the data were compared between each other in the same location to determine possible differences from various products and detect potential emission sources. Afterwards, the data obtained from different literature sources and thus different measurements were compared to distinguish variations between locations and possible trends of flame retardants concentrations in the region of Europe were investigated. In addition, the PBDEs and PCBs indoor air concentrations were used as indications of human exposure and possible health effects. This was done by comparing the measured data with toxicological studies and threshold limit values describing the connectivity of various flame retardants levels to certain disease or generally ill health. Those toxicological data were obtained from the literature and described later on. Finally, conclusions were completed on the current presence of PBDEs and PCBs in European indoor air and channels for mitigation of their health effects were proposed.

3 Results and discussion

PBDE congeners in different indoor locations were observed in the UK and Belgium (Harrad *et al* [10]). Samples taken from the University of Birmingham seem to contain high amount of congeners of PBDE in dust compared to the same congeners found in other cities in UK. In addition, the PBDE 209 congener found in all indoor locations was the highest among the other congeners for both the UK and Belgium. Its value was exceptionally high (mean values range: 2,177 – 45,000 ng/g & maximum: 520,000 ng/g). The reason for this is its high use during manufacturing of products since it has high stability and low-degree-burning properties. PBDE 209 is a DecaBDE and possesses these aforementioned properties due to its fully-brominated diphenyl ether (Darnerud *et al* [1]).

The content of brominated hydrocarbon flame retardants in air samples from a plant recycling electronics goods, a factory assembling printed circuit boards, a computer repair facility, offices equipped with computers and outdoor air have been measured (Sjodin *et al* [11]). Elevated values of the higher PBDEs, i.e. PBDE 183 and 209 as well as of 1,2-bis(2,4,6-tribromophenoxy)-ethane and tetrabromobisphenol A were detected in all the studied areas in the recycling plant (max: 44 – 200 ng/m^3). During the recycling process release of PBDEs and other flame retardant chemicals is observed, since those flame retardants were found at much lower concentrations inside a room where were being assembled circuit boards and in an office with computers (max. of PBDE 209 was 0.32 ng/m^3).

Measurements of PBDEs and PCBs were acquired in air in 31 homes, 33 offices, 25 cars, and 3 public microenvironments (Harrad *et al* [12]). Cars found to be the most contaminated microenvironment for ΣBDE (average) 0.709 ng/m^3, but the least for ΣPCB (average) 1.391 ng/m^3.

The high levels of PBDEs in internet cafes/computer rooms (mean: 0.17 ng/m^3), in public buildings/offices (mean: 2.25 ng/m^3) and in computers/electronics shops (mean: 0.117 ng/m^3) could be attributed to the presence and/or usage of electronic devices, such as personal computers, monitors, televisions, etc. (Mandalakis *et al* [13]). Lower PBDE levels were observed in furniture stores (mean: 0.012 ng/m^3), houses (mean: 0.009 ng/m^3) and in a laboratory (mean: 0.003 ng/m^3). The relatively limited operation of electronic devices might be the reason for the slightly lower PBDE levels in computers/electronics shops compared to offices, internet cafes/computer rooms where electronic devices are in use.

By comparing the sum of six of the most frequently and reliably monitored congeners (Σ_6PBDE, sum of BDE 28, 47, 99, 100, 154 and 153), results show that workplaces in UK (mean: 0.166 – 2.787 ng/m^3) and Sweden (mean: 1.231 ng/m^3) contain higher Σ_6PBDE levels than workplaces in Greece (mean: 0.047 ng/m^3) (Mandalakis *et al* [13]). Moreover, the levels of PBDEs measured in Greek homes (mean: 0.007 ng/m^3) were among the lowest reported in the literature due to the possible low-level usage of PBDE-containing material in residential homes in Greece and/or due to enhanced natural air ventilation (opening of windows) of houses in Greece due to warm climatic conditions.

Measurements of polychlorinated biphenyls (PCBs) were acquired in air in 31 homes, 33 offices, 25 cars, and 3 public microenvironments (Harrad *et al* [12]). The high ΣPCBs concentrations in public microenvironments (MEs) (mean: 30.73 ng/m^3) and offices (mean: 18.15 ng/m^3) were possibly due to the high content of PCB-containing materials that are used inside those buildings. Inversely, lower PCBs concentrations were detected at homes in Birmingham (mean: 2.82 ng/m^3) and in cars (mean: 1.39 ng/m^3).

PCB-containing materials were found to still release PCBs in homes since the indoor PCB concentrations were constantly higher than outdoors, approximately by a factor of 2–50 circa (Menichini *et al* [14]). The indoor concentrations of Σ_{62} PCBs varied between 6.5 and 33 ng/m^3. The windows were closed during the measurements for almost all the sampling time; thus, the contribution of the indoor sources was close to the highest achievable under these experimental conditions. No apparent association was originated linking concentrations and height of the floor: the high/low floor ratios for both Σ_{62} PCBs (sum of: PCBs17, 18, 28, 30, 41, 44, 47+48, 49, 52, 64, 66+80, 70, 74, 77, 81, 85, 87, 91, 95, 97, 99, 100, 101, 105, 110, 114, 118, 123, 128, 135, 136, 137, 138+163, 141, 146, 149, 151, 153, 155, 156, 157, 167, 170, 171, 172, 174, 176, 177, 180, 183, 187,

189, 194, 195, 200, 201, 202, 203+196) and Σ_6 PCBs (sum of: PCBs 28, 52, 101, 138 (+coeluting 163), 153, 180) were close to one at three buildings and 0.3–0.5 in the fourth one.

The PCB concentrations at most of the 181 public buildings in the state of Schleswig-Holstein were below the German guideline-level that has been set at 300 ng/m^3 (Heinzow *et al* [15]). Buildings reinforced with concrete, erected between 1968 and 1972, had elevated PCB concentrations in indoor air. Contrarily, the lowest levels were found in brickwork-constructed buildings. High concentration of the Σ_{15}PCB (sum of: PCBs 28, 52, 101, 138, 153, 180, 8, 18, 22, 31, 44, 49, 70, 99, 118) was marked (maximum: 1071 ng/m^3), generating doubts about its potential health effects.

Lower chlorinated PCBs were the prevailing congeners in a contaminated school building in Nuremberg, Germany (Liebl *et al* [16]). The concentrations of PCB 28, 52, 101 ranged between 4-600, 38-2300, 3-1100 ng/m^3 respectively. Concentrations of the three higher chlorinated indicator congeners were less than 80 ng/m^3. Total concentrations of all PCB congeners (sum of indicator congeners times 5) ranged between 690-20800 ng/m^3 (median 2044 ng/m^3). These levels were much higher than those in the previous study environments in Europe and it would be very important to distinguish the sources of materials that release such high levels of PCBs.

Total PCB levels in indoor air up to 4200 ng/m^3 were observed in various industrial and public buildings in Switzerland emitted from joint sealings. In an industrial building contaminated with PCB (former production of transformers), total PCB levels up to 13000 ng/m^3 were detected. Public buildings had lower PCBs than industrial buildings but in a few of those values close to the maximum value that have been imposed by a range of countries were reached. Depending on the quantity of joint sealings in a room, on the PCB type and content in the materials, and on the air exchange rate, total PCB concentrations in indoor air up to several 1000 ng/m^3 may take place. (Kohler *et al* [17])

3.1 Health effects of PBDEs and PCBs

PBDEs have a propensity to bio-accumulate and have similar chemical structure to thyroid hormones and polychlorinated biphenyls (PCBs). Bio-accumulation of PBDE congeners in human tissue and other biota could cause disruption of thyroid hormones, neurobehavioral deficits and possibly cancer (McDonald [3]). Hydroxylated metabolites of PBDEs structurally resemble thyroid hormones and therefore compete for binding to thyroid hormone receptors and transporter proteins and thus bind to thyroid hormone transport protein (i.e. transthyretin) (Meerts *et al* [18]) and to the thyroid hormone receptors TR-α1 and TR-β (Marsh *et al* [19]). Studies at newborn mice point out that PBDEs like PCBs, cause learning and motor deficits as well as brain development that worsen by ageing (Eriksson *et al* [20]). Workers exposed to polybrominated biphenyls and polybromodiphenyl ethers, e.g., DeBDE, during manufacture revealed a higher than normal prevalence of primary hypothyroidism and a noteworthy decline in sensory and fibula motor velocities (WHO [21]).The fully-brominated deca-BDE has been accused for inducing carcinogenicity in male and female rats [22]. High levels of BDE-47 in adipose tissue in persons has been correlated to non-Hodgkin lymphoma (NHL) disease (Hardell *et al.* [23]). Skin irritation is also caused from PBDE exposure (WHO [21]).

Inversely to the PBDEs, PCBs have been classified from the International Agency for Research on Cancer (IARC) for their carcinogenicity. Limited evidence for carcinogenicity has been found in humans and sufficient in animals. Overall summary evaluation of carcinogenic risk to humans is Group 2A. Therefore, the agent is probably carcinogenic to humans (IARC [24]). PCBs akin to PBDEs are toxic substances, stable in the environment with propensity to bio-accumulate. As a result they are present in the aquatic as well as in the terrestrial food chain (Jensen [25]). The health effects attributable to PCBs exposure include: (a) cancer, including increases in liver cancer and malignant melanoma; (b) immune effects such as suppress immune responses and decrease host resistance; (c) reproductive effects, including possible reduction of birth weight, conception rates and live birth rates as well as reduced sperm counts; (d) neurological effects including possible deficits in visual recognition, short-term memory and learning and possible peripheral neuropathy; (e) endocrine effects, such as exert effects on thyroid hormone levels and suppress thyroid hormone receptor (TR)-mediated transcription ; (f) elevations in blood pressure, serum triglyceride and serum cholesterol, chlorance, eye and skin irritation, headache, dizziness, depression, increased eye discharge, increased sweating at the palms and feeling of weakness, hyperpigmentation of skin and mucous membranes. (Faroon *et al* [26])

3.2 Threshold or safe level/restrictions

PentaBDE and octaBDE are candidates for inclusion in the United Nation's Stockholm Convention on Persistent Organic Pollutants (POPs). DecaBDE, remains in use today in North America, but in the EU it was banned on 1 April 2008 by the European Court of Justice.

The National Institute of Occupational Health (NIOSH) of the US considers chlorodiphenyl containing 54% chlorine to be a potential occupational carcinogen (Aroclor 1254). NIOSH usually recommends that occupational exposures to carcinogens should be limited to the lowest feasible concentration. The recommended exposure limit for 10 hours time-weighted average is set to 0.001 mg/m^3 (or 1000 ng/m^3) (NIOSH [27]). The guideline-level of total PCB indoor air concentrations in buildings in Germany has been set at 300 ng/m^3 (tolerable concentration) and 3000 ng/m^3 (action level) [28]. In addition, based on a tolerable daily intake (TDI) of 1 μg of PCB per kg body weight per day, a tentative guideline value (maximum tolerable concentration) for PCB in indoor air of 6000 ng/m^3 was communicated by the Swiss authorities. This limit was set for buildings such as schools or offices, where people spend an average of 8 hours per day. For buildings such as residences, where permanent exposure can be assumed (24 h per day), the maximum tolerable concentration is 2000 ng/m^3 [29].

4 Conclusions

High concentrations of PBDE congeners 99 and 209 were found in various locations in the UK and also in Sweden in a recycling plant. Those high concentrations are attributed to particularly stringent fire safety regulations existing in the UK and to the fact that Deca-PBDE (PBDE 209) had not been banned in the EU during these measurements. PBDEs in Athens were higher in places were computers or electrical devices were in use than places with electrical devices switched-off. Moreover, comparison among diverse places in Europe showed that dismantling halls, PC rooms in Sweden and offices in UK had the highest Σ_6BDEs (sum of BDE 28, 47, 99, 100, 154 and 153) compared to other indoor environments.

The indoor air concentration of PCBs is independent from the height of the apartments and is mainly related to indoor materials containing PCBs. High PCB concentrations in indoor air were detected in buildings reinforced with concrete. Lower PCB levels were found at brickwork constructed buildings. Very high PCB concentrations were monitored in a school in Nuremberg, Germany, where the maximum values of the congeners PCB 28, 52 and 101 were higher than the proposed tolerable concentration limit (300 ng/m^3).

Limited data on the human health effects of PBDEs exist in literature. PBDEs seem to act as thyroid disruptors and as skin irritants in humans. Carcinogenic, reproductive and neurobehavioral diseases have been observed in animal studies as reviewed in the work of (Darnerud *et al* [30]). Inversely PCBs have been accountable for serious diseases including cancer, endocrine disruption, neurological and reproductive diseases and dermal effects. Food, especially from the aquatic environment as well as mother's milk seems to increase the human risk in such diseases. In addition environments where high air concentrations of HFRs exist should be avoided.

4.1 Suggestions

Recycling plants, schools and hospitals should be designed in such a way, so they do not impose risk for occupational environment. Increase of the ventilation inside those buildings, and removal of products or materials that are possible sources of PBDEs and PCBs could provide healthier exposed air. The exposure to flame retardants is not simply a national problem, since products are used in countries other than the place of manufacture, whereas migration of such chemicals through air is inevitable. Therefore harmonised guidelines of PBDEs and PCBs are needed with the co-operation of all the countries. Accidents of additions of PBDEs and PCBs to either food or animal food have to be strictly avoided. Methods of decontamination of aquatic environments will decrease the current levels of HFRs. Alternatives to flame retardants are needed. Those alternatives should have a satisfactory protection against fire but at the same time they have to be environmental friendly as well as not causing health problems. Some include magnesium dioxide and mixtures of it with antimony oxides, boron, melamine, melamine salts, silicon dioxide, and silicones as well as a newer class of materials known as "nano-additives" such as layered clay minerals [31].

References

[1] Darnerud, P.O., G.S. Eriksen, T. Jóhannesson, P.B. Larsen, and M. Viluksela, *Polybrominated Diphenyl Ethers: Occurrence, Dietary Exposure, and Toxicology.* Environmental Health Perspectives, **109**, 2001.

[2] Kemmlein, S., O. Hahn, and O. Jann, *Emissions of organophosphate and brominated flame retardants from selected consumer products and building materials.* Atmospheric Environment, **37**, 5485-5493, 2003.

[3] McDonald, T.A., *A perspective on the potential health risks of PBDEs.* Chemosphere, **46**, 745-755, 2002.

[4] Pakalin, S., T. Cole, J. Steinkellner, R. Nicolas, C. Tissier, S. Munn, and S. Eisenreich, *Review on production processes of Decabromodiphenyl ether (DecaBDE) used in polymeric applications in electrical and electronic equipment and assessment of the availability of potential alternatives to DecaBDE.* EC, IHCP, European Chemicals Bureau, EUR 22693 EN, 2007.

[5] *Polychlorinated biphenyls: Human health aspects.* WHO, Concise International Chemical Assessment Document 55,

[6] WHO, Polychlorinated Biphenyls and Terphenyls. Environmental Health Criteria 2, WHO, Geneva, Switzerland, p. 85, 1976

[7] *Polychlorinated Biphenyls and Terphenyls.* IPCS, Environmental Health Criteria 140,

[8] Commission Directive 2004/73/EC, Official Journal of the European Union, 2004

[9] Directive 1999/45/EC of the European Parliament and of the Council, Official Journal of the European Communities, 1999

[10] Harrad, S., C. Ibarra, M. Diamond, L. Melymuk, M. Robson, J. Douwes, L. Roosens, A.C. Dirtu, and A. Covaci, *Polybrominated diphenyl ethers in domestic indoor dust from Canada, New Zealand, United Kingdom and United States.* Environment International, 2007 in press.

[11] Sjodin, A., H. Carlsson, K. Thuresson, S. Sjolin, A. Bergman, and C. Ostman, *Flame Retardants in Indoor Air at an Electronics Recycling Plant and at Other Work Environments.* Environ. Sci. Technol., **35**, 448-454, 2001.

[12] Harrad, S., S. Hazrati, and C. Ibarra, *Concentrations of Polychlorinated Biphenyls in Indoor Air and Polybrominated Diphenyl Ethers in Indoor Air and Dust in Birmingham, United Kingdom: Implications for Human Exposure.* Environ. Sci. Technol., **40**, 4633-4638, 2006.

[13] Mandalakis, M., V. Atsarou, and E.G. Stephanou, *Airborne PBDEs in specialized occupational settings, houses and outdoor urban areas in Greece.* Environmental Pollution, in press.

[14] Menichini, E., N. Iacovella, F. Monfredini, and L. Turrio-Baldassarri, *Relationships between indoor and outdoor air pollution by carcinogenic PAHs and PCBs.* Atmospheric Environment, **41**, 9518-9529, 2007.

[15] Heinzow, B., S. Mohr, G. Ostendorp, M. Kerst, and W. Korner, *PCB and dioxin-like PCB in indoor air of public buildings contaminated with different PCB sources – deriving toxicity equivalent concentrations from standard PCB congeners.* Chemosphere, **67** 1746-1753, 2007.

www.witpress.com, ISSN 1743-3541 (on-line)

[16] Liebl, B., T. Schettgen, G. Kerscher, H.-C. Broding, A. Otto, J. Angerer, and H. Drexler, *Evidence for increased internal exposure to lower chlorinated polychlorinated biphenyls (PCBs) in pupils attending a contaminated school.* Int. J. Hyg. Environ. Health, **207** 315 - 324, 2004.
[17] Kohler, M., M. Zennegg, and R. Waeber, *Coplanar Polychlorinated Biphenyls (PCB) in Indoor Air.* Environ. Sci. Technol., **36**, 4735-4740, 2002.
[18] Meerts, I.A.T.M., J.J. van Zanden, E.A.C. Luijks, I. van Leeuwen-Bol, G. Marsh, E. Jakobsson, A. Bergman, and A. Brouwer, *Potent competitive interactions of some brominated flame retardants and related compounds with human transthyretin in vitro.* Toxicol. Sci., **56**, 95-104, 2000.
[19] Marsh, G., A. Bergman, L.G. Bladh, M. Gillner, and E. Jakobsson, *Synthesis of p-hydroxybromodiphenyl ethers and binding to the thyroid receptor.* Organohalogen Compounds, **37**, 305-308, 1998.
[20] Eriksson, P., E. Jakobsson, and A. Fredriksson, *Developmental neurotoxicity of brominated flame-retardants polybrominated diphenyl ethers and tetrabromo-bis-phenol A.* Organohalogen Compounds, **35**, 375-377, 1998.
[21] WHO, Environ Health Criteria 162: Brominated Diphenyl Ethers. Available from http://www.inchem.org/documents/ehc/ehc/ehc175.htm as of March 19, 2003, 1994
[22] *National Toxicology Program (NTP), Toxicology and carcinogenesis studies of decabromodiphenyl oxide (CAS No. 1163-19-5) in F344/N rats and B6C3F1 mice (feed studies).* US Department of Health and Human Services, NTP Technical Report 309, NIH Publication No. 86-2565, 1986.
[23] Hardell, L., et al, Oncol. Res., **10** 429-32, 1998.
[24] IARC, Monographs on the Evaluation of the Carcinogenic Risk of Chemicals to Man. Geneva: World Health Organization, International Agency for Research on Cancer, Multivolume work 71, p. S7 1972-present (1987)
[25] Jensen, S., *Report of a new chemical hazard.* New. Sci., **32**, 612-623, 1966.
[26] Faroon, O.M., L.S. Keith, C. Smith-Simon, and C.T. De Rosa, *Polychlorinated biphenyls: Human health aspects.* WHO, Geneva, 2003.
[27] NIOSH, Pocket Guide to Chemical Hazards, Washington, D.C. U.S. Government Printing Office, DHHS (NIOSH) Publication No. 97-140, 64, 1997
[28] PCB-Guideline/PCB-Richtlinie, Berlin, Germany, ARGE BAU, Mitteilungen des Deutschen Instituts fur Bautechnik 2/1995, 1995
[29] Swiss Federal Office of Public Health. BAG Bulletin, Bern, Switzerland, 464-465, 2001.
[30] Darnerud, P.O., S. Atuma, M. Aune, S. Cnattingius, M.-L. Wernroth, and A. Wicklund-Glynn, *Polybrominated diphenyl ethers (PBDEs) in breast milk from primiparous women in Uppsala County, Sweden.* Organohalogen Compounds, **35**, 411-414, 1998.
[31] *http://pubs.acs.org/subscribe/journals/esthag-w/2007/sept/tech/kb_flameretard.html.*

Section 7
Global and regional studies

Improved modelling experiment for elevated PM_{10} and $PM_{2.5}$ concentrations in Europe with MM5-CMAQ and WRF/CHEM

R. San José[1], J. L. Pérez[1], J. L. Morant[1], F. Prieto[2] & R. M. González[3]
[1]*Environmental Software and Modelling Group, Computer Science School, Technical University of Madrid (UPM), Spain*
[2]*Department of Ecology, Building of Sciences, University of Alcalá,*
[3]*Department of Meteorology and Geophysics, Faculty of Physics, Complutense University of Madrid, Spain*

Abstract

The application of the MM5-CMAQ model (PSU/NCAR/EPA, US) to simulate the high concentrations in PM_{10} and $PM_{2.5}$ during a winter episode (2003) in Central Europe has been performed. The selected period is January, 15 – April, 6, 2003. Values of daily mean concentrations of up to 75 μgm^{-3} are found on average from several monitoring stations in Northern Germany. Additionally, the WRF/CHEM (NOAA, US) model has been applied. In this contribution we have performed additional simulations to improve the results obtained in our contribution (San José et al. (2008)). We have run again both models but with changes in emission inventory and turbulence scheme for MM5-CMAQ. In the case of WRF/CHEM many more changes have been performed: Lin et al. (1983) microphysics scheme has been substituted by WSM 5-class single moment microphysics scheme (Hong et al. 2004); Goddard radiation scheme has been substituted by Dudhia radiation scheme and FTUV photolysis model has been substituted by J-FAST photolysis model. The results improve substantially the PM_{10} and $PM_{2.5}$ patterns in both models. The correlation coefficient for PM_{10} for 80 days simulation period and for daily averages has been increased up to 0.851 and in the case of $PM_{2.5}$, it has been increased up to 0.674.

Keywords: emissions, PM_{10} and $PM_{2.5}$, air quality models, air particles.

www.witpress.com, ISSN 1743-3541 (on-line)
doi:10.2495/AIR090341

$PM_{2.5}$ concentrations was developed after Feb. 1 until Feb. 15. During this period of time, Central Europe was under the influence of a high-pressure system coming from Russia through Poland and Southern Scandinavia. In the northern part of Germany, we found south-easterly winds and stable conditions with low winds. These meteorological conditions brought daily PM_{10} concentrations at about 40 μgm^{-3}. The second peak was characterized by a sharp gradient on PM_{10} concentrations after Feb. 15 and until March, 7. This episode reached daily PM_{10} concentrations up to 70 μgm^{-3}. The meteorological conditions on March, 2 (peak values) were characterized by a wind rotation composed of south-westerly winds from Poland over the north of Germany and north-westerly and Western winds in the Central part of Germany. Finally a third peak with values of about 65 μgm^{-3} on March, 27 started on March, 20 ending on April, 5. 2003 was having a similar structure and causes than the second one. The observational data used to compare with the modelling results is referred in San José et al. [25].

3 Emission data

In both models, we have applied the TNO emissions [17] as area and point sources with a geographical resolution of 0.125° latitude by 0.25° longitude and covering all Europe. The emission totals by SNAP activity sectors and countries agree with the baseline scenario for the Clean Air for Europe (CAFE) program [18]. This database gives the PM_{10} and $PM_{2.5}$ emission for the primary particle emissions. We also took from CAFE the PM splitting sub-groups, height distribution and the breakdown of the annual emissions into hourly emissions. The $PM_{2.5}$ fraction of the particle emissions was split into an unspecified fraction, elemental carbon (EC) and primary organic carbon (OC). The EC fraction of the $PM_{2.5}$ emissions for the different SNAP sectors was taken from [19]. For the OC fraction, the method proposed by [20] is applied as follows: an average OC/EC emission ratio of two was used for all sectors, i.e. the OC fraction were set as twice the EC fractions, except if the sum of the two fractions exceed the unity. In this case ($f_{EC} > 0.33$), f_{OC} was set as: $f_{OC} = 1 - f_{EC}$. With this prepared input, the WRF/CHEM and CMAQ took the information as it is. The hourly emissions are derived using sector-dependent, monthly, daily and hourly emission factors as used in the EURODELTA (http://aqm.jrc.it/eurodelta/) exercise. The differences with [25] simulations for MM5-CMAQ are established as follows: Albania, Croatia, Bosnia and Serbia use the Bulgaria daily factors; Turkey uses the Hungary daily factors; Belarus, Moldavia, Ukraine and Russia use the Romania daily factors; Germany use the Federal Republic of Germany daily factors; Czech Republic uses the Slovakia monthly factors. The VOC to TOC factor is 1.14. In case of WRF/CHEM the changes are the same than for MM5-CMAQ but the VOC to TOC factor in the VOC splitting scheme is changed to 3.2.

4 MM5-CMAQ and WRF-CHEM architectures and configurations

MM5 was set up with two domains: a mother domain with 60 x 60 grid cells with 90 km spatial resolution and 23 vertical layers and 61x61 grid cells with 30 km spatial resolution with 23 vertical layers. The central point is set at 50.0 N and 10.0 E. The model is run with Lambert Conformal Conical projection. The CMAQ domain is slightly smaller following the CMAQ architecture rules. We use reanalysis T62 (209 km) datasets as 6-hour boundary conditions for MM5 with 28 vertical sigma levels and nudging with meteorological observations for the mother domain. We run MM5 with two-way nesting capability. We use the Kain-Fritsch 2 cumulus parameterization scheme, the MRF PBL scheme, Schultz microphysics scheme and Noah land-surface model. In CMAQ we use clean boundary profiles for initial conditions, Yamartino advection scheme, ACM2 for vertical diffusion, EBI solver and the aqueous/cloud chemistry with CB05 chemical scheme. Since our mother domain includes significant areas outside of Europe (North of Africa), we have used EDGAR emission inventory with EMIMO 2.0 emission model approach to fill those grid cells with hourly emission data. The VOC emissions are treated by SPECIATE Version 4.0 (EPA, USA) and for the lumping of the chemical species, we have used the [24] procedure, for 16 different groups. We use our BIOEMI scheme for biogenic emission modeling. The classical, Atkin, Accumulation and Coarse modes are used (MADE/SORGAM modal approach). In WRF/CHEM simulation we have used only one domain with 30 km spatial resolution similar to the MM5. We have used the Lin et al. (1983) scheme for the microphysics, Yamartino scheme for the boundary layer parameterization and [23] for the biogenic emissions. The MOSAIC sectional approach is used with 4 modes for particle modeling.

5 Changes in model configurations

In the case of MM5-CMAQ the changes in the model simulations compared with the report of [25] affect only the emissions (as explained above) on the Kz (eddy diffusivity coefficient). The option to use the so-called KZMIN as detailed in CMAQ code is applied. If KZMIN is activated the Kz coefficient is calculated by:

$$Kz = KzL + (KzU - KzL) * UFRAC \tag{1}$$

where Kz is the eddy diffusivity in m^2s^{-1}. KZL is 0.5 (lowest) and KzU is 2.0 (highest). The UFRAC represents the percentage (rage 0-1) of urban landuse in the grid cell.

In the case of WRF/CHEM, the changes affect the microphysics scheme, substituting the [26] scheme by the WSM (WRF single moment) 5-class microphysics scheme [27]. 5 represents the number of water species predicted by the scheme. The Goddard/NASA radiation scheme is substituted by the Dudhia radiation scheme [28]. The FTUV photolysis rate [29] model is substituted by the FAST-J scheme [30].

4 MM5-CMAQ and WRF-CHEM architectures and configurations

MM5 was set up with two domains: a mother domain with 60 x 60 grid cells with 90 km spatial resolution and 23 vertical layers and 61x61 grid cells with 30 km spatial resolution with 23 vertical layers. The central point is set at 50.0 N and 10.0 E. The model is run with Lambert Conformal Conical projection. The CMAQ domain is slightly smaller following the CMAQ architecture rules. We use reanalysis T62 (209 km) datasets as 6-hour boundary conditions for MM5 with 28 vertical sigma levels and nudging with meteorological observations for the mother domain. We run MM5 with two-way nesting capability. We use the Kain-Fritsch 2 cumulus parameterization scheme, the MRF PBL scheme, Schultz microphysics scheme and Noah land-surface model. In CMAQ we use clean boundary profiles for initial conditions, Yamartino advection scheme, ACM2 for vertical diffusion, EBI solver and the aqueous/cloud chemistry with CB05 chemical scheme. Since our mother domain includes significant areas outside of Europe (North of Africa), we have used EDGAR emission inventory with EMIMO 2.0 emission model approach to fill those grid cells with hourly emission data. The VOC emissions are treated by SPECIATE Version 4.0 (EPA, USA) and for the lumping of the chemical species, we have used the [24] procedure, for 16 different groups. We use our BIOEMI scheme for biogenic emission modeling. The classical, Atkin, Accumulation and Coarse modes are used (MADE/SORGAM modal approach). In WRF/CHEM simulation we have used only one domain with 30 km spatial resolution similar to the MM5. We have used the Lin et al. (1983) scheme for the microphysics, Yamartino scheme for the boundary layer parameterization and [23] for the biogenic emissions. The MOSAIC sectional approach is used with 4 modes for particle modeling.

5 Changes in model configurations

In the case of MM5-CMAQ the changes in the model simulations compared with the report of [25] affect only the emissions (as explained above) on the Kz (eddy diffusivity coefficient). The option to use the so-called KZMIN as detailed in CMAQ code is applied. If KZMIN is activated the Kz coefficient is calculated by:

$$Kz = KzL + (KzU - KzL) * UFRAC \tag{1}$$

where Kz is the eddy diffusivity in m^2s^{-1}. KZL is 0.5 (lowest) and KzU is 2.0 (highest). The UFRAC represents the percentage (rage 0-1) of urban landuse in the grid cell.

In the case of WRF/CHEM, the changes affect the microphysics scheme, substituting the [26] scheme by the WSM (WRF single moment) 5-class microphysics scheme [27]. 5 represents the number of water species predicted by the scheme. The Goddard/NASA radiation scheme is substituted by the Dudhia radiation scheme [28]. The FTUV photolysis rate [29] model is substituted by the FAST-J scheme [30].

6 Model results

The comparison between daily average values (averaged over all monitoring stations) of PM_{10} concentrations and modeled values has been performed with several statistical tools such as: Calculated mean/Observed mean; Calculated STD/Observed STD; bias; squared correlation coefficient (R2); RMSE/Observed mean (Root Mean Squared Error); percentage within +/- 50% and number of data sets. Figure 1 shows the comparison between PM_{10} observed averaged daily values and the modeled values by MM5-CMAQ. The results show that for MM5-CMAQ, the new configuration related to emission data and eddy diffusivity improves the correlation coefficient from 0.828 to 0.851 but the pattern show a substantial improvement with the central peak much closer to the observed data. Figure 2 shows the comparison between observed and modeled average daily data for the episode with the new configuration for the WRF/CHEM model. The results show a much better correlation coefficient going from 0.782 to 0.852 with the new configuration. Figures 3 and 4 show similar results for $PM_{2.5}$. In case of MM5-CMAQ the improvement is from 0.608 to 0.674 and for WRF/CHEM the change is from 0.760 to 0.759. These results show that the new configuration is substantially better than the previous one. New experiments are needed to determine the impact of emissions and the eddy diffusivity respectively.

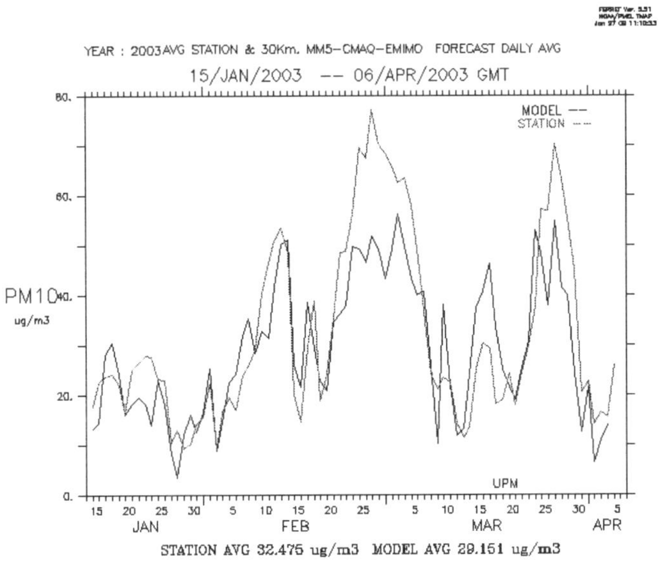

Figure 1: Comparison between daily averaged observed PM_{10} concentrations and model results produced by MM5-CMAQ. The model gets closer to the maximum peak compared with the previous simulation in [25].

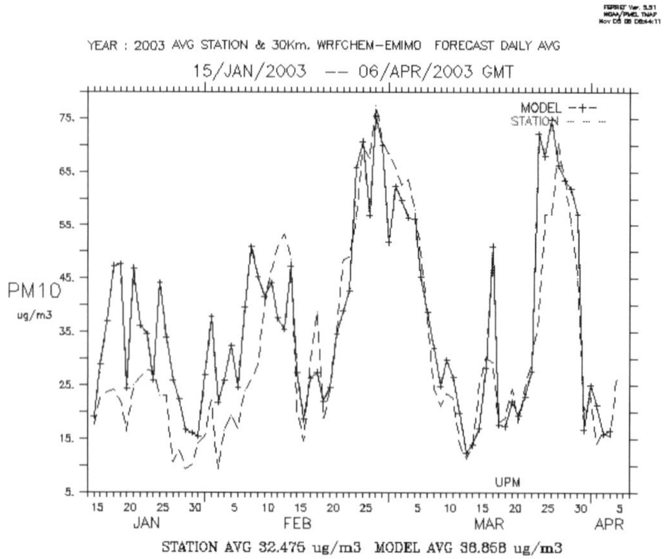

Figure 2: Comparison between daily averaged observed PM_{10} concentrations and model results produced by WRF/CHEM. The model captures even better than in the previous simulation [25] the magnitude of the PM_{10} peaks.

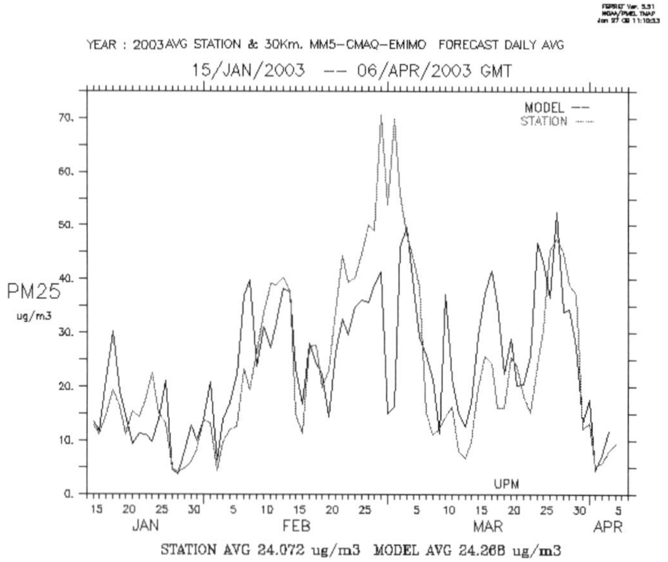

Figure 3: Comparison between daily averaged observed $PM_{2.5}$ concentrations and model results produced by MM5-CMAQ. The model gets closer to the simulation performed in [25].

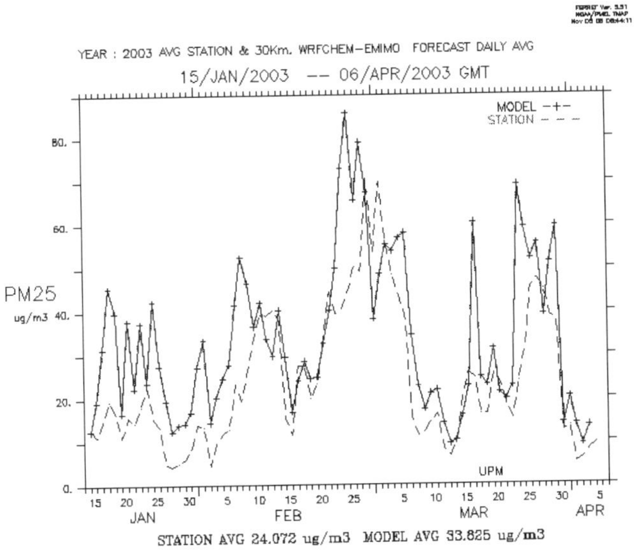

Figure 4: Comparison between daily averaged observed $PM_{2.5}$ concentrations and model results produced by WRF/CHEM. The model overestimates a little bit the observed data but the correlation coefficient gets a light improvement (up to 0.759).

7 Conclusions

We implemented and re-ran two different models (MM5-CMAQ and WRF-CHEM) for the same episode over the northern part of Germany during the winter period of 2003 (Jan. 15-Apr. 5, 2003). The comparison between these simulations and those performed in [25] produce the following results: we have improved substantially the correlation coefficients for the daily averages when comparing observed and modelled data for both models. The WRF/CHEM continue to show better results than MM5-CMAQ but the peaks for PM_{10} and $PM_{2.5}$ for MM5-CMAQ are getting closer to the observed peaks. The patterns for MM5-CMAQ have improved substantially compared with the results obtained in [25]. New experiments are necessary to determine the impact of eddy diffusivity and emission inventory on the new results.

Acknowledgements

We would like to thank Dr. Peter Builtjes (TNO, The Netherlands) for his initial guidance and suggestion for this experiment and also COST 728 project (EU) where the inter-comparison experiment was proposed. The authors thankfully acknowledge the computer resources, technical expertise and assistance provided by the Centro de Supercomputación y Visualización de Madrid (CeSVIMa) and the Spanish Supercomputing Network.

References

[1] Collins W.J., D.S. Stevenson, C.E. Johnson and R.G. Derwent, Tropospheric ozone in a global scale 3D Lagrangian model and its response to NOx emission controls, *J. Atmos. Chem.* **86** (1997), 223-274.

[2] Derwent R., and M. Jenkin, Hydrocarbons and the long-range transport of ozone and PAN across Europe, *Atmospheric Environment* **8** (1991), 1661-1678.

[3] Gardner R.K., K. Adams, T. Cook, F. Deidewig, S. Ernedal, R. Falk, E. Fleuti, E. Herms, C. Johnson, M. Lecht, D. Lee, M. Leech, D. Lister, B. Masse, M. Metcalfe, P. Newton, A. Schmidt, C Vandenberg. and R. van Drimmelen, The ANCAT/EC global inventory of NOx emissions from aircraft, *Atmospheric Environment* **31** (1997), 1751-1766.

[4] Gery M.W., G.Z. Whitten, J.P. Killus and M.C. Dodge, A photochemical kinetics mechanism for urban and regional scale computer modelling, *Journal of Geophysical Research* **94** (1989), D10, 12925-12956.

[5] Grell, G.A., J. Dudhia and D.R. Stauffer, A description of the Fifth-Generation Penn State/NCAR Mesoscale Model (MM5), NCAR/TN- 398+ STR. *NCAR Technical Note*, 1994.

[6] Jacobson M.Z. and R.P. Turco, SMVGEAR: A sparse-matrix, vectorized GEAR code for atmospheric models, *Atmospheric Environment* **28**(1994), 2, 273-284.

[7] Langner J., R. Bergstrom and K. Pleijel, European scale modeling of sulfur, oxidized nitrogen and photochemical oxidants. Model development and evaluation for the 1994 growing season, *SMHI report RMK No. 82*, Swedish Met. And Hydrol. Inst., SE-601 76 Norrkoping, Sweden, (1998).

[8] Roemer M., G. Boersen, P. Builtjes and P. Esser, *The Budget of Ozone and Precursors over Europe Calculated with the LOTOS Model.* TNO publication P96/004, Apeldoorn, The Netherlands, 1996.

[9] San José R., L. Rodriguez, J. Moreno, M. Palacios, M.A. Sanz and M. Delgado, Eulerian and photochemical modelling over Madrid area in a mesoscale context, *Air Pollution II, Vol I., Computer Simulation, Computational Mechanics Publications, Ed. Baldasano, Brebbia, Power and Zannetti.*, 1994, 209-217.

[10] San José R., J. Cortés, J. Moreno, J.F. Prieto and R.M. González, Ozone modelling over a large city by using a mesoscale Eulerian model: Madrid case study, *Development and Application of Computer Techniques to Environmental Studies, Computational Mechanics Publications, Ed. Zannetti and Brebbia,* 1996, 309-319.

[11] San José, R., J.F. Prieto, N. Castellanos and J.M. Arranz, Sensitivity study of dry deposition fluxes in ANA air quality model over Madrid mesoscale area, *Measurements and Modelling in Environmental Pollution,* Ed. San José and Brebbia, 1997, 119-130.

[12] Schmidt H., C. Derognat, R. Vautard and M. Beekmann, A comparison of simulated and observed ozone mixing ratios for the summer 1998 in Western Europe, *Atmospheric Environment* **35** (2001), 6277-6297.

www.witpress.com, ISSN 1743-3541 (on-line)

[13] Stockwell W., F. Kirchner, M. Kuhn and S. Seefeld, A new mechanism for regional atmospheric chemistry modeling, *J. Geophys. Res.* **102** (1977), 25847-25879.

[14] Walcek C., Minor flux adjustment near mixing ration extremes for simplified yet highly accurate monotonic calculation of tracer advection, *J. Geophys. Res.* **105** (2000), 9335-9348.

[15] Janjic, Z. I., J. P. Gerrity, Jr. and S. Nickovic, 2001: An Alternative Approach to Nonhydrostatic Modeling. Monthly Weather Review, Vol. 129, 1164-1178.

[16] Byun, D.W., J. Young, G. Gipson, J. Godowitch, F. Binkowsky, S. Roselle, B. Benjey, J. Pleim, J.K.S. Ching, J. Novak, C. Coats, T. Odman, A. Hanna, K. Alapaty, R. Mathur, J. McHenry, U. Shankar, S. Fine, A. Xiu, and C. Lang. 1998. *Description of the Models-3 Community Multiscale Air Quality (CMAQ) model.* Proceedings of the American Meteorological Society 78th Annual Meeting Phoenix, AZ, Jan. 11-16, 264-268.

[17] Visscherdijk, A. and H. Denier van der Gon, 2005. Gridded European anthropogenic emission data for NOx, SO2, NMVOC, NH3, CO, PM10, PM2.5 and CH4 for the year 2000. TNO-report B&O-AR, 2005/106.

[18] Amann, M., Bertok, I., Cofala, J., Gyarfas, F., Heyes, C., Klimon, Z., 2005. Baseline Scenarios for the Clean Air for Europe (CAFE) Programme. Final Report, International Institute for Applied Systems Analysis, Schlossplatz 1, A-2361 Laxenburg, Austria.

[19] Schaap, M., H. Denier van der Gon, A. Visschedijk, M. van Loon, H. ten Brink, F. Dentener, J. Putaud, B. Guillaume, C. Liousse, P. Builtjes, 2004a. Anthropogenic Black Carbon and Fine Aerosol Distribution over Europe, J. Geophys. Res., 109, D18207, doi:10.1029/2003JD004330.

[20] Beekmann, M., Kerschbaumer, A., Reimer, E., Stern, R., Möller, D., 2007. PM Measurement Campaign HOVERT in the Greater Berlin area: model evaluation with chemically specified observations for a one year period. Atmos. Chem. Phys. 7, 55-68.

[21] Spindler, G., K. Mueller, E. Brueggemann, T. Gnauk, H. Herrmann, 2004. Long-term size-segregated characterization of PM10, PM2.5, and PM1 at the IfT research station Melpitz downwind of Leipzig (Germany) using high and low-volume filter samplers. Atmospheric Environment 38, 5333–5347.

[22] Putaud, J., F. Raesa, R. Van Dingenen, E. Bruggemann, M. Facchini, S. Decesari, S. Fuzzi, R. Gehrig, C. Hueglin, P. Laj, G. Lorbeer, W. Maenhaut, N. Mihalopoulos, K. Mueller, X. Querol, S. Rodriguez, J. Schneider, G. Spindler, H. ten Brink, K. Torseth, A. Wiedensohler, 2004. A European aerosol phenomenology - 2: chemical characteristics of particulate matter at kerbside, urban, rural and background sites in Europe. Atmospheric Environment 38, 2579–2595.

[23] Guenther et al., 1995 A. Guenther, C.N. Hewitt, D. Erickson, R. Fall, C. Geron, T. Graedel, P. Harley, L. Klinger, M. Lerdau, W.A. McKay, T. Pierce, B. Scholes, R. Steinbrecher, R. Tallamraju, J. Taylor and P.

Zimmerman, A global model of natural volatile organic compound emissions, *Journal of Geophysical Research* **100** (1995), pp. 8873–8892.

[24] Carter, W. P. L. (2007): “Development of the SAPRC-07 Chemical Mechanism and Updated Ozone Reactivity Scales,” Final report to the California Air Resources Board Contract No. 03-318. August. Available at http://www.cert.ucr.edu/~carter/SAPRC.

[25] San José, R., J.L. Pérez, J.L. Morant and R.M. González (2008): “Elevated PM10 and PM2.5 concentrations in Europe: a model experiment with MM5-CMAQ and WRF/CHEM” WIT Transactions on Ecology and the Environment, Vol. 116, pp. 3-12. ISSN: 1743-3541 (on-line).

[26] Lin, Y.-L., R. D. Rarley, and H. D. Orville, 1983: Bulk parameterization of the snow field in a cloud model. *J. Appl. Meteor*,, 22, 1065-1092.

[27] Hong, S-.Y., J. Dudhia, S.-H. Chen, 2004: A revised approach to ice-microphysical processes for the bulk parameterization of cloud and precipitation., *Mon. Wea. Rev.,* 132, 103-120.

[28] J. Dudhia. A non-hydrostatic version of the Penn State-NCAR mesoscale model: validation tests and simulation of an Atlantic cyclone and cold front. *Monthly Weather Review*, 121:1493–1513, 1993.

[29] Tie, X. X., Madronich, S., Walters, S., Y., Z. R., Rasch, P., and Collins, W.: Effect of clouds on 5 photolysis and oxidants in the troposphere, J. Geophys. Res., 108(D20), 4642.

[30] Wild, O., Zhu, X., and Prather, M. J.: Fast-J: Accurate simulation of in- and below-cloud photolysis in Global Chemical Models, J. Atmos. Chem., 37, 245–282.

Monitoring of PM_{10} air pollution in small settlements close to opencast mines in the North-Bohemian Brown Coal Basin

S. Hykyšová[1] & J. Brejcha[2]
[1]*Environmental Centre Most and Kralupy, Brown Coal Research Institute, j.s.c., Czech Republic*
[2]*Accredited Testing Laboratory, Brown Coal Research Institute, j.s.c., Czech Republic*

Abstract

In the past, due to air pollution, the North-Bohemian Brown Coal Basin used to belong to the famous zone called the "Black Triangle" which also covered the lower parts of Silesia and Saxony. The air pollution was for the greatest part caused by large industrial sources where its impact on the forest ecosystems gained a cross-border character. At the beginning of 1990s, after reduction measures in industry and the implementation of stricter environmental laws, the air condition started to improve rapidly. Due to extensive opencast brown coal mining and the presence of large combustion resources the area is still classed as a region with a poor air quality mainly due to increased concentrations of dust particles. In recent years the increasing effect of local combustion sources on air pollution may be seen. Nevertheless the air pollution in small settlements has not been mapped sufficiently. The article is based on the data collected from mobile measurements of air purity executed within the remit of the subgroup "emissions-immissions" of the research project "Research of physical and chemical features affected by coal mining, its use and impact on environment in the North-Western Bohemia region". The measurement was performed in the selected communities close to the opencast mines in the period 2004–2007. The aim was to evaluate the immissions status in the small settlements selected from the North-Bohemian region, to appraise the predicative value of collected results and prepare a methodology concept of mobile measurements to describe the local immission situation. This article extends the information gained from mobile measurement and the data from the stationary measuring stations and also takes into account significant sources of pollution such as coal power plants in the North-Bohemian Brown Coal Basin. The results collected during the series of measurements in the course of heating and non-heating seasons are simultaneously compared with the data regarding the manner of heating and type of the fuel used for heating in the communities.

Keywords: air pollution, immissions, emissions, local combustion furnaces, air pollution monitoring, dust particles, PM_{10}, coal opencast mining, Black Triangle.

WIT Transactions on Ecology and the Environment, Vol 123, © 2009 WIT Press
www.witpress.com, ISSN 1743-3541 (on-line)
doi:10.2495/AIR090351

1 Introduction

Continuous measurement of air pollution is generally focused on residential areas with high population density. Sufficient long-term data are available from stationary measuring stations from which we can determine the air quality, predict long-term development trends and consequently design the measures to improve the air pollution condition.

The air pollution condition of small communities up to 10 thousand inhabitants where almost 46,8% of the CR population lives [3], has not been mapped yet. The work of Kotlík *et al.* is the first attempt to map this condition in the Czech Republic [5]. This work confirmed the theory that the immission pollution in small settlements is, with regard to air pollution due to suspended particles, heavy metals and polyaromatic hydrocarbons, often comparable with the pollution in cities with much higher and continuous effect of emissions from transport. High levels of solid fuels (mainly brown coal) for heating in local combustion furnaces contribute to air pollution in small communities.

The aim of this work is to map the air pollution from PM_{10} suspended particles in small communities of the North-Bohemian Brown Coal Basin. The air quality in this region, which used to be a part of the so-called Black Triangle, is distinctively affected by the presence of large combustion facilities and opencast brown coal mines. Moreover the topographic conditions of the bottom of the Krušné Hory Mountains create conditions for more frequent occurrences of inversion weather with unfavourable conditions for the dispersion of pollutants.

2 Characteristics of the region

The so called Black Triangle, which at the end of the last century reached Northern Bohemia, lower part of Silesia and Saxony, gained its name due to the extreme air pollution which affected the condition of forests and also human health. Opencast mines of the North-Bohemian Brown Coal Basin ten years ago covered an area of almost 300 square kilometres. In the period from 1860 up to now 3.85 billion tonnes of brown coal have been excavated. Coal mining required the removal of 106 communities including the royal town of Most and the emigration of 90 thousand inhabitants [4].

At the beginning of 1990s the quality of the environment started to improve partially due to the decline of industrial production and the limitation on brown coal and bunker oil consumption in large sources of air pollution. That was thanks to the implementation of direct measures to decrease emissions and also end technologies.

In spite of these environmental measures the region of the North-Bohemian Brown Coal Basin still belongs to the areas with deteriorated air quality, mainly due to dust particles pollution. Mainly large energy sources, vast opencast brown-coal mines, refineries and petrochemical works affect the environment pollution. In fig. 1 apart from the measurement locations, which shall be discussed in the methodology chapter, there are large stationary sources

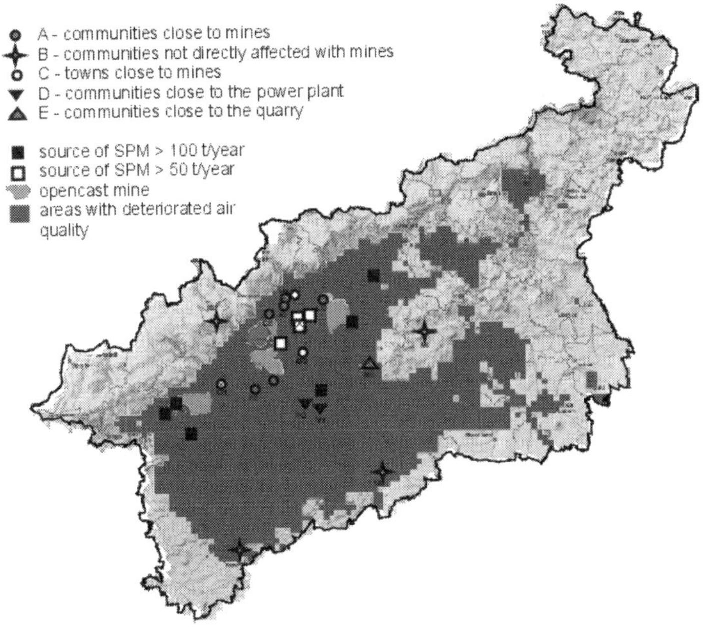

Figure 1: The area of interest with the denoted areas with deteriorated air quality, pollution sources and measurement locations.

contributing the most significantly to the emissions of suspended particulate matters denoted. We should not forget local sources such as individual heating and transport, of course. The latter sources, as opposed to the large and medium ones, show the growth of emissions due to motor transport but also insufficient legislation in the control and fines of the local combustion furnace operators.

3 Trend of PM_{10} air pollution in the North-Bohemian Brown Coal Basin

3.1 Progress of emissions from SPM and immissions of PM_{10}

In the North-Bohemian Brown Coal Basin (and also in other districts of the Czech Republic) the immission limit for the suspended particles PM_{10}, for which zero tolerance applies from 1.1.2005, has been exceeded. The level of air pollution with PM_{10} is illustrated in fig. 2. Since 2000 there has been stagnancy or even an increase in solid pollutant emissions in certain areas of the North-Bohemian Brown Coal Basin. The North-Bohemian Brown Coal Basin region is burdened with the specific emissions of solid pollutants three times more than the average in the Czech Republic.

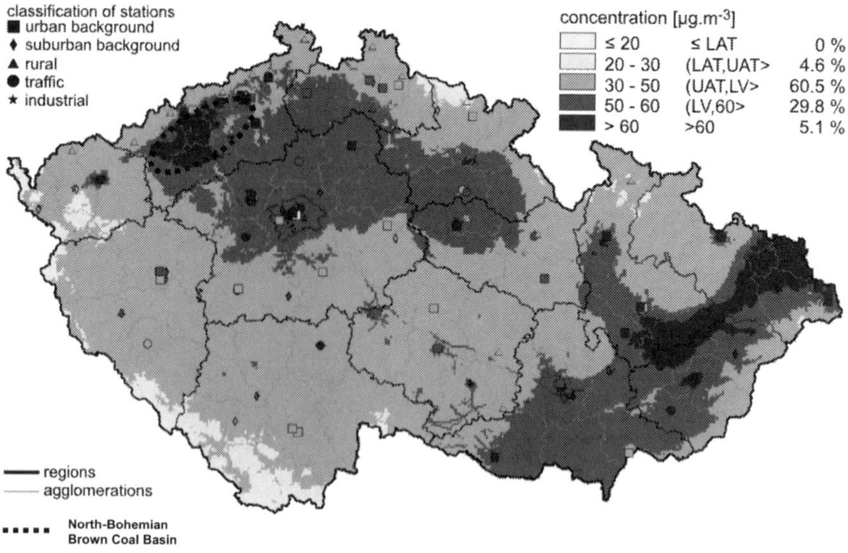

Figure 2: Field of the 36th highest 24-hour concentration of PM_{10} in the CR – year 2005. *Source: Czech Hydrometeorological Institute.*

3.2 Contribution of source groups on air pollution PM_{10}

Local combustion furnaces contribute to air pollution with the suspended particles PM_{10} in the Czech Republic for the most part (pursuant to the Czech legislation the sources which belong to the category of small sources of pollution denoted as REZZO 3). The contribution of these sources has been increasing since the 1990s. It is mainly caused by the inhabitants reverting to the cheaper type of heating with solid fuels mainly to brown coal. The conditions of combustion in local combustion furnaces do not enable them to achieve the efficient burning and cleaning of flue gases as in the industrial sources. Moreover there is often waste combustion. In the Czech Republic local combustion furnaces contribute to air pollution with suspended particles from 38%, to the pollution with polyaromatic hydrocarbons even from 66% [11]. The other new member states of EU show similar contribution while the contribution of the local combustion furnaces of the original member states is lower [1].

3.3 Progress of brown coal mining and overburden disposal

The presence of opencast mines has a negative impact on air pollution with dust particles in the North-Bohemian Brown Coal Basin. Dust nuisance is caused by lots of processes related to coal mining. The sources of dust nuisance in the opencast mines can be classified as passive (temporary coal disposal site, eroded slopes and dumps) and active sources of emissions (coal preparation plants, excavators, conveyor belts and other mining facilities) [6]. Most dust nuisance comes from the overburden disposal. That is because of the higher volume of overburden rock involved compared to coal mining but also different consistency of both materials. We should not forget the geological characteristics of overburden [7]. Coal and overburden mining progress in the North-Bohemian Brown Coal Basin is shown in fig. 3. In the monitored period 2004 – 2007 the annual coal and overburden exploitation was ca on the same level – 38 million tonnes of coal and 115 million m^3 of overburden.

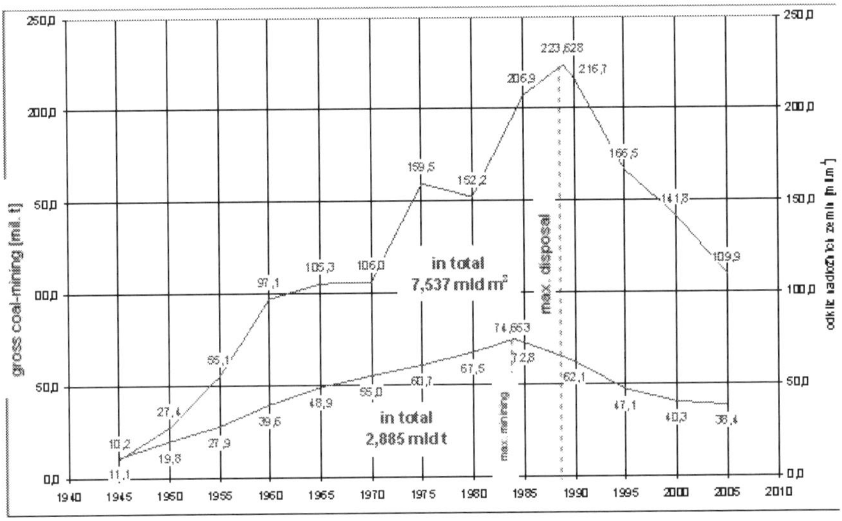

Figure 3: Progress of brown coal mining and overburden rocks disposal in the North-Bohemian Brown Coal Basin. *Source: Chytka, Valášek, 2009.*

4 Methodology

4.1 Groups qualification

Mobile measurement focused on the immission condition monitoring in the small settlements of the North-Bohemian Brown Coal Basin was performed in the series in the years 2004–2007 at 10 measuring locations. Data from 8 measuring locations were used for this work complemented with the data from the same number of measuring locations where continuous measurement within the network of automated immission monitoring of the Czech Hydrometeorological Institute is provided [10].

The locations were divided according to the distance from the opencast mine, the type of development and contribution of large combustion sources on pollution with PM_{10} [9] into five categories – see table 1. Then the contribution of solid fuels on heating in local combustion furnaces was found out at the communities [8]. It is a percentage of number of people using coal or wood for heating. The data do not relate to the total number of inhabitants but only that group of inhabitants which does not use central heating.

As the source operating conditions vary with the changing seasons of the year and also different climate and dispersal conditions, the results were classified not only from the total overview and the view of time trend development but also with respect to the seasons. The evaluation results were divided into three groups corresponding the following periods:

- *Heating period (ca from 15. 10. to 15. 3.)* – mainly with increased output of the all category sources operations from small sources (local combustion furnaces) to large sources (heating plants, power plants). Climate and dispersion conditions often burden the pollutants dispersion.
- *Non-heating period (ca from 15. 5. to 15. 9.)* – with prevailing low output of large and medium size sources and also small sources mainly local combustion furnaces. Climate and dispersion conditions, (qualified only to a certain limit), creation of poor dispersion conditions. Higher summer temperatures can contribute to the increase of dust nuisance from opencast mines, construction sites and increased secondary dust nuisance.
- *Transition period (ca from 15. 9. to 15. 10. and from 15. 3. to 15. 5.)* – with varying output of large and medium and also small sources. Climate and dispersion conditions causing frequent occurrence of temperature inversions and fog thereby worsening the conditions for pollutants dispersion.

4.2 Mobile measurement

The measurement of suspended particles was performed with a continuous β-dust-meter FH 62 I R with heated probes and pre-separator of particles bigger than 10 μm PM_{10} and ESM Eberline. The draw-offs of dust for indicative definition of metals in the air were executed (Ni, Be, Cd, As, Hg and Pb). Other basic pollutants were also monitored. This article is focused on the dust particles PM_{10} and selected heavy metals.

Meteorological unit THIES was used for measuring wind speed and direction, the temperature of air, relative air humidity and barometric pressure. The metering device was located in the measuring vehicle Mercedes 711D.

Table 1: Review of measuring locations.

Community	Estate character	Distance from M, D/ P * in km	No. of measurements 2004 - 2007	No. of people using solid fuels for heating	Contribution of power plants on PM_{10} immissions in % (average – 2004 – 2006)
A - Communities close to mines					
Ce	community	< 1/ > 5	105	274	9 - 17
St	community	< 1/ > 5	41	93	23 - 32
Pe	community	< 5/ < 10	37	8	23 – 32
Lo	community	< 2/ < 10	continuous	1081	8 - 15
Jo	estate housing	< 5/ > 5	20	N	9 - 17
Ha	estate housing	< 5/ <10	22	N	9 - 17
B - Communities non-affected directly with mines					
Mi	community	> 20/ > 10	34	283	20 -28
Ru	- **	> 5/ < 15	continuous	-	32 - 40
Sm	community	> 30/ >30	continuous	55	10 - 18
Sj	community	> 20/ < 25	continuous	0	10 - 18
C – Towns close to mines					
Mo	town	< 5/ < 5	35	290	13 - 27
Ch	town	< 5/ < 10	continuous	555	27 - 35
Li	town	< 5/ < 10	continuous	584	10 - 18
D – Communities close to power plants					
Bl	community	< 10/ < 5	continuous	23	30 - 38
Vy	community	> 10/ < 5	continuous	146	30 - 38
E – Community close to quarry					
Me	community	< 1***/ > 10	34	253	32 - 40

Legend: * M – opencast brown coal mine, D – mine dump, P – power plant; ** - measuring locations places in the mountains out of the urban area, *** - distance from the quarry, N - non-detected

4.3 Basic data files

Data files with short-term 3-minute values were collected from mobile measurements at the measuring locations. These short-term values were analysed for each measuring location and converted into the following output qualities:

- average concentrations from the single measurements
- average concentration from the whole block of measurements
- average concentration in the periods selected according to the operating regime of energy sources (heating, transition and non-heating periods)
- direction characteristics (concentration rosette) - average concentrations from the selected ranges of wind direction in the appraised period
- climate characteristics - average concentrations by the selected range of climate conditions in the appraisal period.

From the files of daily values for the classified periods an arithmetic mean, standard deviation and the number of values, median or geometric average, or quantile corresponding the allowed number of surplus a year according to the specification of the immission limit were calculated or analyzed.

The direction contributions were classified only from the files with 3-minute values where we can expect close dependency between the measured concentrations and wind direction.

5 Measurement results and their evaluation

The results of PM_{10} concentration measurements in the settlement groups monitored are illustrated in fig. 4. Based on these it is obvious that the highest concentrations are seen in the E group, the community closest to the quarry (< 1 km), where also the highest contribution of pollution from the power plants was found and also the highest contribution of solid fuels to the heating in local combustion furnaces (86%). The second highest PM_{10} concentration was found in group A – communities close to the opencast brown coal mines with the exception of non-heating period when higher concentration was found in group D – communities close to the power plant. The communities far from the mines displayed the lowest PM_{10} concentration in all the periods whereas the contribution of power plants on PM_{10} pollution was about the same for both the groups.

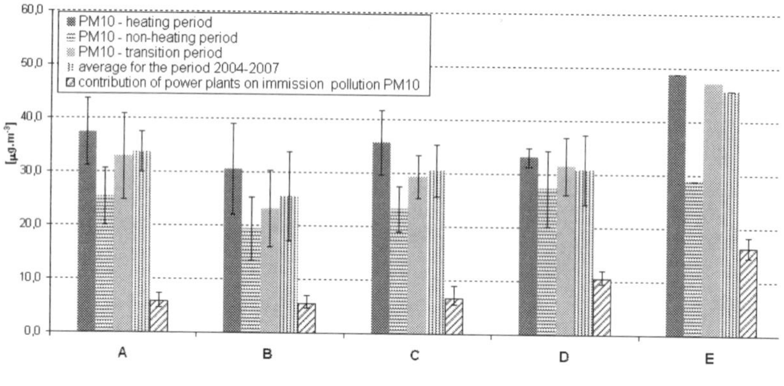

Figure 4: Average concentration of PM_{10} in the settlements monitored.

It is also obvious from the results that the values have a significantly seasonal character with the highest PM_{10} values in the heating period and increased values in the transition period. The ratio of PM_{10} values in the heating and non-heating periods was greatly different in some groups. The E group displayed the most distinctive difference where the values in the heating period reached 169% of values found in the non-heating period. Compared to the group D, only 121% growth of values was found in the heating period.

The results from the small settlements were compared where we managed to find out the details of the fuel types used for heating in local combustion furnaces. It is clear from fig. 5 that for the small settlements from groups B, D and E (communities unaffected directly by opencast mines, communities close to the power plants and quarry) we can state that heating using solid fuels impacts PM_{10} pollution in these communities during the heating and transition period which corresponds with the work of Kotlík *et al.* [5]. Solid fuel combustion in local combustion furnaces, of course, contributes to the pollution and in the communities close to the opencast mines, but in the monitored group the PM_{10} values did not reflect this relationship. The contribution of large stationary sources and opencast mines is significantly reflected and also variable characteristics of wind direction due to the topographic conditions at the bottom of the Krušné Hory mountains.

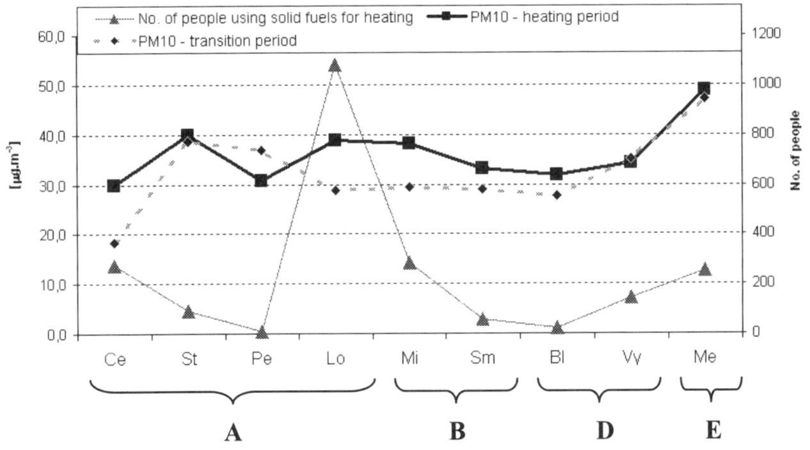

Figure 5: Comparison of PM_{10} concentrations in heating and transition periods related to solid fuel heating.

Fig. 6 compares the concentration rosettes of PM_{10} found at the selected stations during the heating, non-heating and transition periods. Also here a significant seasonality of the measured values is conspicuous not only with respect to different concentration levels but also in relation to various shape of rosette. It is obvious from the concentration rosettes that in summer and winter other sources dominate air pollution. This group is more apparent at the community from group B, so the community is here non-affected with the mine and the community from group E, close to a small quarry. Both these communities are also typical with a high number of people using solid fuels for heating.

Since 2000 the complaints from citizens about air pollution were also monitored. These were accepted through a free hot line of the Environmental Centre Most which has been working as an information centre about environment in the Euroregion Krušnohoří-Erzgebirge. Most of the complaints related to the increased emissions from local combustion furnaces mainly from small settlements. The callers often pointed out illegal waste combustion in the local combustion furnaces which has been, for the time being, very difficult to prosecute. Since 2000 ca. 8000 questions were asked of which ca 80% related to the air quality which proves that the citizens of the North-Bohemian Brown Coal Basin considered the problem of air pollution a priority.

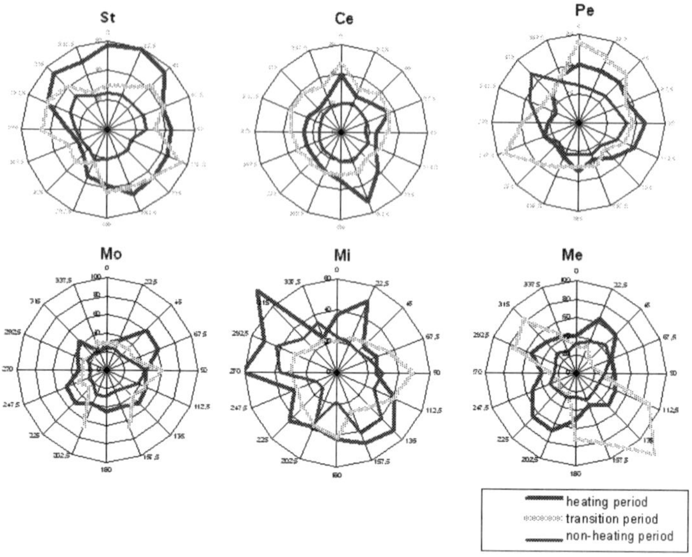

Figure 6: Comparison of directional characteristics PM_{10} in the periods.

6 Conclusion

Statistical analyses of PM_{10} concentrations in the North-Bohemian Brown Coal Basin gathered in the period 2004 – 2007 shows that in the winter (heating) period there is a significant increase in the concentration of substances generated or released in the combustion processes. It is a consequence of significantly more intense operation of energy sources and usually more frequent occurrences of poor dispersion conditions. Local combustion furnaces for solid fuels are proved markedly poor for small settlements on the local scale. In the case of villages situated close to opencast mines it is more problematic to define the exact contribution of the pollution sources. In the case of small settlements not exposed to the effect of opencast mines, significantly larger inter-season variations in PM_{10} values are noted. The results also warn of significant PM_{10} pollution in communities close to a smaller quarry.

The results of this work also led to the conclusion related to the methodology of stratified measurement. That cannot be, in the sense of standard procedure, generally used for all cases, defined mainly in such an exposed region as North-Western Bohemia. Therefore a stratified measurement methodology was proposed for an additional measurements of the results gained from the measurement networks of Brown Coal Research Institute situated at the edges of municipal development in the surroundings of opencast mines. The stratified measurement will be focused on the evaluation of immission situation from the perspective of suspended particles PM_{10} concentrations at the places located a maximum of ca 2 km from the stationary locations, i.e. in the area where continuously measured local meteorological data from the measurement network are valid.

Acknowledgements

This article was prepared within the works on the state task of the Ministry of Education, Youth and Physical Education of the Czech Republic "Research of physical and chemical features affected by coal mining, its use and impact on environment in the North-Western Bohemia region".

References

[1] Amann, M. et al. *Health risks of ozone from long-range trans-boundary air pollution*. Copenhagen: WHO Regional Office for Europe, 2008. 93 s., XIII. Available from WWW: <http://www.euro.who.int/Document /E91843.pdf>. ISBN 978 92 890 4289 5.

[2] Brejcha, J. Summary of data from mobile immission?? measurements in selected small settlements of the North-Bohemian region In *Brown Coal Bulletin*. Most : Research Institute for Brown Coal j.s.c., 2009. pp. 25-36. ISSN 1213-1660.

[3] Halásek, J. Small Lexicon of Communities in the CR 2008 [online]. 15.12.2008 [cit. 2009-04-15]. Available from WWW: <http://www.czso.cz/csu/2008edicniplan.nsf/p/1302-08>.

[4] Goeckeler, S., Reeve, H. *Libkovice: Zdař Bůh*. 1. edition. Prague : DIVUS, 1997. 184 p. ISBN 80-902379-1-6.

[5] Kotlík, B. et al.: Air duality in Czech villages – reasons and ideas for corrections. Air protection, 2006, No. 4, p. 5-8.

[6] Matějíček, L. et al. Spatio-Temporal Modelling of Dust Transport over Surface. In *Sensors*. [s.l.] : Molecular Diversity Preservation International, 2008. s. 3830-3847. Available from WWW: <http://www.mdpi.com /1424-8220/8/6/3830>. ISSN 1424-8220.

[7] Reynolds, L. et al (2003) Toxicity of airborne dust generated by opencast coal mining, *Mineralogical Magazine*, April 2003, Vol. 67, 2, pp.141-152

[8] *Census of people, houses and flats 2001* [online]. Czech Statistic Office, c2005 [cit. 2009-02-20]. Available from WWW: <http://www.czso.cz /sldb/sldb2001.nsf/openciselnik?openform&:cz042>.

[9] *ČEZ Group - Power plants and the environment : Contribution of immission concentration* [online]. ČEZ a.s., c2009 [cit. 2009-03-16].

Available from WWW: <http://www.cez.cz/cs/energie-a-zivotni-prostredi/zivotni-prostredi/sledovani-parametru-pro-ochranu-ovzdusi/podil-na-imisni-koncentraci.html>.

[10] *Columnar Catalogue* [online]. Czech Hydrometeorological Institute, c2006, 30.06.2006 [cit. 2009-02-20]. Available from WWW: <http://www.chmi.cz/uoco/isko/tab_roc/tab_roc.html>.

[11] *Air Pollution Has No Borders* [online]. version 1.0.1. Prague : Ministry of Environment CR, 1997 [cit. 2009-04-16]. Available from WWW: <http://www.env.cz/vytapeni>.

PAH concentrations and seasonal variation in PM_{10} in the industrial area of an Italian provincial town

M. Rotatori, E. Guerriero, S. Mosca, F. Olivieri, G. Rossetti, M. Bianchini & G. Tramontana
Institute for Atmospheric Pollution, CNR, Italy

Abstract

A comprehensive campaign was performed from November 2006 to September 2007 in four sites in the industrial area of an Italian provincial town in order to fully understand the state and characteristics of the air pollution. As part of the work, 13 PAHs in PM_{10} aerosols were chemically characterized in this study by GC/MS. The main aim of this study was identifying the seasonal variation of PAHs concentration and the distribution pattern associated with PM_{10}.
Keywords: PM10, PAHs, seasonal variation, GC/MS, BaP.

1 Introduction

Investigation of the origin of air pollution, its major sources and the importance of pollution from distant regions became increasingly important due to new limits that were set for PM_{10} concentrations in EC-Directive 99/30/CE [1]. The Italian Ministry of Environment, as a response to the first daughter directive on air quality proposed by the European Commission (96/62/CE [2]), published the DM n. 60/2002 [3] in order to assess PM_{10} pollution and monitor the ambient air quality in Italy. The new directive proposed the measurement of PM_{10} concentrations and set limit values to be reached towards health protection. However, it needs to be considered that the chemical composition of particles can also induce health related effects. Polycyclic aromatic hydrocarbons (PAHs) are one of the several hundred organic compounds that have been identified in the particulate matter. PAHs are emitted into the atmosphere by the incomplete combustion of organic materials and fossil fuels such as automobile fuel

WIT Transactions on Ecology and the Environment, Vol 123, © 2009 WIT Press
www.witpress.com, ISSN 1743-3541 (on-line)
doi:10.2495/AIR090361

combustion, industrial combustion, wood and coal burning etc. PAHs in the atmosphere can be present either in the gaseous phase or adsorbed on atmospheric particles depending on their respective chemical and physical properties. Scientific research has proved that many higher molecular weight PAHs with four or more aromatic rings are carcinogenic and they are mostly attached to the particle phase in the atmosphere [4].

The International Agency for Research in Cancer (IARC) has classified a number of individual PAH compounds as probable human carcinogens and a number of "common mixtures of substances" that include PAH compounds as carcinogenic to humans. Despite this possible carcinogenic character of PAHs, a Directive relative to PAH in air was only recently established. The target value for PAH in air shall be assessed for the level of benzo(a)pyrene (BaP): 1 ng/m^3, for the total content in the PM_{10} fraction averaged over a calendar year (Directive 2004/107/EC, implemented in Italy in 2007 [5]). With this Directive, it is expected that each Member State shall take all necessary measures to ensure that, as from 31 December 2012, the concentration of BaPs do not exceed this target value. In addition, each Member State shall monitor other relevant PAHs at a limited number of measurement sites to assess the contribution of BaP in ambient air.

2 Experimental

2.1 Sampling

The study was performed in the industrial area of an Italian provincial town where sources of anthropogenic emissions surrounding the sampling sites included industrial chemical plants (production of H_2SO_4, synthesis of TiO_2, an electrical energy power plant, a water-treatment system, an incinerator). The environmental survey was designed to monitor the impact caused by the release of air pollutants from industrial plants on the surrounding territory, with particular reference to compounds containing sulphur and nitrogen, which are the most significant for their effects on the population's health and ecosystems.

The work provided the monitoring of ambient air and industrial emissions by measuring the concentration of conventional and unconventional atmospheric pollutants, both in the gaseous phase and in particulate. Sampling was carried out on a monthly basis by means of 52 passive samplers to measure SO_2, NO_x, NO_2, O_3, BTX, H_2S and four active samplers for the simultaneous measurement of PM_{10} in different locations on a daily basis. Along with the particulate, PAHs, metals, anions and cations were also determined. By the means of specific models of spatial interpolation, it was possible to estimate the concentration of SO_2, NO_x, NO_2, O_3, BTX and H_2S in sites where there were no samplers, allowing the establishment of maps of the concentration of pollutants and to identify in these maps which are the areas most subject to high concentrations.

In this study, only the results on particulate PAHs are presented.

The map in Fig. 1 shows the location of the selected 52 sites and the four active sampler sites (P1-P4). Site 1 (P1) is in a residential area; Site 2 (P2) is situated in the seaside resort of the municipality of Scarlino; Site 3 (P3) is situated in a peripheral industrial area; Site 4 (P4) is situated in an almost rural area.

Air samples were collected in four 30-days campaigns from November 2006 to September 2007 using a Tecora Skypost PM HV air monitor, provided with a PM_{10} sampling head, which collects airborne particles on a 47mm 2µm pore size Teflon filter. Samples were collected over 24 hour periods (sampling time beginning at midnight) at a flow rate of 38 L/min, yielding sampled air volumes of ~ $55m^3$ per day. Among the filters, a part corresponding to a week's sampling was used for the analysis of PAHs.

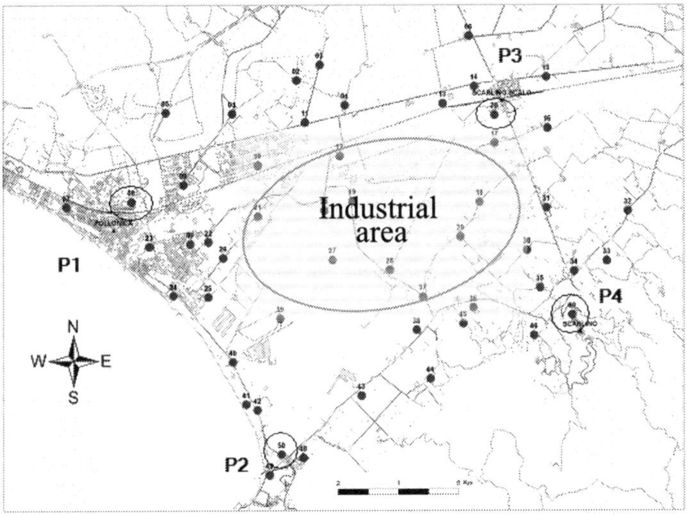

Figure 1: Map of the investigated area showing the position of sampling sites.

2.2 Extraction and analysis

After sampling, the filters were removed from the sampler and stored in a dark and closed chamber at a controlled temperature for 24h. Then they were weighed using a microbalance in order to obtain the mass of PM_{10} collected.

Each filter was then extracted for PAHs analysis as follows: it was put in an amber vial, with the addition of perpetuated-PAH surrogate standard and extracted with 3ml of DCM:Acetone (4:1) in an ultrasonic bath for 30 min, twice. The extract was transferred to a clean vial, concentrated to about 100µl by a gentle flow of nitrogen and cleaned using solid-liquid chromatography on silica gel. Each sample was placed on the top of the column and hexane was used to elute the n-alkanes fraction and toluene to collect the fraction containing PAHs.

The analysis was performed using a Thermo Trace GC ultra gas chromatograph coupled to a DSQ mass spectrometer detector (electron ionization mode and quadrupole analyser) fitted with a fused silica capillary column (Restek Rxi^{TM}-50 30m x 0.25mm ID x 0.25 μm df) working in SIM mode. Helium was used as carrier gas at a constant flow of 1ml/min and transfer line was heated at 280°C. All injections were in PTV mode with an injection volume of 1μl.

The oven temperature was programmed as follows: 40°C (held 1 min), 30°C/min till 150°C (held 1 min), 8°C/min till 190°C (held 5 min), 6°C/min till 308°C (held 29 min).

The quantifications were performed by using an internal standard method for the following PAHs: phenantrene (PHE, 3-ring), anthracene (AN, 3-ring), fluoranthene (FA, 4-ring), pyrene (PY, 4-ring), benzo[a]anthracene (BaA, 4-ring), chrysene (CHR, 4-ring), benzo[b]fluoranthene (BbFA, 4-ring), benzo[j]fluoranthene (BjFA, 4-ring), benzo[k]fluoranthene (BkFA, 4-ring), benzo[a]pyrene (BaP, 5-ring), dibenzo(ah)anthracene (DBahA, 5-ring), indeno(1,2,3-cd)pyrene (IP, 6-ring), benzo(ghi)perylene (BghiP, 6-ring).

3 Results and discussion

3.1 PM_{10}

Table 1 reports the mean PM_{10} concentrations and ranges at the four sites in each season. The Italian air quality health guideline for PM_{10} is a 24-h average concentration of 50 μg/m^3 [3].

The mean daily concentration ranged from 17 to 37 μg/m^3. The highest values were found in the third campaign (23-29 May 2007) at each site. It could be due to the fact that the investigated area is a seaside resort and a major contribution to the concentration is due to inorganic components (salt, sulphates, metals). Moreover, the campaign included a weekend and it could agree with the beginning of holidays. This hypothesis is enhanced by the fact that the highest daily mean value in site P4 (rural area, far away from the sea) is lower than the maximum values of the other sites.

Table 1: The mean, minimum and maximum values of PM_{10} (μg/m^3).

	15-22 Nov 2006			17-23 Feb 2007			23-29 May 2007			5-12 Sep 2007		
	Mean	Min	Max	Mean	Min	Max	Mean	Min	Max	Mean	Min	Max
P1	25	21	32	18	15	21	35	20	49	35	18	56
P2	22	16	28	31	16	60	37	26	48	26	19	33
P3	21	15	28	19	14	23	33	16	50	27	13	44
P4	21	12	28	17	14	20	32	20	38	25	16	35

3.2 Particulate PAH

Tables 2–5 show mean, minimum and maximum concentrations for each PAH (in ng/m^3) over the whole period and in each season and at each site.

The concentration of particulate PAHs exerted distinct seasonal variation presenting a maximum during winter (17–23 February) and a minimum during spring (23–29 May). In winter (mean temperature: 10°C), the atmospheric condition is rather stagnant due to the frequent occurrence of low inversion layers. Under these conditions, photo-degradation and pollutant dispersion are least, resulting in the accumulation and more gas-to-particle conversion of PAHs. In spring (mean temperature: 21°C), higher temperature and higher mixing height could lead to the shift of PAHs from the particle to the gas phase with increased dispersion of pollutants [6–8].

BaP is considered one of the most powerful mutagens and often used as a general indicator of PAHs and regarded by the World Health Organization (WHO) as a good index for whole PAH carcinogenicity.

BaP, the classical chemical carcinogen, was detected in all samples with average daily values of 0.02–0.46 ng/m^3. These values are all below the annual average value of 1 ng/m^3 that the Italian regulation indicates as a quality objective.

Table 2: The mean, minimum and maximum values of PAHs in P1 site (ng/m^3).

	15-22 nov 2006			17-23 feb 2007			23-29 may 2007			5-12 sep 2007		
	Mean	Min	Max	Mean	Min	Max	Mean	Min	Max	Mean	Min	Max
PHE	0.07	0.04	0.12	0.23	0.14	0.32	0.32	0.28	0.39	0.18	0.06	0.24
AN	0.02	0.01	0.04	0.08	0.01	0.19	0.26	0.24	0.30	0.10	0.08	0.14
FA	0.09	0.04	0.14	0.05	0.02	0.10	0.12	0.08	0.18	0.21	0.16	0.27
PY	0.09	0.04	0.14	0.06	0.02	0.10	0.13	0.10	0.19	0.33	0.28	0.37
BaA	0.09	0.03	0.14	0.14	0.04	0.31	0.06	0.05	0.06	0.14	0.06	0.24
CHR	0.16	0.08	0.21	0.21	0.07	0.43	0.08	0.05	0.11	0.27	0.21	0.35
BbFA	0.21	0.10	0.27	0.33	0.21	0.58	0.08	0.05	0.13	0.28	0.19	0.35
BjFA	0.06	0.03	0.12	0.16	0.11	0.29	0.04	0.02	0.06	0.13	0.10	0.17
BkFA	0.09	0.04	0.18	0.20	0.12	0.31	0.06	0.04	0.08	0.19	0.16	0.25
BaP	0.14	0.04	0.25	0.34	0.27	0.46	0.03	0.01	0.06	0.22	0.15	0.35
IP	0.14	0.06	0.23	0.39	0.24	0.58	0.05	0.02	0.09	0.28	0.21	0.35
DBAhA	0.05	0.02	0.07	0.13	0.08	0.19	0.04	0.02	0.07	0.09	0.05	0.12
BghiP	0.18	0.10	0.30	0.47	0.33	0.64	0.08	0.06	0.12	0.27	0.24	0.30
ΣPAHs	1.38	0.76	1.80	2.79	1.77	3.95	1.34	1.05	1.69	2.70	2.45	2.92

Table 3: The mean, minimum and maximum values of PAHs in P2 site (ng/m^3).

	15-22 nov 2006			17-23 feb 2007			23-29 may 2007			5-12 sep 2007		
	mean	min	max	mean	min	max	mean	min	max	mean	min	max
PHE	0.07	0.04	0.10	0.16	0.13	0.19	0.34	0.30	0.37	0.56	0.11	1.07
AN	0.02	0.01	0.03	0.26	0.15	0.35	0.26	0.25	0.26	0.10	0.08	0.14
FA	0.10	0.06	0.14	0.10	0.04	0.14	0.21	0.15	0.26	0.56	0.00	1.08
PY	0.13	0.09	0.18	0.10	0.05	0.13	0.22	0.13	0.29	0.98	0.20	1.58
BaA	0.10	0.04	0.20	0.12	0.08	0.15	0.08	0.06	0.11	0.43	0.24	0.61
CHR	0.14	0.09	0.18	0.15	0.07	0.23	0.12	0.07	0.17	0.77	0.55	1.13
BbFA	0.20	0.14	0.36	0.38	0.25	0.52	0.10	0.05	0.18	0.63	0.50	0.98
BjFA	0.10	0.06	0.18	0.19	0.13	0.26	0.04	0.02	0.07	0.25	0.11	0.57
BkFA	0.17	0.10	0.31	0.17	0.14	0.20	0.07	0.04	0.12	0.33	0.24	0.52
BaP	0.19	0.10	0.36	0.26	0.17	0.41	0.06	0.03	0.10	0.40	0.30	0.61
IP	0.19	0.14	0.26	0.27	0.19	0.36	0.05	0.03	0.09	0.35	0.00	0.57
DBAhA	0.05	0.03	0.09	0.13	0.08	0.17	0.04	0.02	0.07	0.04	0.02	0.06
BghiP	0.28	0.17	0.40	0.29	0.12	0.44	0.10	0.05	0.14	0.80	0.50	1.31
ΣPAHs	1.74	1.24	2.41	2.54	2.00	3.09	1.70	1.36	2.15	6.24	3.67	8.36

Table 4: The mean, minimum and maximum values of PAHs in P3 site (ng/m^3).

	15-22 nov 2006			17-23 feb 2007			23-29 may 2007			5-12 sep 2007		
	mean	min	max	mean	min	max	mean	min	max	mean	min	max
PHE	0.07	0.05	0.10	0.12	0.04	0.33	0.34	0.31	0.45	0.18	0.08	0.29
AN	0.02	0.01	0.03	0.04	0.00	0.14	0.32	0.27	0.51	0.10	0.08	0.14
FA	0.09	0.04	0.13	0.06	0.03	0.13	0.19	0.15	0.26	0.26	0.22	0.34
PY	0.11	0.06	0.16	0.05	0.03	0.08	0.18	0.10	0.29	0.30	0.15	0.44
BaA	0.07	0.03	0.18	0.13	0.07	0.22	0.13	0.06	0.18	0.08	0.00	0.12
CHR	0.20	0.12	0.31	0.23	0.14	0.27	0.08	0.04	0.16	0.27	0.20	0.33
BbFA	0.40	0.31	0.57	0.52	0.36	0.64	0.07	0.05	0.09	0.27	0.19	0.36
BjFA	0.23	0.16	0.32	0.26	0.18	0.32	0.05	0.03	0.07	0.10	0.05	0.18
BkFA	0.28	0.18	0.39	0.21	0.12	0.31	0.06	0.04	0.11	0.17	0.09	0.29
BaP	0.24	0.14	0.38	0.46	0.26	0.69	0.03	0.01	0.05	0.15	0.09	0.20
IP	0.29	0.21	0.34	0.35	0.24	0.46	0.03	0.02	0.05	0.18	0.10	0.28
DBAhA	0.21	0.16	0.28	0.12	0.07	0.15	0.02	0.02	0.03	0.04	0.02	0.06
BghiP	0.38	0.13	0.66	0.43	0.31	0.64	0.07	0.05	0.13	0.27	0.15	0.36
ΣPAHs	2.58	2.05	3.23	2.99	2.00	3.86	1.56	1.36	2.23	2.40	1.99	2.90

Table 5: The mean, minimum and maximum values of PAHs in P3 site (ng/m^3).

	15-22 nov 2006			17-23 feb 2007			23-29 may 2007			5-12 sep 2007		
	mean	min	max	mean	min	max	mean	min	max	mean	min	max
PHE	0.05	0.03	0.07	0.16	0.13	0.18	0.32	0.29	0.39	0.14	0.10	0.21
AN	0.02	0.00	0.04	0.11	0.06	0.17	0.26	0.24	0.29	0.10	0.08	0.14
FA	0.05	0.02	0.10	0.05	0.03	0.07	0.11	0.07	0.19	0.19	0.13	0.27
PY	0.06	0.02	0.09	0.04	0.02	0.06	0.13	0.09	0.21	0.24	0.16	0.27
BaA	0.05	0.01	0.12	0.14	0.02	0.28	0.05	0.05	0.07	0.09	0.05	0.17
CHR	0.10	0.05	0.17	0.14	0.04	0.20	0.07	0.04	0.12	0.24	0.14	0.33
BbFA	0.14	0.11	0.17	0.19	0.17	0.23	0.07	0.04	0.11	0.24	0.16	0.32
BjFA	0.06	0.02	0.08	0.10	0.09	0.11	0.03	0.02	0.05	0.08	0.04	0.13
BkFA	0.11	0.07	0.18	0.20	0.16	0.29	0.05	0.04	0.06	0.09	0.05	0.13
BaP	0.10	0.07	0.14	0.23	0.13	0.32	0.02	0.01	0.04	0.13	0.10	0.18
IP	0.12	0.08	0.16	0.40	0.27	0.61	0.03	0.01	0.04	0.14	0.10	0.18
DBAhA	0.02	0.01	0.03	0.18	0.12	0.30	0.02	0.01	0.03	0.04	0.02	0.06
BghiP	0.18	0.05	0.27	0.41	0.19	0.78	0.07	0.04	0.10	0.19	0.16	0.24
ΣPAHs	1.05	0.78	1.30	2.38	1.69	3.34	1.22	1.03	1.67	1.93	1.44	2.33

4 Conclusions

In this study, concentration of thirteen ambient air polycyclic aromatic hydrocarbons were identified and quantified in the PM_{10} during four seasons at four sampling sites. Results showed the concentrations of ΣPAHs ranged from 0.76 to 8.36 ng/m^3, with the average ΣPAHs measured highest in winter and lowest in summer. The relative proportions of individual compounds could reflect changes in PAHs emissions sources and/or atmospheric conditions. Daily levels of BaP, the most investigated PAH and often used as an indicator of total PAHs, are in the range of 0.01-0.69 ng/m^3, definitively below the annual average value of 1 ng/m^3 that the Italian regulation indicates as a quality objective.

References

[1] Directive 1999/30/CE of 22 April 1999 relating to limit values for sulphur dioxide, nitrogen dioxide and oxides of nitrogen, particulate matter and lead in ambient air.

[2] Council Directive 96/62/CE of 27 September of 1996 on ambient air quality assessment management.
[3] Ministerial Decree No. 60/2002 implementing Directive 1999/30/CE of 22 April 1999 relating to limit values for sulphur dioxide, nitrogen dioxide and oxides of nitrogen, particulate matter and lead in ambient air.
[4] Chetwittayachan, T., Shimazaki, D. & Yamamoto, K., A comparison of temporal variation of particle-bound polycyclic aromatic hydrocarbons (pPAHs) concentration in different urban environments: Tokyo, Japan, and Bangkok, Thailand. *Atmos. Environ.* **36**, 2027–2037, 2002.
[5] Legislative Decree No. 152/07 implementing Directive 2004/107/EC of the European Parliament and of the Council relating to arsenic, cadmium, mercury, nickel and polycyclic aromatic hydrocarbons in ambient air.
[6] Sharma, H., Jain, V.K. & Khan, Z.H., Characterization and source identification of polycyclic aromatic hydrocarbons (PAHs) in the urban environment of Delhi. *Chemosphere* **66**, 302–310, 2007.
[7] Sienra, M., Rosazza N.G. & Préndez, M., Polycyclic aromatic hydrocarbons and their molecular diagnostic ratios in urban atmospheric respirable particulate matter. *Atmos. Res.* **75**, 267-281, 2005.
[8] Tham, Y.W.F., Takeda, K. & Sakugawa H., Polycyclic aromatic hydrocarbons (PAHs) associated with atmospheric particles in Higashi Hiroshima, Japan: Influence of meteorological conditions and seasonal variations. *Atmos. Res.* **88**, 224-233, 2008.

Section 8
Pollution effects and reduction

Influence of CO_2 on the corrosion behaviour of 13Cr martensitic stainless steel AISI 420 and low-alloyed steel AISI 4140 exposed to saline aquifer water environment

A. Pfennig[1] & A. Kranzmann[2]
[1]*FHTW University of Applied Sciences Berlin, Germany*
[2]*BAM Federal Institute of Materials Research and Testing, Germany*

Abstract

The CCS technique involves the compression of emission gasses in deep geological layers. To guarantee the safety of the site, CO_2-corrosion of the injection pipe steels has to be given special attention when engineering CCS-sites. To get to know the corrosion behaviour samples of the heat treated steel AISI 4140, 42CrMo4, used for casing, and the martensitic stainless injection-pipe steel AISI 420, X46Cr13 were kept at T=60°C and p=1-60 bar for 700 h-8000 h in a CO_2-saturated synthetic aquifer environment similar to the geological CCS-site at Ketzin, Germany. The isothermal corrosion behaviour obtained by mass gain of the steels in the gas phase, the liquid phase and the intermediate phase gives surface corrosion rates around 0.1 to 0.8 mm/year. Severe pit corrosion with pit heights around 4.5 mm are only located on the AISI 420 steel. Main phase of the continuous complicated multi-layered carbonate/oxide structure is siderite $FeCO_3$ in both types of steel.
Keywords: steel, pipeline, corrosion, carbonate layer, CCS, CO_2-injection, CO_2-storage.

1 Introduction

In the oil and gas production carbon dioxide corrosion may easily cause failure of pipelines [1–7] and this problem will become an issue when emission gasses are compressed from combustion processes into deep geological layers (CCS Carbon Capture and Storage) [8, 9]. Generally steels applied in pipeline

WIT Transactions on Ecology and the Environment, Vol 123, © 2009 WIT Press
www.witpress.com, ISSN 1743-3541 (on-line)
doi:10.2495/AIR090371

technology precipitate slow growing passivating $FeCO_3$-layers (siderite) [10–12, 26, 30, 31]. First CO_2 is dissolved to build a corrosive environment. Because the solubility of $FeCO_3$ in water is low (p_{Ksp} = 10.54 at 25°C [12, 13] a siderite corrosion layer grows on the alloy surface as a result of the anodic iron dissolution. In geothermal energy production the CO_2-corrosion is sensitively dependent on alloy composition, environmental conditions like temperature, CO_2 partial pressure, flow conditions and protective corrosion scales [10–23]. Engineering the geological CCS-site Ketzin, Germany, the first on-shore CO_2-storage (CO_2-SINK) no experience of the corrosion behaviour of the steels and therefore of the necessity to monitor the site was available for the aquifer water T=60°C / p=80 bar [24, 25] 40°C to 60°C). 60°C is a critical temperature region well known for severe corrosion processes [4, 6, 7, 19, 20, 26–29].

This work was carried out to predict the reliability of the on-shore CCS site at Ketzin, Germany and to get a better understanding of the corrosion behaviour of steels used for CO_2-injections.

2 Materials and methods

Exposure tests were carried out using samples made of thermal treated specimen of AISI 4140 (1%Cr) and AISI 420 (13%Cr) with 8 mm thickness and 20 mm width and 50 mm length. A hole of 3.9 mm diameter was used for sample positioning. The surfaces were activated by grinding with SiC-Paper down to 120 μm under water. Samples of each base metal were positioned within the vapour phase, the intermediate phase with a liquid/vapour boundary and within the liquid phase. The brine (as known to be similar to the Stuttgart Aquifer [32]: Ca^{2+}: 1760 mg/L, K^{2+}: 430 mg/L, Mg^{2+}: 1270 mg/L, Na^{2+}: 90,100 mg/L, Cl^-: 143,300 mg/L, SO_4^{2-}: 3600 mg/L, HCO^{3-}: 40 mg/L) was synthesized in a strictly orderly way to avoid precipitation of salts and carbonates. Flow control (2 NL/h) was done by a capillary meter GDX600_man by QCAL Messtechnik GmbH, München. The heat treatment of the samples between 700 h to 8000 h was disposed in a chamber kiln according to the conditions at the geological site at Ketzin/Germany at 60°C at 60 bar in an autoclave system and for reference at ambient pressure as well. X-ray diffraction was carried out in a URD-6 (Seifert-FPM) with CoKα-radiation with an automatic slit adjustment, step 0.03 and count 5 sec. Phase analysis was performed by matching peak positions automatically with PDF-2 (2005) powder patterns. Mainly structures that were likely to precipitate from the steels were chosen of the ICSD and refined to fit the raw-data-files using POWDERCELL 2.4 [33] and AUTOQUAN® by Seifert FPM.

Then the samples were embedded in a cold resin (Epoxicure, Buehler), cut and polished first with SiC-Paper from 180 μm to 1200 μm under water and then finished with diamond paste 6 μm and 1 μm. Different light optical and electron microscopy techniques were performed on specimens to investigate the layer structures and morphology of samples 60°C/700 h, 60°C/2000 h and 60°C/4000 h.

3 Results and discussion

3.1 Kinetics

Figure 1 illustrates the isothermal oxidation behaviour of the alloys AISI 420 X46Cr13 and AISI 4140 42CrMo-4 at 60°C/ambient pressure characterized by mass gain according to DIN 905 part 1-4.

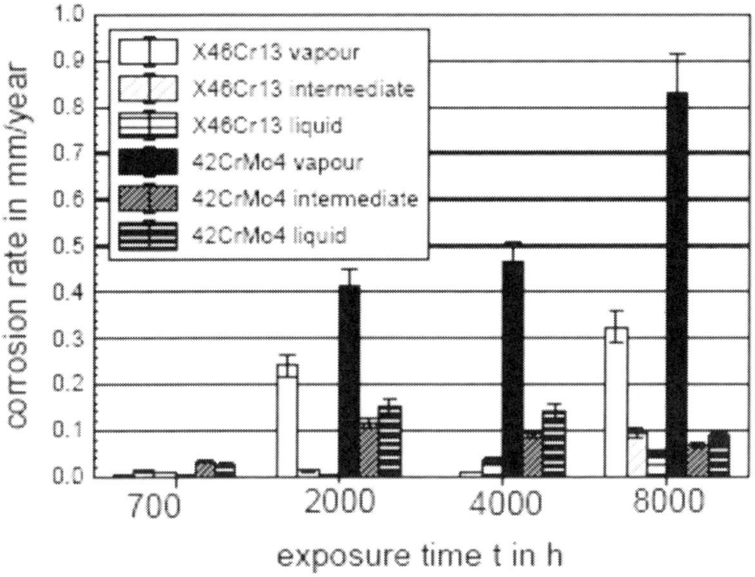

Figure 1: Corrosion rate as a function of heat treatment of the alloys X46Cr13 and 42CrMo-4 (60°C / CO_2 saturated brine / ambient pressure).

The greatest corrosion rates are found within the vapour phase, the lowest in the liquid phase. The high corrosion rates within the vapour phase are mainly due to the better access of CO_2 towards the sample surface in the water saturated CO_2-vapour phase. Samples in the intermediate phase show typical corrosion scale of the media they were exposed to but neither enhancement nor reduction of the corrosion rates were found. The greatest increase of the corrosion rates up to 2000 h is correlated to a passivating layer of siderite in CO_2-atmosphere. In general the corrosion rate increases with increasing CO_2 partial pressure [2], but in the presence of iron carbonate precipitates the corrosion rate may even decrease. This is the reason for the lower corrosion rates at longer exposure times with exception of the low Cr steel in the vapour phase. After 1 year of exposure (8000 h) there is an increase, which shows an extended reaction time and gives evidence of a change in mechanism.

After 700 h at 60 bar the results of corrosion rates are in good agreement with results at ambient pressure.

www.witpress.com, ISSN 1743-3541 (on-line)

Pitting and shallow pit corrosion are only observed on the surface of the sample of X46Cr13 kept in the liquid phase with a maximum penetration depth of 4.6 mm after one year. This is nearly half of the pipe wall thickness (Figure 2). The time of heat treatment has little to no influence on the penetration depth of the pits, but significant influence on the number of counts [34–36].

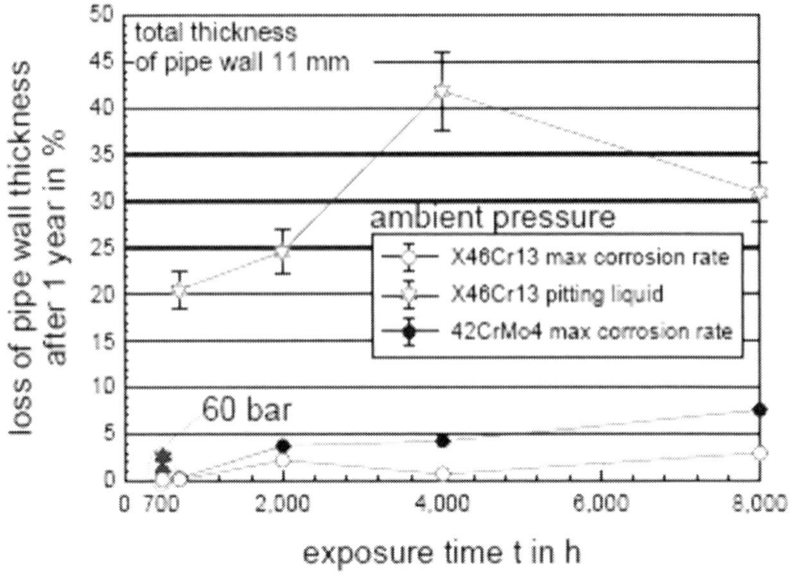

Figure 2: Corrosion rate as a function of heat treatment in CO2 saturated brine of the alloys X46Cr13 and 42CrMo-4 at 60°C at ambient pressure and at 60 bar.

3.2 Microstructure

3.2.1 AISI 4140, 42CrMo-4

Already after 700 h at ambient pressure as well as at 60 bar the 1% Cr alloy (AISI 4140) kept in the vapour phase shows a characteristic duplex corrosion layer (Figure 3). This comprises the outer corrosion layer, mainly consisting of siderite $FeCO_3$ and goethite α-FeOOH, and the inner layer mainly composed of siderite and spinel phase. Both layers contain mackinawite FeS and akaganeite $Fe_8O_8(OH)_8Cl_{1.34}$. The total layer thickness varies from 20 μm to 130 μm at ambient pressure and around 60 μm to 90 μm at 60 bar, where the inner layer shows elliptical islands of 10 μ to 30 μm. Samples in the intermediate phase show shallow pits with a depth about 50 μm and a width of 200 μm.

After 2000 h in the vapour phase the thickness of the outer layer various significantly but can rise up to 1.5 mm in the vapour phase and after 4000 h even a small inner layer is connected to 3-4 mm outer layer. Mainly consisting of the inner layer samples in the liquid phase have an inner layer altitude around 30 μm – 150 μm which grows in depth from 2000 h to 4000 h of heat treatment. Liquid phase samples do not show a typical inner corrosion layer.

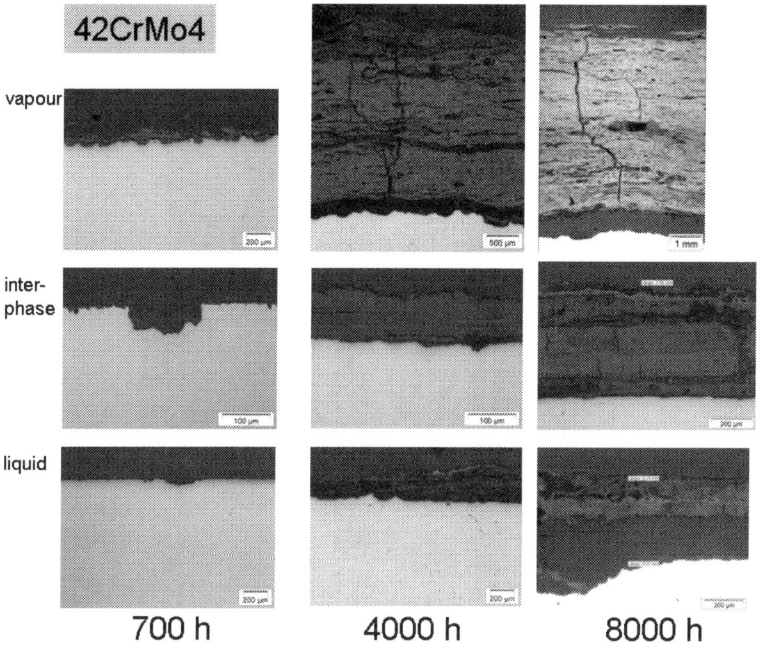

Figure 3: Cross section micrographs of 42CrMo-4 after heat treatment in CO_2 saturated brine at ambient pressure in saline aquifer water.

3.2.2 AISI 420, X46Cr13

As with the 1%Cr steel the 13% Cr steel X46Cr 13 shows surface corrosion with a typical duplex layer formation. Also the thickness of the corrosion layer in the vapour phase is much greater than in the intermediate and especially in the liquid phase. After 700 h at 60 bar and at ambient pressure there are islands of outer corrosion layers in the vapour and the intermediate phase that grow to 70 μm after 2000 h and even to 1.5 mm and a 500 μm inner layer after 4000 h while the liquid phase shows little to no surface corrosion (Figure 4).

While the outer layer mainly consists of siderite $FeCO_3$, goethite α-FeOOH and akaganeite $Fe_8O_8(OH)_8Cl_{1.34}$, the inner layer is composed of siderite, goethite and spinel phase. Mackinawite was not analysed. The carbides within the inner layer follow the stochiometric formula of me23C6 and are most likely manganese carbides.

The complicated multi-layered carbonate/oxide structure reveals siderite $FeCO_3$, with small amounts of Ca [10, 29] is goethite α-FeOOH, mackinawite FeS and spinelphases of various compositions as the main phases (Figures 5 and 6). Lepidocrocite γ-FeOOH and akaganeite $Fe_8O_8(OH)_8Cl_{1.34}$ are minor phases. Although carbides are not expected at these conditions Mn23C6 or some other not clearly stated composition may precipitate as followed from thermodynamic calculations (Software FACTSAGE®). In CO_2-environment precipitates on carbon steel may also consist of Fe_3C [12].

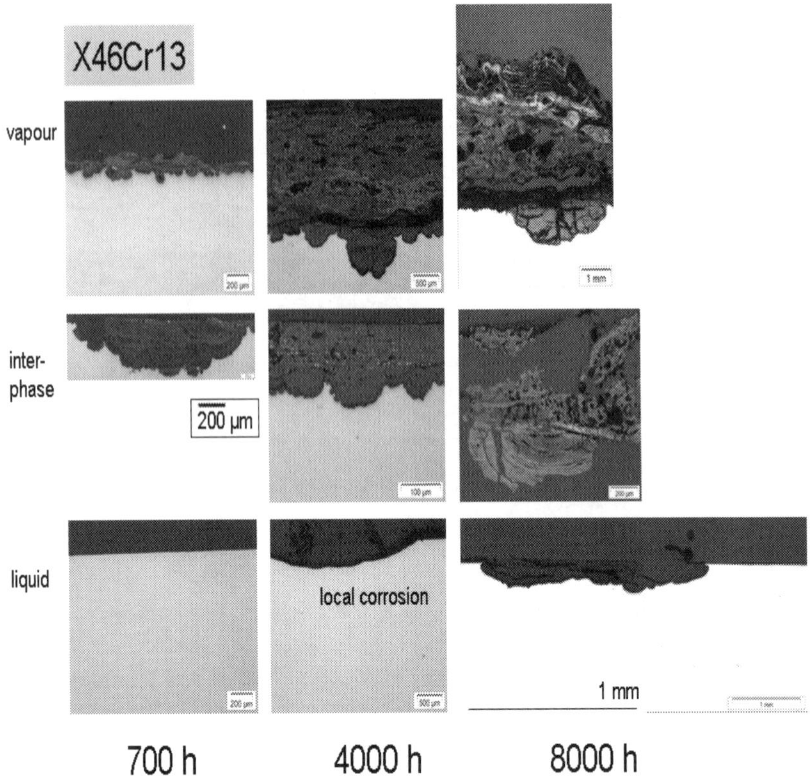

Figure 4: Cross section micrographs of X46Cr13 after heat treatment in CO_2 saturated brine at ambient pressure in saline aquifer water

The formation of the scale in geothermal water takes place in 2 steps as described in detail by Pfennig and Kranzmann [34, 35]. The first step may be attributed to the formation of Fe[II] compounds $FeOH_2$ [13]. The second step corresponds to the formation of a magnetite spinel type with Cr content in the 13% Cr steel and goethite and to the formation of siderite $FeCO_3$. Mackinawite FeS forms due to the saturation of the brine with H_2S and akaganeite $Fe_8O_8(OH)_8Cl_{1.34}$ due to the high salt content of the brine. Iron does not lead to a corrosion resistant stable oxygen film in O_2-free brine saturated with CO_2 at the presence of H_2S. Already little amounts of H_2S in geothermal water cause the change in mechanism of the iron corrosion in the H_2O-CO_2-system [13]. The strong adsorption of sulphide anions blocks the development of a protective oxide film. Therefore predominating phases are carbonates $FeCO_3$, hydroxides FeOOH and sulphides FeS. The further phase study including TEM-technique and the investigation of the combining reaction mechanism is a topic of future research project.

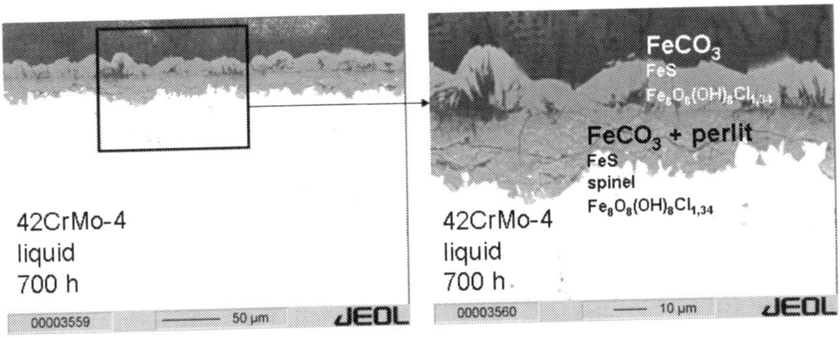

Figure 5: Cross section micrograph of 42CrMo-4 after 700 h of heat treatment in CO_2 saturated brine at ambient pressure in saline aquifer water.

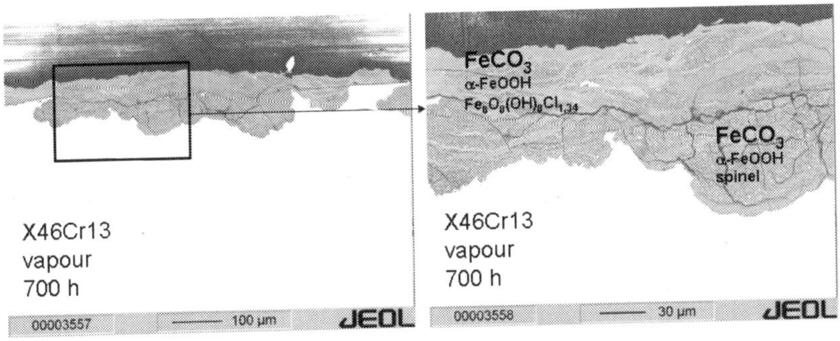

Figure 6: Cross section micrographs of X46Cr13 after 700 h of heat treatment in CO_2 saturated brine at ambient pressure in saline aquifer water.

4 Conclusion

The saturation of a geothermal brine with CO_2 leads to near linear corrosion rates for a 1% Cr (42CrMo-4) and 13%Cr (X46Cr13) steel. Highest surface corrosion rates are 0.8 mm/year (1% Cr) and 0.3 mm/year (13% Cr) in the vapour phase. The average intermediate and liquid corrosion rate for both type of steel is 0.1 mm/year. Severe pit corrosion with pit heights around 4.5 mm is only located on the X46Cr13 steel kept in the liquid where the gas flow and pressure is low. Main phase of the continuous scale is $FeCO_3$ siderite and FeS mackinawite in both types of steel. A complicated multi-layered carbonate/oxide structure reveals goethite α-FeOOH, lepidocrocite γ-FeOOH, spinelphases of various compositions and akaganeite $Fe_8O_8(OH)_8Cl_{1.34}$. In the high chromium-bearing alloy carbides ($Mn_{23}C_6$) are found. Since it is not clear whether the scales are protecting and therefore adhesive to the wall one should be aware of parts of the scale falling down into the injection pipe part. Following from these corrosion rates the steels used for injecting technical CO_2 in Ketzin will withstand at least a 1 year period of injection without a need of replacement. if the tensile stresses can be reduced to a minimum and the decrease in pressure along the pipe wall thickness is kept low.

References

[1] L.S. Moiseeva and O.D. Kuksina, Material Science and Corrosion Protection – Predicting the Dorrosion Aggressiveness of CO_2-Containing Media in Oil and Gas Wells, Chemical and Petroleum Engineering, Vol. 36, No. 5-6, (307 – 311) 2000

[2] S. Nešić, "Key issues related to modelling of internal corrosion of oil and gas pipelines – A review", Corrosion Science 49 (2007) 4308–4338

[3] J.D. Drugli, T. Rogne, M. Svenning, S. Axelsen, The effect of buffered solutions in corrosion testing of alloyed 13% Cr martensitic steels for mildly sour applications, in: Proceedings of Corrosion/99, NACE International, Houston, TX, 1999, paper 99-586

[4] D.S. Carvalho, C.J.B. Joia, O.R. Mattos, Corrosion rate of iron and iron-chromium alloys in CO_2-medium, Corrosion Science 47 (2005) 2974-2986

[5] B.R. Linter, G.T Burstein, Reactions of pipeline steels in carbon dioxide solutions, Corrosion Science 41 (1999) 117-139

[6] M. Seiersten, Material selection for separation, transportation and disposal of CO_2, Corrosion paper no. 01042 (2001)

[7] R. Nyborg, Controlling Internal Corrosion in Oil and Gas Pipelines, Business Briefing: Exploration & Production: The Oil & Gas Review, issue 2 (2005) 70-74

[8] D.C. Thomas, Carbon Dioxide Capture for Storage in Deep Geologic Formations – Results from CO_2 Capture Project, Volume 1: Capture and Separation of Carbon Dioxide form Combustion Sources, CO_2 Capture Project, Elsevier Ltd UK 2005, ISBN 0080445748

[9] A.G. Reyes, W.J. Trompetter, K. Britten, J. Searle, Mineral deposits in the Rotokawa geothermal pipelines, New Zealand, Journal of Volcanology and Geothermal Research 119 (2002) 215-239

[10] Z.D. Cui, S.L. Wu, S.L. Zhu, X.J. Yang, Study on corrosion properties of pipelines in simulated produced water saturated with supercritical CO_2, Applied Surface Science 252 (2006) 2368-2374

[11] D.A. Lopez, W.H. Schreiner, S.R. de Sánchez, S.N. Simison, The influence of carbon steel microstructure on corrosion layers An XRS and SEM characterization, Applied Surface Science 207 (2003) 69-85

[12] D.A. Lopez, T. Perez, S.N. Simison, The influence of microstructure and chemical composition of carbon and low alloy steels in CO_2 corrosion. A state-of-the art appraisal, Materials and Design 24 (2003) 561-575

[13] J. Banaś, U. Lelek-Borkowska, B. Mazurkiewicz, W. Solarski, Effect of CO2 and H2S on the composition and stability of passive film on iron alloy in geothermal water, Electrochimica Acta 52 (2007) 5704-5714

[14] I. Thorbjörnsson, Corrosion fatigue testing of eight different steels in an Icelandic geothermal environment, Materials and Design Vol.16 No. 2 (1995) 97-102

[15] C. Bohne, Mikrostruktur, Eigenspannungs-zustand und Korrosionsbeständigkeit des kurzzeitlaserwärmebehandelten hochstickstofflegierten Werkzeugstahls X30CrMoN15 1 Dissertation, Technische Universität Berlin 2000

[16] H. Gräfen, D. Kuron, Lochkorrosion an nichtrostenden Stählen, Materials and Corrosion 47, (1996), S.16

[17] http://www.ews-steel.com/pdf/vortrag_drvabe_06-11.pdf

[18] M. Kemp, A. van Bennekom, F.P.A. Robinson, Evaluation of the corrosion and mechanical properties of a range of experimental Cr-Mn stainless steels; Materials Science and Engineering A199 (1995) 183-194

[19] A. Tahara, T. Shinohara, Influence of the alloy element on corrosion morphology of the low alloy steels exposed to atmospheric environments, Corrosion Science 47 (2005) 2589-2598

[20] B. Brown, S. R. Parakala, S. Nesic, CO_2 corrosion in the presence of trace amounts of H_2S, Corrosion, paper no. 04736 (2004) 1-28

[21] R.M. Moreira, C.V. Franco, C.J.B.M. Joia, S. Giordana, O.R. Mattos, The effects of temperature and hydrodynamics on the CO_2 corrosion of 13Cr and 13Cr5Ni2Mo stainless steels in the presence of free acetic acid, Corrosion Science 46 (2004) 2987-3003

[22] T. Okazawa, T. Kobayashi, M. Ueda, T. Kushida, Development of super 13Cr stainless steel for CO_2 environment containing a small amount of H_2S, in: Proceedings of Corrosion/93, NACE Intern

[23] M. Ueda, A. Ikeda, Effect of Microstructure and Cr Content in Steel on CO_2 Corrosion, NACE Corrosion, paper no. 13 (1996) 1-6

[24] GeoForschungszentrum Potsdam, CO_2-SINK – drilling project, description of the project PART 1 (2006) 1-39

[25] http://www.co2sink.org/techinfo/drilling.htm

[26] H. Inaba, M. Kimura, H. Yokokawa, „ An analysis of the corrosion resistance of low chromium-steel in a wet CO_2-environment by the use of an electrochemical potential diagram, Corrosion Science 38 (1996) 1449-1461

[27] JFE Steel Corporation – Pipes and Tubes – OCTG, http://www.jfesteel.co.jp/en/products/pipes/octg/index.html

[28] S.L Wu, Z.D. Cui, G.X. Zhao, M.L. Yan, S.L. Zhu, X.J. Yang, EIS study of the surface film on the surface of carbon steel form supercritical carbon dioxide corrosion", Applied Surface Science 228 (2004) 17-25

[29] T. Kamimura, M. Stratmann, "The influence of chromium on the atmospheric corrosion of steel", Corrosion Science 43 (2001) 429-447

[30] J. Carew, S. Akashah, "Prediction techniques for materials performance in the crude oil production system", Modelling Simulation, Material Science Engineering 2 (1994) 371-382

[31] S.L Wu, Z.D. Cui, F. He, Z.Q. Bai, S.L. Zhu, X.J. Yang, Characterization of the surface film formed from carbon dioxide corrosion on N80 steel, Materials Letters 58 (2004) 1076-1081

[32] A. Förster, B. Norden, K. Zinck-Jørgensen, P. Frykman, J. Kulenkampff, E. Spangenberg, J. Erzinger, M. Zimmer, J. Kopp, G. Borm, C. Juhlin, C. Cosma, S. Hurter, 2006, Baseline characterization of the CO2SINK geological storage site at Ketzin, Germany: Environmental Geosciences, V. 13, No. 3 (September 2006), pp. 145-161.

[33] SW. Kraus and G. Nolze, POWDER CELL – a program for the representation and manipulation of crystal structures and calculation of the resulting X-ray powder patterns, J. Appl. Cryst. (1996), 29, 301-303

[34] Pfennig, A., Kranzmann, A., "Influence of CO_2 on the corrosion behaviour of 13Cr martensitic stainless steel AISI 420 and low alloyed steel AISI 4140 exposed to saline aquifer water environment", Intl. Conference on Environment 2008 ICENV 2008, Penang, Malaysia, December 15th to 17th, 2008

[35] Pfennig, A., Kranzmann, A., "Effects of Saline Aquifere Water on the Corrosion Behaviour of Injection Pipe Steels 1.4034 and 1.7225 during Exposure to CO_2 Environment", Green House Gas Emission Reduction Technologies GHGT9 Conference, Washington DC, USA, November 16th to 20th, 2008

[36] Pfennig, A., Kranzmann, A., "Effect of CO_2 on the stability of steels with 1% and 13% Cr in saline water", Water, Steam and Aqueous Solutions – Advances in Science and Technology of Power Generation, ICPWS XV, 15th, Berlin, September 8-11, 2008

Effects of flattening the stockpile crest and of the presence of buildings on dust emissions from industrial open storage systems

C. Turpin & J. L. Harion
Department of Industrial Energetics, Ecole des Mines de Douai, France

Abstract

On industrial sites, fugitive dust emissions from open storage systems for bulk materials, such as coal or iron, represent a significant part of overall estimated particle emissions and can lead to environmental and health risks. The aeolian erosion process depends strongly on the turbulent flow structure over the stockpiles and is significantly conditioned by the topography of the site, the wind direction and intensity and also the shape of the stockpiles. This paper presents wind flow structures obtained by three-dimensional numerical simulations for various stockpile configurations and for a global industrial site simulation with different wind directions. The first aspect investigated by these simulations is the effect caused by the flattening of the stockpile crest on dust emissions. This study provides information to industrials on the best geometrical pile characteristics of an oblong shape pile configuration in order to limit particles emissions. The second aspect investigated in this study is the wind exposure over stockpiles while considering its environment on an industrial site. This study highlights the influence of the presence of surrounding buildings or stockpiles on the real exposure of granular materials, and will allow a more accurate evaluation of fugitive dust emissions on industrial sites.
Keywords: computational fluid dynamics (CFD), fugitive dust emissions, emission factors, wind erosion, flat-topped stockpiles.

1 Introduction

Investigations [1–3] about aeolian erosion concurred to state that the particles emissions from a pile of granular material strongly depend on the wind flow structure over the pile which depends itself on the topography of the studied site, the

WIT Transactions on Ecology and the Environment, Vol 123, © 2009 WIT Press
www.witpress.com, ISSN 1743-3541 (on-line)
doi:10.2495/AIR090381

wind direction and the pile shape. The understanding of the flow interactions with the pile led to the estimation of fugitive dust emissions. In fact, the knowledge of the near surface velocity allows to estimate the particles quantity likely to take off.

To quantify fugitive dust emissions from stockpiles, most of manufacturers use a methodology based on the determination of an emission factor for each given source. The most common emission factor formulations for wind erosion is proposed by the US EPA (United States Environmental Protection Agency). The EPA emission factor EF for wind-generated particulate emissions is expressed in units of grams per year as follows [4]:

$$EF = k \sum_{i=1}^{N} P_i S_i \tag{1}$$

where k is a particle size multiplier, N the number of disturbances per year, P_i an erosion potential corresponding to the observed (or probable) fastest mile of wind for the ith period between disturbances in gm^{-2} and S_i the pile surface area in m^2.

The erosion potential function for a dry exposed surface is :

$$P = 58(u^* - u_t^*)^2 + 25(u^* - u_t^*) \qquad \text{for } u^* > u_t^* \tag{2}$$

$$P = 0 \qquad \text{for } u^* \leq u_t^*$$

where u^*, the wind friction velocity, is given by $u^* = 0.1u_{10}^+(u_s/u_r)$. u_t^* is the threshold friction velocity (ms^{-1}), u_{10}^+ the fastest mile value collected on a anemometer reference height of $10m$, u_s the wind speed measured at $25cm$ from the piles surface and u_r the wind speed reference measured at a height of $10m$.

For large disturbances of wind, such as example on large storage piles, the emission factor is estimated by the sum of local erosion potentials corresponding to a same value of u_s. The emission factor in then given by :

$$EF = k \sum_{i=1}^{N} \sum_{j=1}^{M} (P_j S_j)_i \tag{3}$$

where M is the number of area parts, S_j is the corresponding surface area. This formulation allows the consideration of the influence of the pile geometry and the wind direction on the velocity distribution over the stockpiles by dividing the pile area into sub-areas of constant degree of wind exposure. The EPA's report provides cartographies, derived from wind tunnel studies, representing the wind exposure of two representative pile shapes [4].

Further studies [5–7], focusing on fugitive dust emissions on industrial sites, introduce a new approach to quantify the wind exposure over a stockpile by using data coming from Computational Fluid Dynamics simulations. Three-dimensional numerical simulations, previously validated against wind tunnel measurements, have been employed to simulate the wind flow structure over a wide variety of pile forms and dimensions that can be found on industrial sites. These simulations are used to extend the number of available wind exposure cartographies of stockpiles. The analysis of the results obtained for a wide variety of stockpile geometries allowed to suggest solutions to limit aeolian erosion. For example, an optimal stockpile aspect ratio [6] was proposed to reduce dust emissions.

This paper presents the results of two different studies carried out using the three-dimensional numerical simulations previously validated by [5]. In the first study, the flow structure over various flat-topped stockpiles was simulated in order to find the best clipping height on an oblong shape stockpile which could lead to a decrease of dust emission rate. This study was initiated from the assessment that stockpile crest is the area most subject to erosion, so it becomes relevant to simulate wind flow over different stockpiles having undergone a crest clipping. In the second part of this paper, the first results of the simulated wind flow structure over a real configuration of industrial site are presented. This study aims to highlight the necessity to take into account the presence of buildings surrounding the storage areas in the estimation of the fugitive dust emissions by wind erosion. Very few studies are dedicated to this type of complex configuration compared to isolated stockpile configurations.

2 Numerical simulations description

2.1 Geometry, mesh and turbulence model definitions

The commercial Computational Fluid Dynamics (CFD) software FLUENT was employed to simulate the wind flow over various pile configurations. Full verification and validation of numerical simulations to experimental results [8] was demonstrated and discussed by [5]. The numerical model was also validated for a full scale height. The dimensions of the calculation domain were defined according to the geometric characteristics of the stockpiles on the site and the incidence angle of the flow on the pile. Various tests were performed to ensure aerodynamical independence of the flow patterns over the tested piles as to the effect of the domain boundaries. Characteristic elements of the mesh are shown on Figure 1. An irregular mesh was applied to follow the shape of the geometry.

A grid sensitive test was realized to ensure that numerical results are independent of the grid. Profiles of velocity, turbulent kinetic energy and specific dissipation rate, which are specifics of an atmospheric boundary layer, were used to define the entry of calculation domain. Symmetry boundary conditions were used for the lateral sides and the upper limit of the domain. The lower boundary was considered as a wall. Turbulence closure model was achieved through application of the two-equation $k - \omega$ SST model [9]. This choice is based on a comparative study between experimental results and numerical results for different closure models and a single pile configuration. These validated numerical simulations can be used for various piles configurations, and their results associated to the formulation 1, 2 and 3 allow the estimation of the erosion potential for the configuration tested.

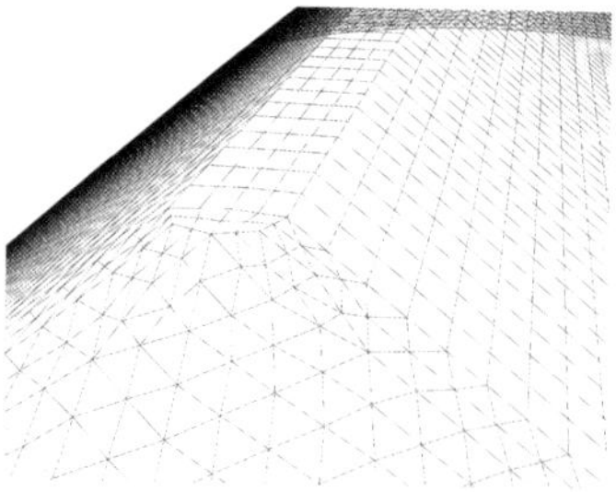

(a)

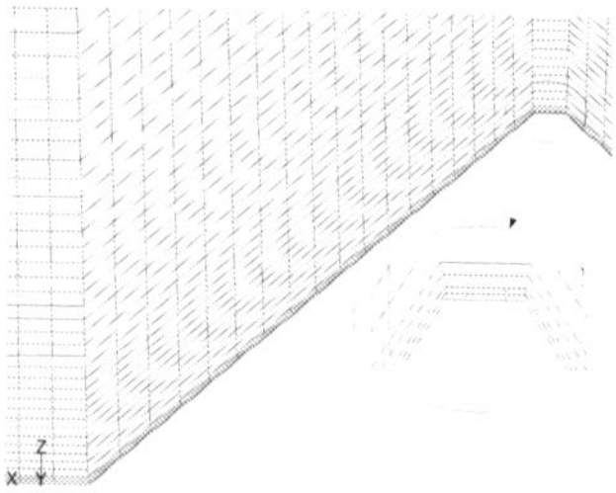

(b)

Figure 1: (a) Top view of the grid over the test 2, (b) transversal cut of the domain representing the grid around the test 2.

2.2 Tested configurations

The numerical model, previously explained, was applied to simulate flow over various flat topped stockpiles heights and over a complete industrial site. The whole simulations were carried out for only one wind speed, $u = 5ms^{-1}$. Indeed it was shown that the normalized wind speed u_s/u_r is independent of the wind speed flow [6]. For each configuration, data of normalized velocity values u_s/u_r, grouped in steps of 0.1 and their corresponding areas were computed at $25cm$ from the piles surface. Results were then integrated in the EPA emission factor formulations to estimate the rates of dust emissions. A coal material was taken as a reference for the calculations, its threshold friction velocity was determined from a wind tunnel experiment and averaged about $u_t^* = 0.35ms^{-1}$.

2.2.1 Flat topped piles

Different configurations with various pile heights and wind conditions were simulated. The modeled piles had a constant volume, shape (oblong), and a constant side slope angle (37°), but different clipping heights. The characteristics of the pile without clipping correspond to the pile configuration determined by [6] as the one having the weakest erosion potential for various wind directions. Geometric properties of the tested configurations are reported in Table 1. The clipping heights applied to the stockpile ranges from 0 to $3m$ and varies by step of $0.5m$. The wind flow structure over two stockpile configurations was studied for four wind incidence angles : $\theta = 0°, 30°, 60°$ and $90°$. Only one configuration of flat topped pile (test 4) was compared to the sharp crested pile configuration in order to limit the number of calculations.

2.2.2 Complete industrial site

The previously validated three-dimensional numerical simulations method was also used to simulate the wind flow over coal stockpiles on a power plant in

Table 1: Geometric characteristics of the tested piles.

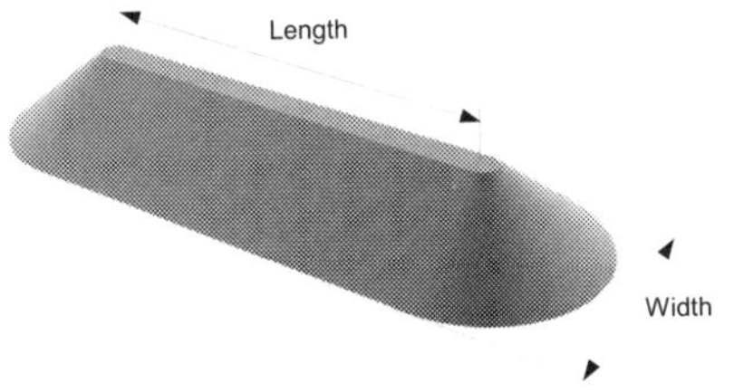

Test	θ (°)	*Volume* (m^3)	*Pile angle* (°)	*Height* (m)	*Length* (m)	*Width* (m)
1, 8, 10, 12	90, 60, 30,0	31477	38	16	74,5	41.000
2	90	31477	38	15,5	74,5	41.016
3	90	31477	38	15	74,5	41.060
4, 9, 11, 13	90, 60, 30, 0	31477	38	14,5	74,5	41.140
5	90	31477	38	14	74,5	41.256
6	90	31477	38	13,5	74,5	41.414
7	90	31477	38	13	74,5	41.616

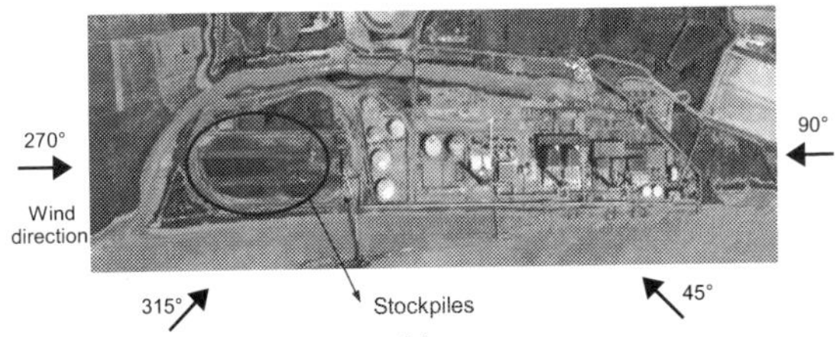

(a)

θ	Complete site	Site with only the stockpiles
270	test 1A	test 1B
315	test 2A	test 2B
45	test 3A	test 3B
90	test 4A	test 4B

(b)

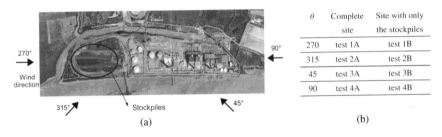

Figure 2: (a) Top view of the industrial site, (b) Test names for the different configurations.

real conditions. Two configurations were tested for various wind flow directions: $\theta = 270°, 315°, 45°$ and $90°$. The first configuration represents the site with only the stockpiles. The second one represents completely the site with the surrounding buildings. The various configurations tested are presented on Figure 2 and in Table 1.

3 Results and discussion

3.1 Flat topped piles

Qualitatively, the overall results are consistent with literature and experimental data [1, 10–12]: flow deceleration at the base of the pile, flow acceleration up the

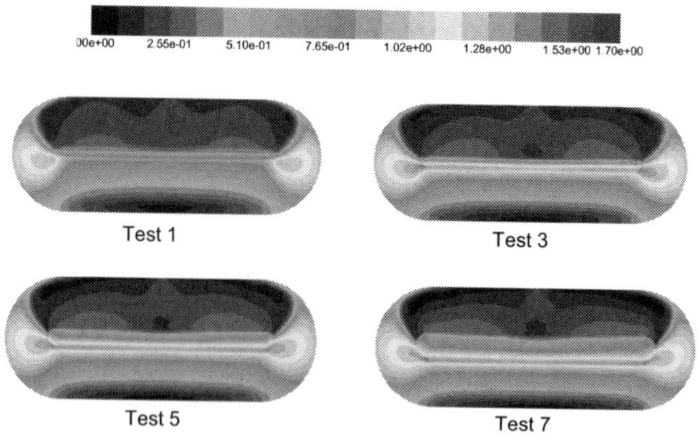

Figure 3: Top view of the normalized velocities contours u_s/u_r at $25cm$ from the pile surface for the tests 1, 3, 5 and 7.

windward slope and toward the sides, followed by a region of flow separation on the crest, and finally a low velocity region of reversed flow in the lee side.

The pile geometry effect on the wind speed near the pile surface is shown in Figure 3 through the contours of the normalized velocity u_s/u_r. On the windward slope of the stockpiles, where the flow is accelerated, the normalized velocity u_s/u_r increases progressively from the toe to the top of the stockpile. Maximum u_s/u_r value are then recorded at the top, and on the lateral sides of the stockpiles, what implies a high erosion potential in these zones. When the clipping height increases, the ratio u_s/u_r is increasing, this happens mostly on the top of the flat-topped stockpiles (Figure 3). Figures 4(a) and (b) show the velocity contours for the tests 1 and 4. They show that the pile geometry near the crest has a very strong influence on the flow detachment. Figures 4(b) and (d) show, for flat topped pile configuration, a strong low pressure field, and much larger velocity values near the crest. Simulations predict a strongly negative pressure gradient on the first ridge of the flat topped pile in comparison with the sharp crested configuration (Figure 4(d)) . This negative pressure gradient justifies the fact that the flow acceleration up the stoss slope is higher on the flat topped pile configuration than in the sharp crested configuration.

After the flat top, the flow separates from the surface and creates a large recirculation. Figure 3 shows that the flow recirculation in the lee side induces a low velocity region which implies that the erosion process stops.

Figure 5 presents the emission factors calculated by the EPA method for the tests 1 to 7 for different wind speeds. It shows that the evolution of the emission factors for each configuration tested not depend on the wind velocity. For a clipping height between 0 and $1.5m$, the calculated emission factors increase significantly. For a clipping height larger than $1.5m$, the emission factors remain approximately constant in comparison with the test 4. This is due to the fact that the pile's heights

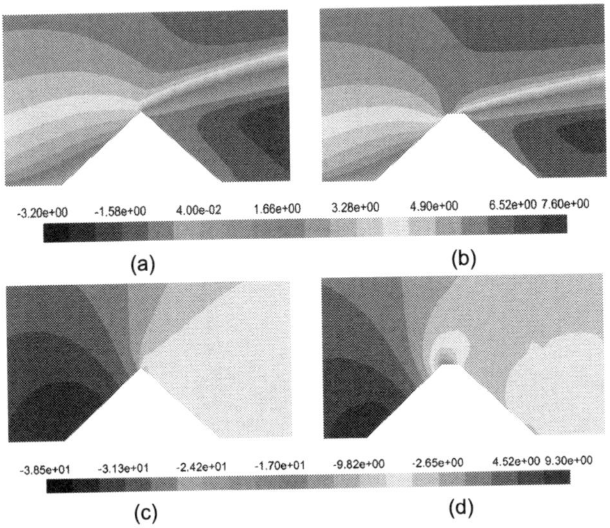

Figure 4: Transversal cut following the wind flow direction of the velocity contours (test 1 (a) and test 4 (b)) and the pressure contours (test 1 (c) and test 4 (d)).

are smaller, and therefore, the stoss slope areas subject to the aeolian erosion are reduced. Nevertheless, for these configurations, the maximum value of normalized velocity increases as the clipping height grows up, in particular at the approach of the stockpile flat top.

The contours of normalized velocity above the pile surfaces for the tests 1, 4 and 8 to 13 for different velocity directions are shown on Figure 6. They reveal that when the wind incidence angle on the stockpiles θ varies from 90° to 0°, the impact zone of the flow on the pile upwind and the recirculation zone on the leeward slope downstream are considerably modified and reduced. As a result, the areas of low normalized velocity become smaller to the advantage of regions subject to the aeolian erosion. These results provide important information on the influence of the wind direction on dust emissions.

The clipping of the stockpiles does not allow to reduce dust emission but, on contrary, increases the dust emission rates. Areas corresponding to erosion are systematically higher for the flat topped configuration. The configuration having the weakest potential erosion, among the tested configurations, is the sharped crest stockpile put perpendicularly to the wind flow.

3.2 Complete industrial site

In order to extend the number of cartographies representing the wind exposure of a stockpile available in the EPA report, three-dimensional numerical simulations

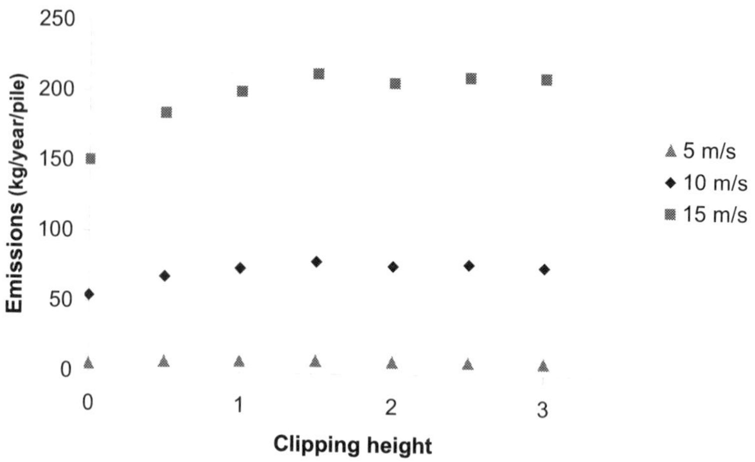

Figure 5: Evolution of the emission factor of a pile as a function of the clipping height for different wind speeds $u_{10}^{+} = 5, 10$ and $15m/s$.

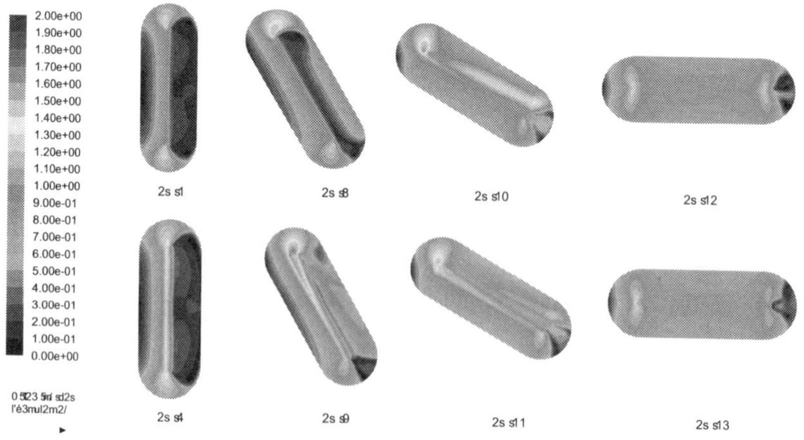

Figure 6: Top view of the contours of normalized velocities u_s/u_r at $25cm$ from the pile surface for the tests 1, 4 and 8 to 13.

have been employed to simulate the wind flow structure over a wide variety of pile geometries and dimensions that can be found on industrial sites [5–7]. In these studies, the wind flow structure was analyzed only for isolated piles configurations without considering their environment on the industrial site. However, the wind erosion process is considerably affected by flow field changes, which are caused by every modifications of the terrain geometry. Nevertheless, till now, at no time the

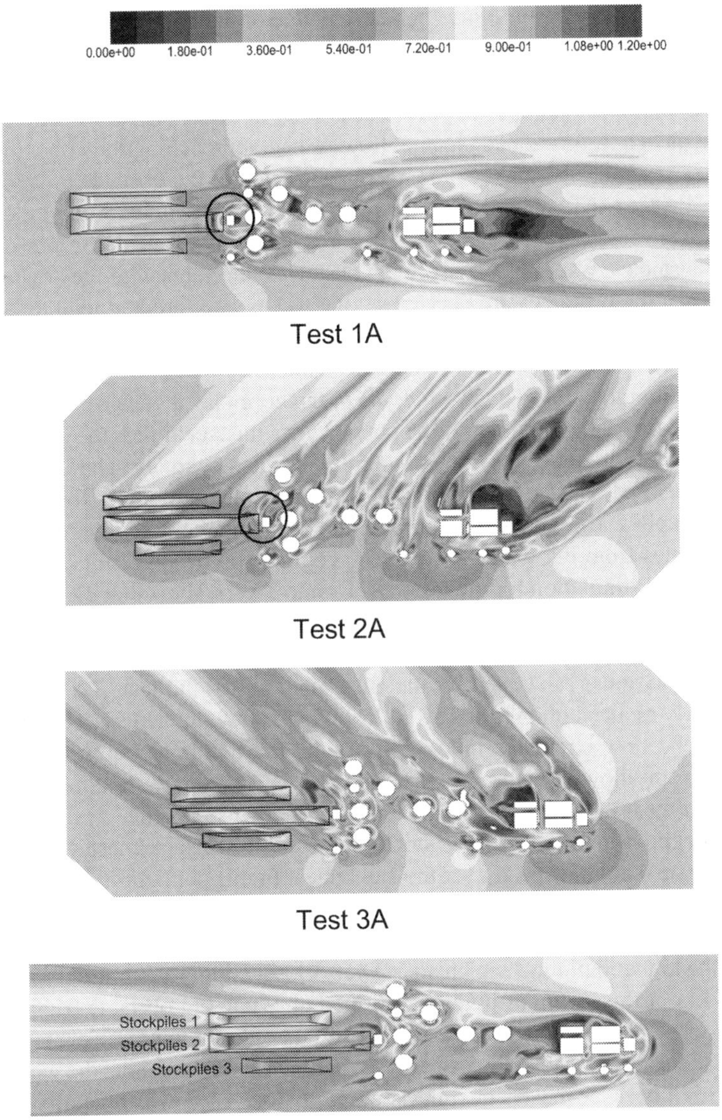

Figure 7: Top view of the normalized velocities contours u_s/u_r at $25cm$ from the surface for the tests 1A, 2A, 3A and 4B.

effect of the buildings surrounding the stockpiles has been taken into account. In this study, two configurations were tested for various wind flow directions: the first one without representing the buildings and the second one with the surrounding buildings in the calculation domain.

In a general manner, buildings behave as huge obstacles for the wind flow and perturb largely the dynamic flow structures over stockpiles. The turbulent flow around isolated obstacles has been extensively studied notably by [13]. From their experimental investigations, they have described in detail the flow past a cubical obstacle placed in developed channel flow. When approaching the cube, the flow is decelerates and separates on the upstream corners of the obstacle creating separation vortexes. The main vortex wraps as a horseshoe vortex around the cube. A broad recirculation zone develops downstream of the obstacle. In presence of several obstacles, wakes of the upstream obstacles interferes with the downstream ones. When the obstacles are very close, the wakes interfere and tend to become similar to the wake of a single body.

The effect of the surroundings buildings on the wind speed around the stockpiles is shown in Figure 7 through the contours of the normalized velocity u_s/u_r. In the test 1A and 2A, where the wind first come across the stockpiles, the buildings do not significantly disturb the flow structure over the stockpiles. The low velocity region generated by the flow deceleration upstream the buildings (for the closest to the stockpiles surrounded on Figure 7) interacts with the stockpiles and protects the stockpiles from the wind erosion. The comparison of the results obtained by the numerical simulation for the tests 1A-1B and 2A-2B show that the presence of the buildings downstream the stockpiles lead to a decrease of the erosion potential of the stockpiles. For the wind directions $\theta = 270°$ and $315°$, the estimation of the dust emissions, calculated by the EPA method, for the tests 1A and 2A are respectively reduced of 14% and 2.5% in comparison with the tests 1B and 2B.

When the flow circulates first of all around the buildings before reaches the stockpiles, the flow structure over the stockpiles is strongly modified. Figure 7 shows, for the test 3A, that the vortexes generate downstream the closest buildings to the stockpiles interfere with the stockpiles. Near the surface, these vortexes tend to accelerate the wind speed over the stockpiles. In this configuration, the erosion potential estimation increases of 21% in comparison with the configuration where the buildings are not simulating (test 3B). In the test 4A, the wind flow first comes across the higher buildings which are set very closely. Figure 7 shows that the flow behave like if this group of buildings formed only one big obstacle (right side of the picture), in fact the flow accelerates toward the sides of this large obstacle and a large wake zone is created downstream. Over the stockpiles 1 and 2 the wind speed is very high because these stockpiles are exactly located downstream the vortexes generated by the buildings. The stockpile 3 is, as for it, located in the wake of the buildings, so it is protected from the wind erosion. On average for the test 4A, the erosion potential of the three stockpiles, calculated by the EPA method, has dramatically (90%) increased by taking into account the presence of the buildings.

For this real configuration of a power plant, the comparison of the results from the two tested configurations (with and without buildings) shows by evidence that the topography of the site exerts large perturbations on the flow structure over the stockpiles. This study shows the necessity to take into account the topography of the industrial site, at least the main buildings and all the significant stockpiles, to enhance the accuracy of the dust emission estimation.

4 Conclusion

In this paper are presented the numerical modeling results of flow structures over various flat-topped stockpiles and over a real configuration of a power plant. The first part of the study consisted in analyzing the flattening effect of the stockpiles crest on dust emissions. This work have revealed that the flattening of the stockpiles crest not allowed to reduce the wind erosion of the stockpiles. The optimal geometrical characteristics of a stockpile, among those tested, minimizing dust emission is a sharped crest stockpile with the main direction perpendicular to the wind flow. In the second part of the study, the presence of the surrounding buildings around the stockpiles was considered for the calculation of the dust emissions rate from the stockpiles. The results of these simulations emphasize the necessity to take into account the topography of the site, at least the main buildings and all the significant stockpiles. In fact, previous studies carried out without regarding the topography of the site led to an inaccurate evaluation of the fugitive dust emissions.

The study reported in this paper improves the understanding of fugitive dust emissions on industrial sites and in mining zones. These findings will allow a more accurate and relevant evaluation of fugitive dust emissions from open storage systems on industrial sites and a better evaluation of its environmental impacts.

Acknowledgements

This work was carried out with the financial support of ArcelorMittal Dunkerque and Fos-sur-Mer (steelworks in France), EDF Research & Development and ADEME (the French Agency for Environment and Energy Management).

References

[1] Neuman, C.M., Lancaster, N. & Nickling, W., Relations between dune morphology, air flow, and sediment flux over reversing dunes. *Sedimentology*, **44**, pp. 1103–1113, 1997.
[2] Lancaster, N., Nickling, W., Neuman, C.M. & Wyatt, V., Sediment flux and airflow on the stoss slope of a barchan dune. *Geomorphology*, **17**, pp. 55–62, 1996.
[3] Parson, D., Wiggs, G., Walker, I., Ferguson, R. & Garvey, B., Numerical modelling of airflow over an idealised transverse dune. *Environmental Modelling and Software*, **19**, pp. 153–162, 2004.
[4] EPA, Update of fugitive dust emissions factors in ap-42. *Midwest Research Institute, Kansas City*, **MRI No. 8985-K**, pp. AP–42 section 11.2, 1988.
[5] Badr, T. & Harion, J., Numerical modelling of flow over stockpiles: implications on dust emissions. *Atmospheric Environment*, **39**, pp. 5579–5584, 2005.
[6] Badr, T. & Harion, J., Effect of aggregate storage piles configuration on dust emissions. *Atmospheric Environment*, **41**, pp. 360–368, 2007.

[7] Torano, J., Rodriguez, R., Diego, I. & Pelegry, J.R.A., Influence of pile shape on wind erosion cfd emission simulation. *Applied mathematical modelling*, **31**, pp. 2487–2502, 2006.
[8] Stunder, B. & Arya, S., Windbreak effectiveness for storage pile fugitive dust control: a wind tunnel study. *Journal of the air pollution control association*, **38**, pp. 135–143, 1988.
[9] Menter, F., Two-equation eddy-viscosity turbulence models for engineering applications. *AIAA Journal*, **32**, pp. 1598–1605, 1994.
[10] Neuman, C.M., Lancaster, N. & Nickling, W., The effect of unsteady winds on sediment transport on the stoss slope of a transverse dune. *Sedimentology*, **47**, pp. 211–226, 2000.
[11] Walker, I. & Nickling, W., Secondary airflow and sediment transport in the lee of a reversing dune. *Earth Surface Processes and Landforms*, **24**, pp. 438–448, 1999.
[12] Walker, I. & Nickling, W., Dynamics of secondary airflow and sediment transport over and in the lee of transverse dunes. *Progress in Physical Geography*, **26**, pp. 47–75, 2002.
[13] Martinuzzi, R. & Tropea, C., The flow around surface mounted prismatic obstacle placed in a fully developed channel flow. *Journal of Fluid Engineering*, **115**, pp. 85–92, 1993.

Synergies between energy efficiency measures and air pollution in Italy

T. Pignatelli[1], M. Bencardino[1], M. Contaldi[2], F. Gracceva[1] & G. Vialetto[1]
[1]*ENEA, National Agency for New Technology, Energy and the Environment, Rome, Italy*
[2]*APAT, Environment Protection and Technical Services Agency, Rome, Italy*

Abstract

Greenhouse gases, such as carbon dioxide, and atmospheric pollutants, such as sulphur, nitrogen oxides and particulate matter, are mainly generated from energy production and consumption. For several years the European Commission has given great importance to the control of atmospheric pollutants, establishing emission limits for LCPs (Large Combustion Plants), as well as national emission ceilings, at Member State level. The development of energy scenarios is the most crucial step in creating emission projections of air pollutants and greenhouse gases from energy sources, and "business as usual" (b.a.u.) energy scenarios are normally considered to elaborate upon reference emission scenarios, on the basis of the abatement measures ruled in the current legislation. However, the b.a.u. energy scenarios could be deeply modified by the implementation of policies and measures aimed at energy saving, so emphasising the synergies existing between Air Quality and Climate Change, and at the same time reducing the abatement costs to ultimately protect human health and the environment. From this point of view, energy efficiency and renewable sources comprise two essential policy instruments to achieve the final objective. In this paper, an alternative energy scenario, taking into account a set of policies for increasing the energy efficiency and the share of renewable sources, is analysed and assessed in terms of decreased air pollutants and carbon dioxide emissions. Synergies and trade offs in air pollution are also analysed through a comparative and quantitative analysis of the reference and alternative scenarios, under the commitment to achieve the reduction objectives. The study has been carried out by the GAINS-Italy model, a national integrated assessment model developed by ENEA in collaboration with the International Institute for Applied Systems Analysis (IIASA) in Laxenburg. The modelling activity is ongoing under the sponsorship of the Italian Ministry for the Environment, the Land and the Sea.
Keywords: air pollution, climate change, integrated assessment modelling, emission scenarios, cost curves, GAINS, GHGs.

1 Introduction

In recent years, in the framework of the European Union as well as in the larger context of the UN-ECE Convention on Long Range Transboundary Air Pollution (CLRTAP), the links between Climate Change and Air Pollution have become more and more evident. Greenhouse Gases (GHGs), such as CO2, and air pollutants, such as Sulphur Oxides (SO2), Nitrogen Oxides (NO2) and Particulate Matter (PM) are essentially generated by the combustion process of fossil fuels. International agreements, such as the Kyoto Protocol and the Gothenburg Protocol, call for a substantive reduction in the emissions of those pollutants in the coming decades in order to meet environmental and human health preservation objectives. As a consequence, the national experts, while developing energy scenarios, have placed increasing attention on including measures suitable to simultaneously address the envisaged reductions both in GHGs and Air Pollutant emissions. For these reasons, the Italian Ministry for the Environment and the Protection of the Land and the Sea has mandated ENEA to explore the potential in emission reduction associated with a number of structural and technical measures that have already been planned to be implemented in coming years. In this analysis, such a potential is then quantified by the comparison with the estimated emissions from a reference scenario, which do not include the measures under assessment. The comparative analysis of this study also allows one to highlight the advantages and disadvantages of the selected measures, with the ultimate objective of providing the policy makers with sufficient elements to define cost-effective policies to combat both Climate Change and Air Pollution.

2 Description of the analysis tool

The analyses reported in this study have been developed by the Integrated Assessment Model GAINS-Italy, the result of a research project jointly carried out by ENEA and IIASA (International Institute for Applied Systems Analysis, Laxenburg, Austria). GAINS-Italy derives directly from the analogous continental model GAINS-Europe (Klassen [1]), which in turn is an extension to the GHGs of the RAINS-Europe model (Amann et al. [2]). GAINS-Italy, similarly to its precursor version RAINS-Italy (Vialetto et al. [6]), computes emissions scenarios, at 5 year intervals, for SO_2, NOx, NH_3, VOCs, PM, Ozone and GHGs, on the basis of data concerning the anthropogenic activities (energy consumptions, industrial production, livestock, agriculture etc.) and a long list of applicable abatement technologies, which the user selects in the so called Control Strategy, according to the implementation of measures due to the Current Legislation (CLE Strategy) or alternative Reduction Strategies. Once emissions are calculated through the use of Atmospheric Transfer Matrices, elaborated by the Eulerian Model AMS (Atmospheric Modelling System, Zanini et al. [3]), representing, in the form of exchange coefficients, the source-receptor relationship between emission sources and the single cell, 20km x 20km, in the calculus domain, the dispersion of the emissions over the territory is calculated and mapped. Then the estimated emission depositions and concentrations are compared with "critical loads" and "critical levels", cell by cell, to have an assessment of the environmental impact in terms of acidification and eutrophication and ground level ozone (EMEP [4]). Moreover, the impact of PM2,5 and Ozone on human health can be also estimated, in terms of "Life Expectancy Reduction" for PM2,5 (Mechler et al. [5]) and "Premature Deaths" for Ozone, on the basis of epidemiological studies and statistical analysis (Amann et al. [2]). Abatement costs can also be evaluated, including investment and operative costs, to compare alternative control strategies and ultimately identifying the most cost-effective solution. The analysis process described is summarized in the flow diagram in Fig. 1. In this study, the analysis is limited to the emission analysis only.

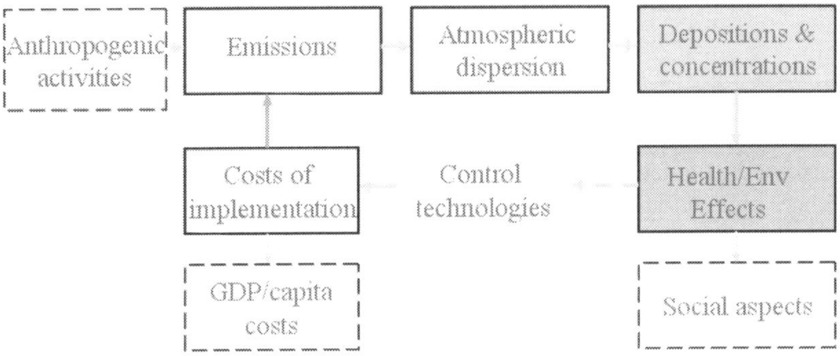

Figure 1: Flow diagram of the Integrated Assessment Approach in GAINS.

In order to better understand the relationship between the input and the output in the GAINS-Italy Model, a simplified schema is reported in Fig. 2

The left and bottom boxes are input data selected by the user. In particular, the energy scenarios are developed exogenously by a suitable energy model called Markal-Italy, and then imported as input in GAINS-Italy, once properly re-formatted and re-aggregated for GAINS. The Control Strategy is the penetration of the abatement technologies, along time, in terms of the percentage of "covered" input data (usually called activity level), for any combination of all the sector, sub-sector, fuel or technology considered in the GAINS-Model, at any year under analysis, e.g. the percentage of input coal, in Power Pants, where Electro-Static Precipitator (ESP) is applied, at years 2000, 2005, 2010 etc.

3 Assumption of the analysis

As mentioned above, the main objective of this study is the estimation of the potential emission reductions associated with the introduction of a series of measures into an alternative energy scenario (here called "ALT"), in comparison

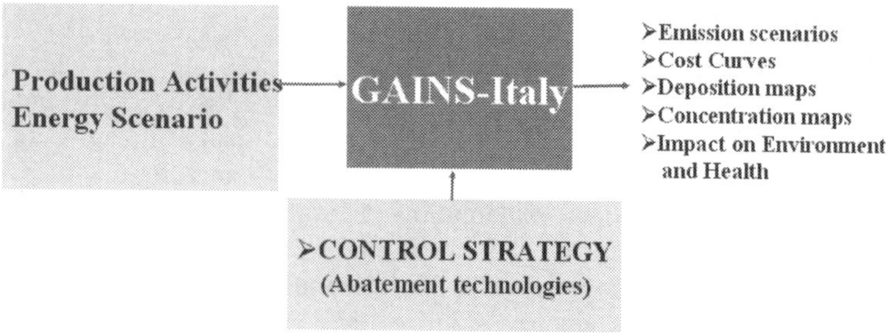

Figure 2: Simplified schema input/output in GAINS-Italy.

Table 1: Additional measures characterizing the ALT scenario with respect the TEND scenario.

INCREASED ELECTRIC GENERATION FROM RENEWABLE ENERGIES (REN)
Increased production of fuel from municipal wastes, up to 1/3 of the total annual production of wastes
1000 Mw installed power at 2015, from photo-voltaic plants
15% of energy demand for heating supplied by remote centralized plants
Increased share of coal (CCS included) in electricity production from 2020
Increased use of low consumption bulbs up to triple of current trend
Electric power: increased share of high energy efficiency engines over 50% of the market, at 2015
House appliances: high efficiency energy appliances doubled with respect the current trend
Thermal solar power: 30% of demand at 2020
Thermal insulation: increased efficiency up to 10% consumption saving
Transport Sector: bio-fuel share increased up to 7,5% at 2020
Transport Sector: Low consumption vehicles share increased up to 80% of the total new vehicles, at 2020
Transport Sector: increased share of gas buses in public transport
Freight Transport: Doubled share with respect the current value

with a reference energy scenario (here called "TEND"). The TEND scenario is considered to be a business as usual (b.a.u.) scenario, where from the point of view of the emission controls implemented, the abatement measures included are those foreseen in the Current Legislation only (therefore the whole emission scenario is referred as CLE). The ALT scenario, instead, includes additional measures with respect to the reference scenario, concerning energy saving, higher energy efficiency and a higher share of renewable energy, which are summarized in Table 1. The emission controls associated with the ALT scenario are the same as in the TEND scenario, so that any difference in emissions is related to structural changes in the energy scenarios only. The measures listed in Table 1 have already been included in the national energy plan and formally adopted, at national level, from the Italian Ministry for Economic Development.

4 Characterization of the scenarios from an energy perspective

In order to better understand the results of this study as differences of the two scenarios analyzed, in terms of emission reduction achievable, it is useful to have a characterization of the two energy scenarios from the perspective of the input to the GAINS-Italy Model. Keeping in mind Fig. 2 and remembering that in this study only the energy content is different, comparing the TEND and the ALT scenarios, it becomes clear that any single change in the energy input is reflected in output, as a change in emissions. In Fig. 3 the two scenarios are compared in terms of global energy consumption. The ALT scenario is characterized by lower global energy consumption, starting from 2010 and further increasing until 2030. The global energy content is also reported in Table 2 in absolute values (PJoule). In Fig. 4, the comparison between the TEND and the ALT scenarios are reported in terms of energy consumption per sector at 2020. The additional measures included in the ALT scenario have effect mainly in the Industry, Residential and Transport sectors as well as in the Power Plant sector, generally reducing energy consumptions in the ALT scenario.

5 Results and discussion

In Figs. 5–8 the results of the comparative analysis are reported. The red line is the emission ceiling, at 2010, established by the EU Directive 2001/81/CE. For SOx and NOx there is no significant difference in terms of emission reductions

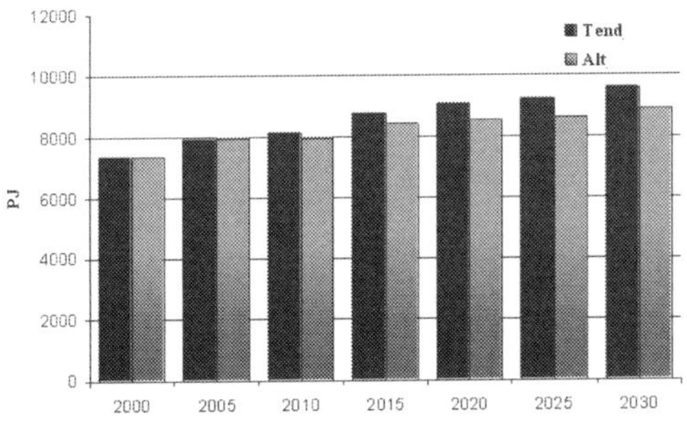

Figure 3: Comparison between the TEND and the ATL scenarios, as global energy content (PJoule).

Total Energy Content		
Tend	Alt	Year
7311	7332	2000
7919	7936	2005
8121	7965	2010
8743	8386	2015
9057	8502	2020
9219	8593	2025
9600	8905	2030

Table 2: Comparison between the TEND and the ATL scenarios (absolute values Joule).

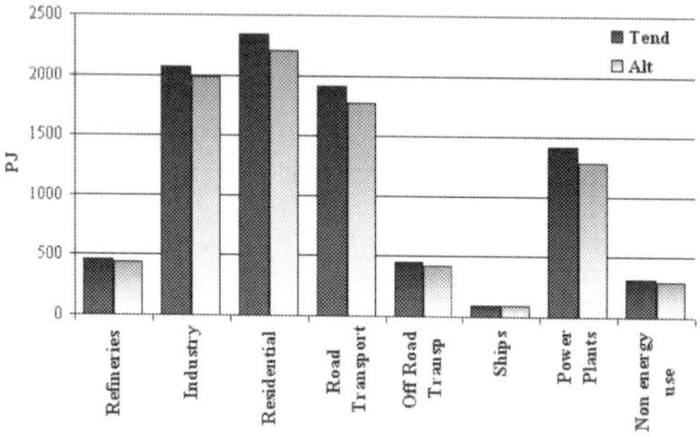

Figure 4: Energy content per sector in the TEND and the ALT scenarios.

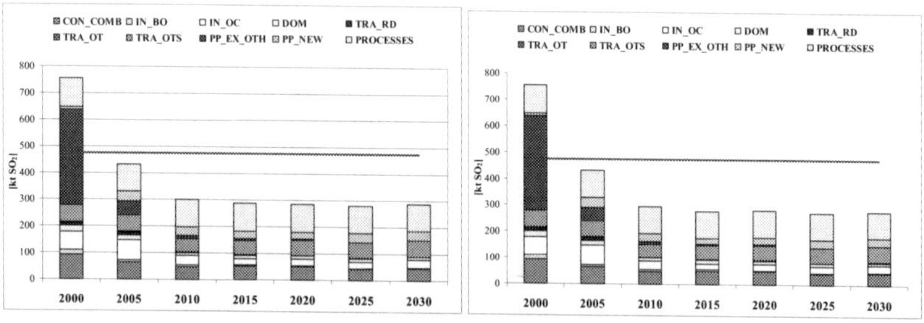

Figure 5: SOx emissions in the TEND (right) and the ALT (left) scenarios, at 2010, by sector, in kt.

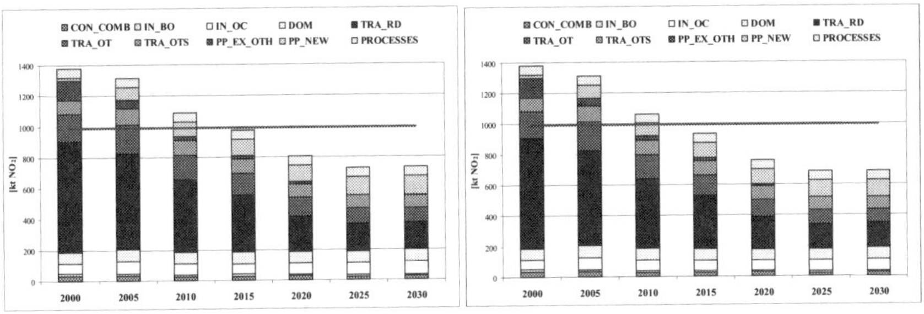

Figure 6: NOx emissions in the TEND (right) and the ALT (left) scenarios, at 2010, by sector, in kt.

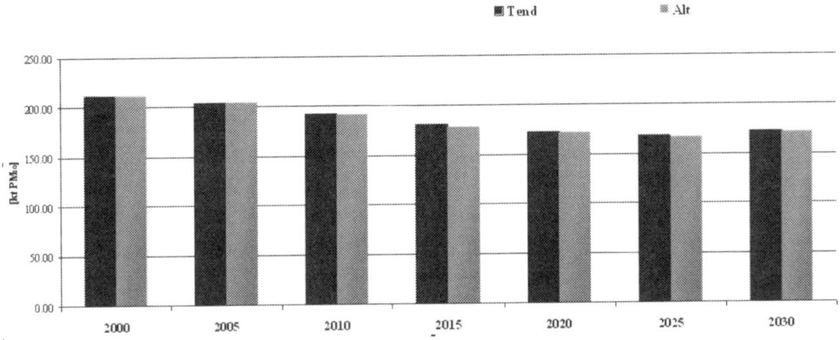

Figure 7: Total PM emissions in the TEND and the ALT scenarios, in kt.

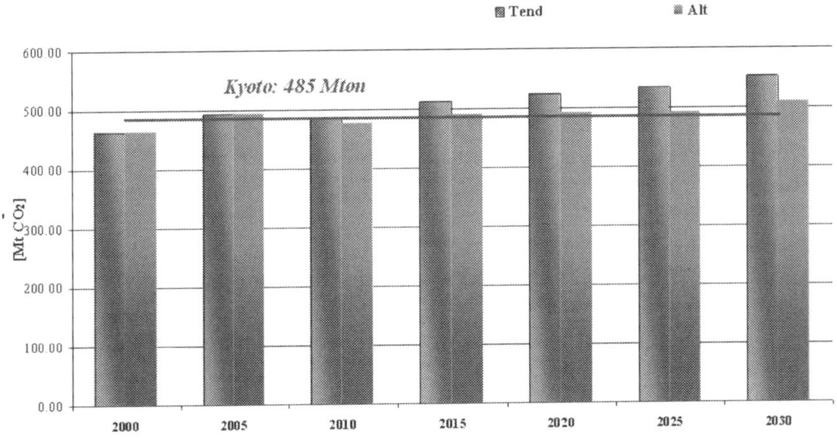

Figure 8: CO2 emission in the TEND and the ALT scenarios, absolute values, in Mt.

and compliance with the Directive, for NOx the ceiling is exceeded and for PM no ceiling is yet established, although the emissions are quite similar in both scenarios. For CO2 emissions a significant reduction is noted after 2015, although at 2010 the achievement of the Kyoto target is quite uncertain, as even in the ALT scenario after 2015 the target is constantly exceeded. Moreover, the Transport sector is identified as the most critical. Therefore, it can be concluded that the measures listed in Table 1, although introducing some benefits in terms of emission reductions, mainly concerning CO_2, do not allow one to achieve compliance with the EU regulation (NOx ceiling) and do not ensure successful fulfilment of the Kyoto target after 2010.

Being as the ultimate perspective of Integrated Assessment Modelling is to provide the policy makers with elements for defining cost-effective policies, the suggestion coming from this study is to further increase efforts in the direction of the already planned measures, especially in the use of renewable energies and improving energy efficiency, so that investment costs in the implementation of the measures are do not vanish by missing environmental and human health protection objectives. It should also be considered that additional costs are associated with incompliance with the EU Directives due to penalties.

References

[1] Klassen G., "The Extension of the RAINS Model to Greenhouse Gases"

[2] Amann M., Cofala J., Heyes C., Klimont Z., Mechler R., Posch M., Schöpp W., "The RAINS model. Documentation of the model approach prepared for the RAINS peer review 2004", February 2004.

[3] Zanini G., Pignatelli T., Monforti F., Vialetto G., Vitali L., Brusasca G., Calori G., Finardi S., Radice P., Silibello C.; The MINNI Project: An Integrated Assessment Modelling System for policy making. In Zerger, A. and Argent, R.M. (eds) MODSIM 2005 International Congress on Modelling and Simulation. Modelling and Simulation Society of Australia and New Zealand, December 2005, pp. 2005-2011. ISBN: 0-9758400-2-9. *http://www.mssanz.org.au/modsim05/papers/zanini.pdf*

[4] EMEP, 2003. Transboundary acidification, eutrophication and ground level ozone in Europe. Part I: Unified EMEP model description, EMEP status Report 1/2003.

[5] Mechler R., Amann M., Shoepp W., "A methodology to estimate changes in statistical life expectancy due to the control of particulate matter air pollution", 2002.

[6] Vialetto, G., Contaldi, M., De Lauretis, R., Lelli, M., Mazzotta, V., Pignatelli, T., 2005. Emission Scenarios of Air Pollutants in Italy using Integrated Assessment Models. Pollution Atmosphérique 185, 71.

Quantification of the effect of both technical and non-technical measures from road transport on Spain's emissions projections

J. M. López[1], J. Lumbreras[2], A. Guijarro[2] & E. Rodriguez[2]
[1]*University Institute of Automobile Research (Instituto de Investigación del Automóvil- INSIA), Carretera de Valencia, Madrid, Spain*
[2]*Department of Chemical & Environmental Engineering, Technical University of Madrid (UPM), José Gutiérrez Abascal, Madrid, Spain*

Abstract

Atmospheric emissions from road transport have increased all around the world since 1990 more rapidly than from other pollution sources. Moreover, they contribute to more than 25% of total emissions in the majority of European Countries. This situation confirms the importance of road transport when complying with emission ceilings (e.g. Kyoto Protocol and National Emissions Ceilings Directive).

A methodology has been developed to evaluate the effect of transport measures on atmospheric emissions (EmiTRANS). Its application to Spain in the horizon of 2020 allows the quantification of the effect of several measures on emission reductions.

This quantification was done through scenario development. Several scenarios were calculated considering technical measures (e.g. vehicle scrapping systems, higher penetration of hybrid and electric vehicles, fuel substitution, etc.) and non-technical measures (mileage reduction, implementation of Low Emission Zones and/or Congestion Charges in main cities, reduction of average speeds, logistical improvements that affect heavy duty vehicle load factors, etc.). The scenarios show the effect of each measure on NO_x, SO_2, CO, PM_{10}, $PM_{2.5}$, VOC, CO_2 and CH_4 emissions. The main conclusion is the necessity to combine both technical and non-technical measures to increase global effectiveness. In the analysis of specific pollutants, there is a great dispersion on reduction effects: technical measures are more effective to reduce air pollutants while non-technical measures are better options to reduce greenhouse effect gases (even though they also reduce air pollutants in a less efficient way).

Keywords: emissions, road transport, air quality, green house gases, methodology, policies and measures.

WIT Transactions on Ecology and the Environment, Vol 123, © 2009 WIT Press
www.witpress.com, ISSN 1743-3541 (on-line)
doi:10.2495/AIR090401

1 Introduction

In recent years the growing traffic demand combined with an increase in exhaust gas emissions is the main reason for permanent deterioration of air quality in urban areas (Lim *et al* [1], Colvile *et al* [2]). In order to reduce emissions, we need to gain precise information about the emission behaviour of motor vehicles. Vehicle exhaust emissions have been the cause of much concern regarding the effects of urban air pollution on human health (Curtis *et al* [3]) and green house gas (GHG) emissions.

The International Energy Agency's (IEA's) World Energy Outlook Reference Case projects that between 2000 and 2030, transport energy use and CO_2 emissions in OECD countries will increase by 50%, despite recent and ongoing policy initiatives intended to dampen this growth.

As an example, CO_2 emissions from road transport in Spain have increased in an 80% during the period 1990-2005 (fig.1). This percentage is higher than the increase in the number of vehicles in the same period. SO_2 emissions show a sharp decline due to reduction in sulphur content of fuels. Nevertheless, N_2O emissions suffered a strong increase but the amount, in terms of CO_2 equivalent, is significantly lower than CO_2 emissions from road transport. CO and VOC experimented a significant reduction, about 55%. NOx emissions were stabilized while $PM_{2.5}$ have only increased a 25%.

In order to facilitate the analysis of this situation, environmental protection authorities are interested in performing emission and air pollution simulation as well as scenario analysis by means of model based simulation systems (Winiwarter *et al* [4]). Traffic flow models provide a promising approach (Schmidt and Schäfer [5], Xia and Shao [6]), including calculations of air pollutant emissions from all transport sectors (Symeomidis *et al* [7]).

This paper presents a methodology to estimate atmospheric emissions from road transport including the development of a tailored software tool. Pollutants considered are those related to current air quality problems in urban areas (SO_2, NMVOC, NO_X and PM) while N_2O and CO_2 as GHG.

2 Methodology

In the present work, we start analysing the factors that have a relevant influence on emissions from road transport. The main parameters that contribute to emissions from road vehicles have been selected from the methodology developed [8], most of which are included in the EMEP/CORINAIR methodology [9].

Then, we have developed a software tool called EmiTRANS, which allows the inclusion of technical and non-technical measures leaded to quantity their influence in emissions reduction.

The purpose of this tool is to obtain emissions from developed scenarios through Copert4 software (Gkatzoflias *et al* [10]) and other outputs that are useful to get conclusions.

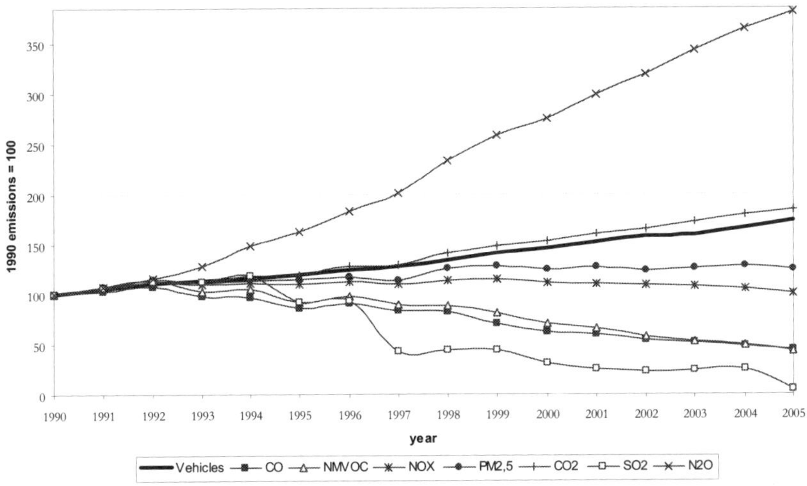

Figure 1: Historic trend of Spanish road transport emissions vs. number of vehicles.

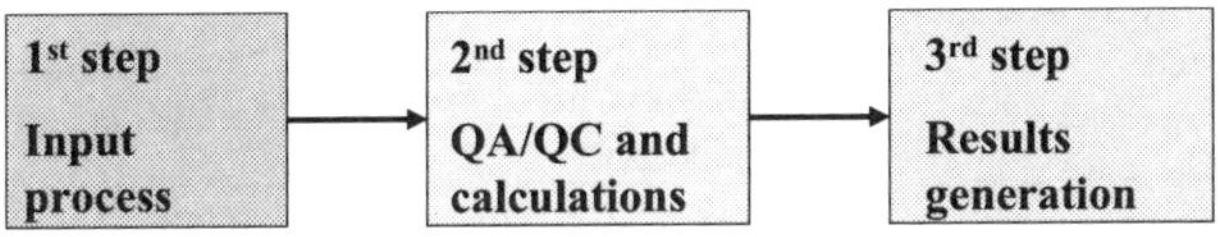

Figure 2: Blocks for running EmiTRANS.

We have also applied EmiTRANS to the case of Spain. This application consists of three blocks, as indicated in fig. 2.

- The first block includes input data. These are divided into six different sectors: passenger cars, light duty vehicles, buses, heavy duty vehicles, mopeds and motorcycles.
- In the second block, implicit variables (those that are not directly used by Copert4, e.g. mileage in units of passenger-km, occupancy rate, load factor, etc.) are transformed into explicit variables (e.g. mileage in veh-km). Afterwards, Quality Assurance/Quality Control (QA/QC) procedures are used (e.g. check that the sum of % is equal to 100, contrast if parameters are in previously assigned range, etc.). Eventually, developed algorithms are applied to obtain outputs.
- In the third block, results are generated according to Copert4 requirements. These outputs are, for instance, number of vehicles for each year by sector, subsector and technology; average speed by type of vehicle; fuel consumption by subsector, etc.

3 Sensitivity analysis

The method has been applied to Spain, carrying out a sensitivity analysis of the factors and using the EmiTRANS tool to develop different scenarios for Spanish road transport emission up to 2020. In order to compare the variation results of the different factors, emissions have been calculated for the road transport sector, no matter the scope of the factor modified (type of vehicle, driving modes, etc.).

The sensitivity analysis has been done according to the changes included in table 1 to identify the influence of several factors in atmospheric emissions.

Table 1: Selection of factors that influence the emissions.

Factor	Sensitivity analyses
Fuel distribution for vehicles	Reference: 46.6% petrol, 53,4% diesel
	30% petrol, 70% diesel
	40% petrol, 60% diesel
	60% petrol, 40% diesel
	70% petrol, 30% diesel
Urban average speed	Reference: 25 km/h
	20 km/h
	22.5 km/h
	27.5 km/h
	30 km/h
Highway average speed	Reference: 105 km/h
	84 km/h
	94.5 km/h
	115.5 km/h
	126 km/h
% of large vehicles	Reference: vehicles with engine cylinder>2 l are 6.2% for petrol and 14.2% for diesel
	Number of large vehicles are tripled
	Number of large vehicles are doubled
	Number of large vehicles are divided by 2
	There are no large vehicles
Number of old passenger cars	Reference: 5,375 M vehicles (26.5%)
	20% substitution by Euro 5 vehicles
	40% substitution by Euro 5 vehicles
	60% substitution by Euro 5 vehicles
	80% substitution by Euro 5 vehicles

Fig. 3 shows the influence on road transport emissions due to slight variations on urban speed for passenger cars. Every pollutant increases when speed is reduced and vice versa. The variation on CO and NMVOC is higher due to inefficient combustion process mainly in spark engines (volumetric efficient has a high influence in urban driving). CO_2 emissions decreased when incrementing the speed due to open throttle condition.

Fig. 4 shows the influence on road transport emissions due to slight variations on highway speed for passenger cars. At the range of reference speed (105 km/h), positive variations increase emissions. For instance, CO_2 emissions experiment a 5% raise when average speed is incremented in a 20%. These results are the consequence of the increasing rolling and drag resistance with the speed. Concerning CO emissions, the enrichment of the mixture at higher speeds causes its large augmentation.

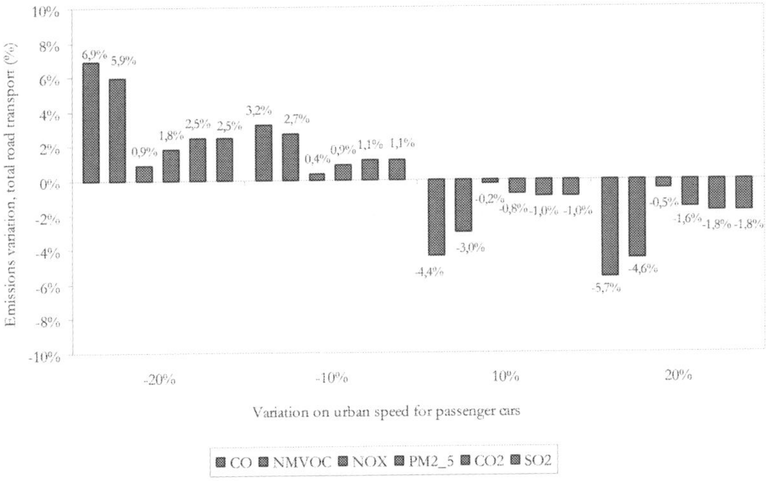

Figure 3: Sensitivity analysis to changes in urban speed. Ref. speed: 25 km/h.

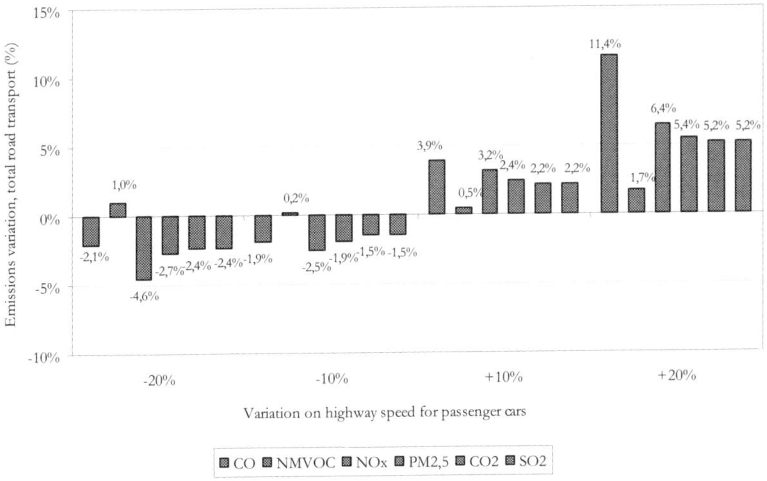

Figure 4: Sensitivity analysis to changes in highway speed. Ref. speed: 105 km/h.

Fig. 5 shows the influence of the load in heavy-duty vehicles on road transport emissions. In this case, mileage per vehicle has been assumed constant for all scenarios; therefore, the amount of tonnes-km of each scenario is different. Considering these hypothesis, NOx and $PM_{2.5}$ emissions increase with the load due to higher torque and fuel injection. CO_2 emissions show a growth of 0.6% when the load achieved 85%. CO and NMVOC are not relevant in diesel engines.

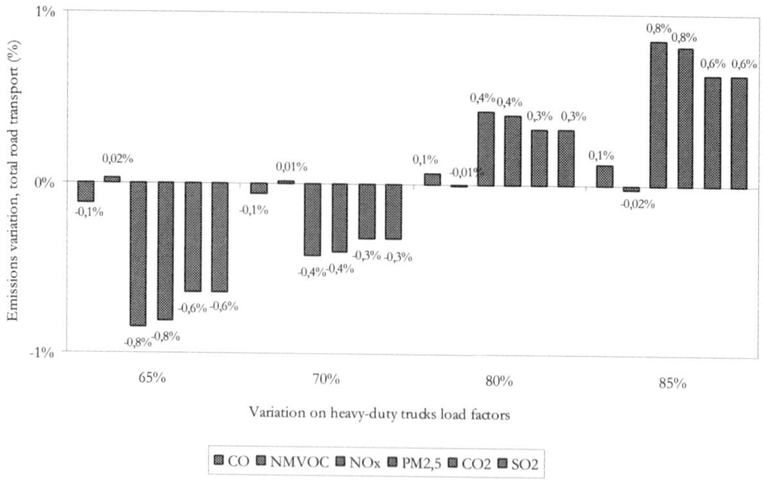

Figure 5: Sensitivity analysis to changes in heavy-duty vehicles load (constant mileage). Ref. load range: 75%.

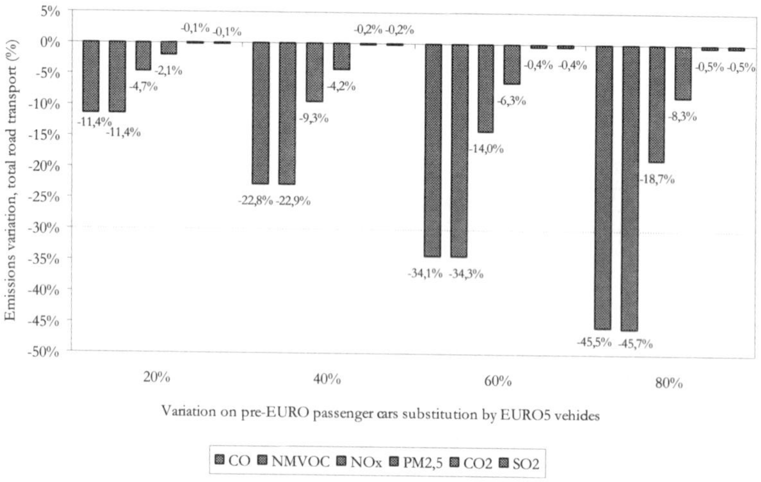

Figure 6: Sensitivity analysis to old vehicles substitution by Euro 5. Ref. case: 26% pre-Euro of passenger car park.

Fig. 6 shows that exhaust technologies have a decisive effect on reduction of every type of road transport pollutant. The high efficiency of three way catalysers decline CO, NMVOC and NOx emissions up to 95% in spark engines. Nevertheless, $PM_{2.5}$ emissions are more difficult to reduce in diesel engines. CO_2 emissions remain almost constant when replacing old vehicles because emission standards did include neither CO_2 limits nor efficiency improvements.

Fig. 7 evidences the proportionality of road transport pollutant emission to mobility variations. Higher mobility of diesel cars is more important to increase NOx and $PM_{2.5}$ in the sector. CO_2 emissions experiment an increment of 6.8% when increasing the mobility a 20%.

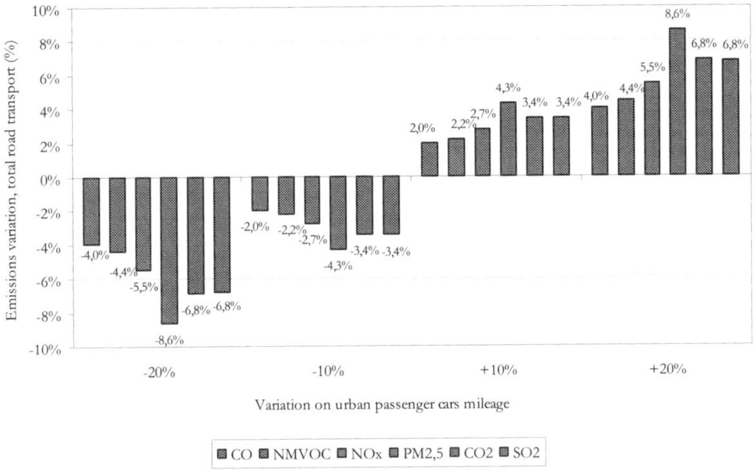

Figure 7: Sensitivity analysis for changes in mobility for passenger cars in urban areas. Baseline: 1070 km/year for gasoline cars and 7954 km/year for diesel cars.

4 Emission scenarios results

Regarding emission projections, the main assumptions for each of the five scenarios are shown in table 2. These assumptions were used to calculate emission projections using EmiTRANS model and Copert4 software as presented in section 2.

Business as Usual (BAU) scenario was defined under the hypothesis of high passenger and freight mobility, non-presence of biofuels and new technologies. The rest scenarios present several improvements related to mobility, penetration of new technologies or biofuel use.

Baseline scenario presents a moderated increment of mobility and includes the effect of Policies and Measures planned by the Spanish Administration. Besides, includes more environmental friendly technologies such as hybrid vehicles, electric vehicles and hydrogen or natural gas vehicles.

Biofuels, New Technologies and Mobility scenarios present improvements in their different fields compare to Baseline scenario.

Fig. 8 displays the results for CO_2 emissions. The largest emissions correspond to the BAU scenario. It does not include any technological measure and the passenger and freight mobility evolve as they did in the past (from 1990-2005). The lowest emissions scenario is the “lower mobility”. This remarks that

Table 2: Selection of factors that influence the emissions.

Scenario	Mobility	Technology in 2020	Power	Biofuels
Business as usual	+4% PC +6% HDV	Same as baseline	Same as baseline	Same as baseline
Baseline	+3.6-0.5% PC +5.1-0.2% HDV	1.4% Electric/H_2 3.2% Hybrid 16% NG urban buses	Petrol: 41%<1,4l; 52% ∈ (1,4l-2l); 7%>2l Diesel: 86%<2l; 14% >2l	2010: 5.83% 2012: 8% 2016-2020: 10%
Technological	Same as baseline	10% Electric/H_2 20% Hybrid 50% NG urb. buses	Same as baseline	Same as baseline
Lower mobility	No mobility increase	Same as baseline	Same as baseline	Same as baseline
Biofuel promotion	Same as baseline	Same as baseline	Same as baseline	2010: 6.88% 2012: 9.5% 2020: 20%

PC: Passenger cars. HDV: Heavy duty vehicles.

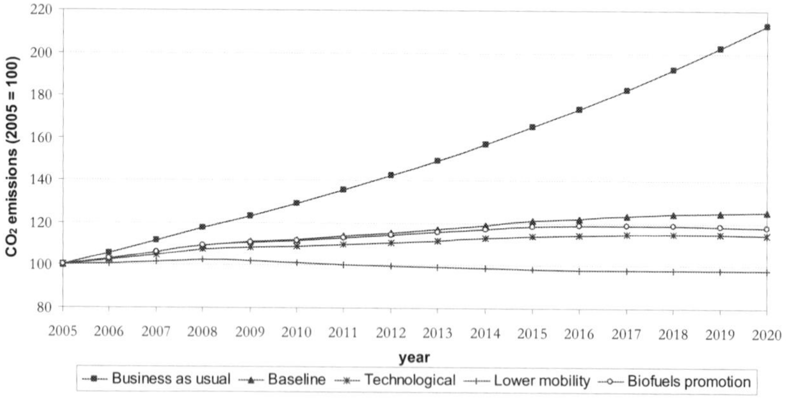

Figure 8: CO_2 emission projections for road transport in Spain.

the most effective measure to reduce CO_2 emissions is mobility cutback. The other scenarios project similar emissions: baseline has higher emissions but not far away from the promotion of biofuels and higher technology penetration. That is, in 2020 emissions under baseline scenario are 25% higher than in 2005 while "technological" and "Biofuel promotion" scenarios only increase a 14 and 18%, respectively.

Emission projections for NOx and $PM_{2.5}$ are shown in figs. 9 and 10. Every measure included in the Baseline scenario related to mobility, new technologies and biofuels yield relevant emission reductions respect to BAU scenario. In 2020, these reductions would be of 54.5% and 41.2% for NOx and $PM_{2.5}$, respectively. The most advantageous scenarios for emission decline, under the assumed hypotheses, are the "lower mobility" followed by "new technologies".

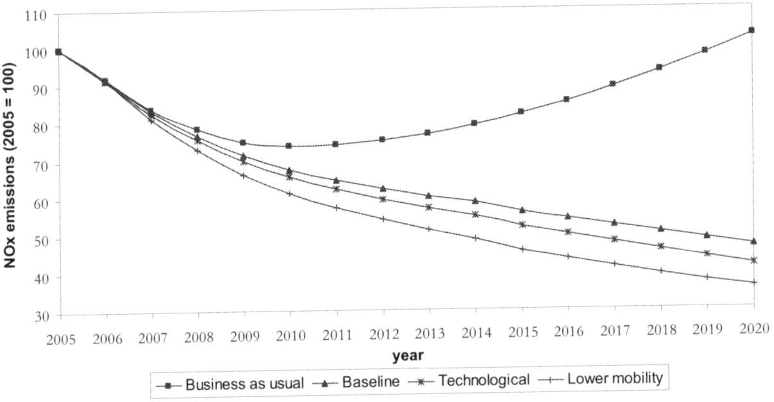

Figure 9: NOx emission projections for road transport in Spain.

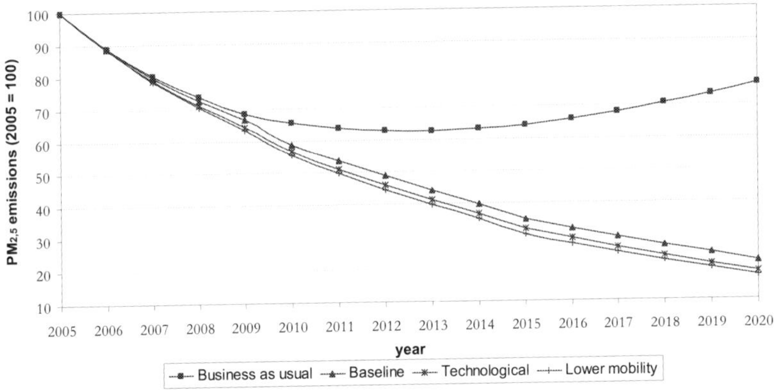

Figure 10: $PM_{2.5}$ emission projections for road transport in Spain.

5 Conclusions

We have developed a model, called EmiTRANS, which is able to estimate the influence of several factors on atmospheric emissions from road transport in a flexible and coherent way. It contributes to incorporate scientific data on decision making process. This methodology has successfully been applied to the case of Spain. It shows the importance of some variables in road transport emissions. According to the sensitivity analyses done, the selected variables (technical and non-technical measures) show clear trends on emissions pollutant. Furthermore, the model is a valuable tool for environmental planning and for the delineation of rational strategy towards the reduction of the atmospheric pollution levels.

This methodology also allows the development of different emission scenarios for future years. The application to Spain for the period 2006-2020 shows that the most effective measures to abate CO_2 emissions are those aimed to reduce passenger and freight mobility while vehicle scrapping systems are also effective to reduce air pollution.

References

[1] Lim, L., Hughes, S. Hellawell, E., Integrated decision support for urban air quality assessment. *Environmental Modelling & Software 20 (2005) 947-954,* 2005.

[2] Colvile, R.N., Hutchinson, E.J., Warrenl, R.F., The transport sector as a source of air pollution (Chapter 6). *Developments in Environmental Sciences*, Volume 1, pp 187-239, 2002.

[3] Curtis, L., Rea, W., Smith-Willis, P., Fenyves, E., Pan, Y., Adverse health effects of outdoor air pollutants. *Environment International* 32, 815–830, 2006.

[4] Winiwarter, W., Sscmidt, G., Environmental software system for emission inventories. *Environmental Modelling & Software 20 (2005) 1469-1477,* 2005.

[5] Schmidt, M., Schäfer, R., An integrated simulation system for traffic induced air pollution. *Environmental Modelling & Software 13(1998) 295-303,* 1998.

[6] Xia, L., Shao, Y., Modelling of traffic flow and air pollution emission with application to Hong Kong Island. *Environmental Modelling & Software 20 (2005) 1175-118,* 2005.

[7] Symeonidis, P., Ziomas, I., Proyou, A., Development of an emission inventory system from transport in Greece. *Environmental Modelling & Software 19(2004) 413-421,* 2004.

[8] Lumbreras, J., Guijarro, A., Lopez, J.M., Rodríguez, E., 2008. Methodology to quantify the effect of policies and measures in emission reductions from road transport. Urban Transport Conference. Wessex Institute of Technology.

[9] European Environment Agency, EMEP/CORINAIR Emission Inventory Guidebook. 2007 update of the third edition. 2007.

[10] Gkatzoflias, D., Kouridis, CH., Ntziachristos, L., Samaras, Z., COPERT 4. Computer programme to calculate emissions from road transport. ETC-ACC (European Topic Centre on Air and Climate Change), 2007.

Author Index